Bayesian M
Selection and
Statistical
Modeling

STATISTICS: Textbooks and Monographs

D. B. Owen

Founding Editor, 1972–1991

STATISTICS: Textbooks and Monographs

Recent Titles

Computer-Aided Econometrics, *edited by David E.A. Giles*

The EM Algorithm and Related Statistical Models, *edited by Michiko Watanabe and Kazunori Yamaguchi*

Multivariate Statistical Analysis, Second Edition, Revised and Expanded, *Narayan C. Giri*

Computational Methods in Statistics and Econometrics, *Hisashi Tanizaki*

Applied Sequential Methodologies: Real-World Examples with Data Analysis, *edited by Nitis Mukhopadhyay, Sujay Datta, and Saibal Chattopadhyay*

Handbook of Beta Distribution and Its Applications, *edited by Arjun K. Gupta and Saralees Nadarajah*

Item Response Theory: Parameter Estimation Techniques, Second Edition, *edited by Frank B. Baker and Seock-Ho Kim*

Statistical Methods in Computer Security, *edited by William W. S. Chen*

Elementary Statistical Quality Control, Second Edition, *John T. Burr*

Data Analysis of Asymmetric Structures, *Takayuki Saito and Hiroshi Yadohisa*

Mathematical Statistics with Applications, *Asha Seth Kapadia, Wenyaw Chan, and Lemuel Moyé*

Advances on Models, Characterizations and Applications, *N. Balakrishnan, I. G. Bairamov, and O. L. Gebizlioglu*

Survey Sampling: Theory and Methods, Second Edition, *Arijit Chaudhuri and Horst Stenger*

Statistical Design of Experiments with Engineering Applications, *Kamel Rekab and Muzaffar Shaikh*

Quality by Experimental Design, Third Edition, *Thomas B. Barker*

Handbook of Parallel Computing and Statistics, *Erricos John Kontoghiorghes*

Statistical Inference Based on Divergence Measures, *Leandro Pardo*

A Kalman Filter Primer, *Randy Eubank*

Introductory Statistical Inference, *Nitis Mukhopadhyay*

Handbook of Statistical Distributions with Applications, *K. Krishnamoorthy*

A Course on Queueing Models, *Joti Lal Jain, Sri Gopal Mohanty, and Walter Böhm*

Univariate and Multivariate General Linear Models: Theory and Applications with SAS, Second Edition, *Kevin Kim and Neil Timm*

Randomization Tests, Fourth Edition, *Eugene S. Edgington and Patrick Onghena*

Design and Analysis of Experiments: Classical and Regression Approaches with SAS, *Leonard C. Onyiah*

Analytical Methods for Risk Management: A Systems Engineering Perspective, *Paul R. Garvey*

Confidence Intervals in Generalized Regression Models, *Esa Uusipaikka*

Introduction to Spatial Econometrics, *James LeSage and R. Kelley Pace*

Acceptance Sampling in Quality Control, *Edward G. Schilling and Dean V. Neubauer*

Applied Statistical Inference with MINITAB®, *Sally A. Lesik*

Nonparametric Statistical Inference, Fifth Edition, *Jean Dickinson Gibbons and Subhabrata Chakraborti*

Bayesian Model Selection and Statistical Modeling, *Tomohiro Ando*

Bayesian Model Selection and Statistical Modeling

Tomohiro Ando

Keio University

Kanagawa, Japan

CRC Press
Taylor & Francis Group
Boca Raton London New York

CRC Press is an imprint of the
Taylor & Francis Group, an **informa** business
A CHAPMAN & HALL BOOK

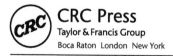

CRC Press
Taylor & Francis Group
6000 Broken Sound Parkway NW, Suite 300
Boca Raton, FL 33487-2742

First issued in paperback 2019

ISBN-13: 978-1-4398-3614-9 (hbk)
ISBN-13: 978-0-367-38397-8 (pbk)

Library of Congress Cataloging-in-Publication Data

Ando, Tomohiro.
Bayesian model selection and statistical modeling / Tomohiro Ando.
p. cm. -- (Statistics, textbooks, and monographs)
Includes bibliographical references and index.
ISBN 978-1-4398-3614-9 (hardcover : alk. paper)
1. Bayesian statistical decision theory. 2. Mathematical statistics. 3. Mathematical models. I. Title. II. Series.

QA279.5.A55 2010
519.5'42--dc22 2010017141

Visit the Taylor & Francis Web site at
http://www.taylorandfrancis.com

and the CRC Press Web site at
http://www.crcpress.com

Contents

Preface

Bayesian model selection is a fundamental part of the Bayesian statistical modeling process. In principle, the Bayesian analysis is straightforward. Specifying the data sampling and prior distributions, a joint probability distribution is used to express the relationships between all the unknowns and the data information. Bayesian inference is implemented based on the posterior distribution, the conditional probability distribution of the unknowns given the data information. The results from the Bayesian posterior inference are then used for the decision making, forecasting, stochastic structure explorations and many other problems. However, the quality of these solutions usually depends on the quality of the constructed Bayesian models. This crucial issue has been realized by researchers and practitioners. Therefore, the Bayesian model selection problems have been extensively investigated.

A default framework for the Bayesian model selection is based on the Bayes factor, which provides the scientific foundations for various fields of natural sciences, social sciences and many other areas of study. From the Bayes factor, Bayesian information criterion (BIC), generalized Bayesian information criterion (GBIC), and various types of Bayesian model selection criteria have been proposed. One of the main objectives of this book is to provide comprehensive explanations of the concepts and derivations of the default framework for the Bayesian model selection, together with a wide range of practical examples of model selection criteria.

The Bayesian inference on a statistical model was previously complex. It is now possible to implement the various types of the Bayesian inference thanks to advances in computing technology and the use of new sampling methods, including Markov chain Monte Carlo (MCMC). Such developments together with the availability of statistical software have facilitated a rapid growth in the utilization of Bayesian statistical modeling through the computer simulations. Nonetheless, model selection is central to all Bayesian statistical modeling. There is a growing need for evaluating the Bayesian models constructed by the simulation methods.

Recent Bayesian model selection studies have been mainly focusing on the evaluation of Bayesian models constructed by the simulation methods. We have seen advances of theoretical development for this area of study. A secondary objective of this book is to give plenty of simulation-based Bayesian model evaluation methods with practical advice. Various types of simulation-based Bayesian model selection criteria are explained, including the numerical calculation of the Bayes factors, the Bayesian predictive information criteria and the deviance information criteria. This book also provides a theoretical basis for the analysis of these criteria.

In addition, Bayesian model averaging is applied to many problems. By averaging over many different Bayesian statistical models, it can incorporate model uncertainty into the solution of the decision problems. In the modeling process, researchers and practitioners generally face a problem: how to specify the weight to average over all models as well as how to calculate the number of models to be combined. Heavily weighting the best fitting models, Bayesian model selection criteria have also played a major role in these issues. The third purpose of book is to cover the model averaging.

R code for several worked examples that appear in the book is available. From the link http://labs.kbs.keio.ac.jp/andotomohiro/Bayesianbook.htm, readers can download the R code to run the programs.

The author would like to acknowledge the many people who contributed to the preparation and completion of this book. In particular, the author would like to acknowledge with his sincere thanks Sadanori Konishi (Kyushu University) and Arnold Zellner (University of Chicago), from whom the author has learned so much about the concepts of Bayesian statistics and statistical modeling. The author would like to thank Ruey Tsay (University of Chicago) for an opportunity to visit Booth School of Business, University of Chicago, where he gained much experience.

The author's ideas on Bayesian statistics and statistical modeling for interdisciplinary studies have been greatly influenced by: Neeraj Bharadwaj (Temple University), Pradeep Chintagunta (University of Chicago), Alan Gelfand (Duke University), John Geweke (University of Iowa), Genshiro Kitagawa (Institute of Statistical Mathematics), Takao Kobayashi (University of Tokyo), Hedibert Lopes (University of Chicago), Teruo Nakatsuma (Keio University), Yasuhiro Omori (University of Tokyo), Nicholas Polson (University of Chicago) and many scholars.

The author is grateful to four anonymous reviewers for comments and suggestions that allowed him to improve the original draft greatly. David Grubbs patiently encouraged and supported the author throughout the final preparation of this book. The author would like to express his sincere thanks to all of these people.

TOMOHIRO ANDO

Chapter 1

Introduction

1.1 Statistical models

The practice of statistical modeling undergoes continual change as a result of both methodological developments and progress in the computer environment. The high-performance computers facilitated widespread advances in the development of statistical modeling theory to capture the underlying nature of a phenomenon. It is evident that the amount of information has been increasing both in size and variety thanks to recent advancement of science technology. With the advancement of computers and the information age, the challenge of understanding vast amounts of complicated data has led to the development of various types of statistical models.

Statistical model, a researchers' view to a phenomenon in our world, provides an useful tool for the description of stochastic system, the prediction, the information extraction, the casual inference, the decision making and so on. Simply speaking, we can regard a statistical model as a simplification of a complex reality. Statistical models are used not only in the social sciences: economics, finance marketing, psychology, sociology and political science, but are also employed in the natural sciences and engineering. Researchers and practitioners in various study fields have been using statistical models extensively.

Mathematically, a statistical model is defined as a set of probability distributions on the sample space (Cox and Hinkley (1974)). We usually consider a parametric family of distributions with densities $\{f(\boldsymbol{x}|\boldsymbol{\theta}); \boldsymbol{\theta} \in \Theta\}$. In this case, a statistical model $f(\boldsymbol{x}|\boldsymbol{\theta})$ parameterized by $\boldsymbol{\theta}$ is thus a parameter set Θ together with a probability distribution function from the sample space to the set of all probability distributions on the sample space (McCullagh (2002)). In developing statistical models, we need to specify the components of a statistical model $f(\boldsymbol{x}|\boldsymbol{\theta})$, a probability distribution function and a parameter value. The next example illustrates a development of statistical models.

Example

Figure 1.1 shows the daily returns of Nikkei 225 index from August 28, 2001 to September 22, 2005 on which the market was open leading to a set of 1,000 samples. The vertical axis is the differences in the logarithm of the daily closing value of Nikkei 225 index and the horizontal axis is the time. The returns y_t are defined as the differences in the logarithm of the daily closing value of Nikkei 225 index $x_t = \{\log(y_t) - \log(y_{t-1})\} \times 100$, where y_t is the closing price on day t. The basic statistics, the sample mean, and the sample standard deviation, are given as $\hat{\mu} = -0.018$, $\hat{\sigma} = 1.930$, respectively.

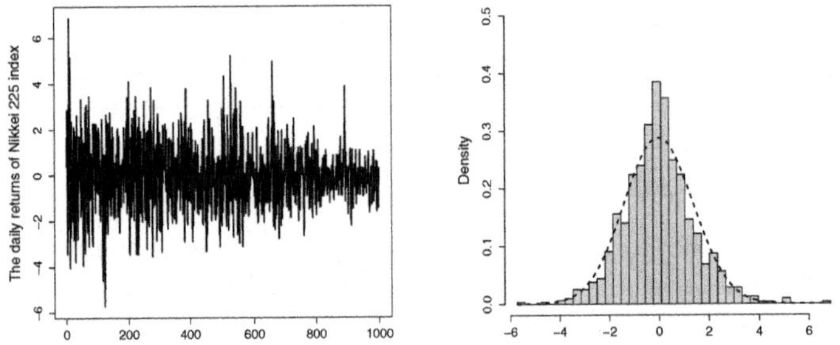

(a) (b)

FIGURE 1.1: (a): Time series plot for Nikkei 225 index return data with sample period from August 28, 2001 to September 22, 2005. (b): The fitted normal density function with the mean $\hat{\mu} = -0.018$ and the standard deviation $\hat{\sigma} = 1.93$. Histograms of Nikkei 225 index return data are also shown.

In analyzing asset return data, however, the summary statistics may not have enough information. To present asset return data, one of the most common ways is to estimate the underlying true structure by fitting a parametric family of statistical models. Here, we consider fitting the normal distribution:

$$f(x|\mu, \sigma^2) = \frac{1}{\sqrt{2\pi\sigma^2}} \exp\left\{ -\frac{(x-\mu)^2}{2\sigma^2} \right\}.$$

After we specify the model, we have to determine unknown parameter values, i.e., μ and σ^2. Although there are various approaches to determine these parameter values, let us simply use the sample mean $\hat{\mu}$ and the sample standard deviation $\hat{\sigma}^2$. Figure 1.1 (b) shows the fitted normal density function $f(x|\hat{\mu}, \hat{\sigma}^2)$ with the mean $\hat{\mu} = -0.018$ and the standard deviation $\hat{\sigma} = 1.930$, respectively. Histograms of data are also shown in Figure 1.1 (b). We can see that the fitted normal density provides a rough approximation of the observations.

In summary, to describe a stochastic system of Nikkei 225 index return data, we employed the normal distribution $f(x|\mu, \sigma^2)$. The constructed statistical model $f(x|\hat{\mu}, \hat{\sigma}^2)$ might allow us to perform a forecasting of future Nikkei 225 index returns.

As shown in the above example, we first specified the probability distribution and then determined the parameter values within the specified model.

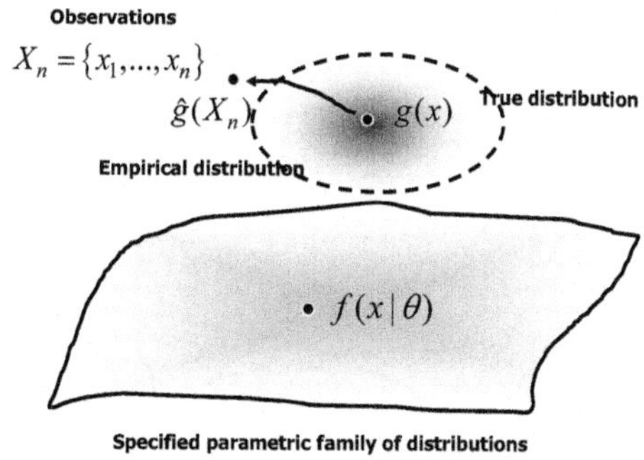

FIGURE 1.2: The relationship between the true model $g(x)$ and the empirical distribution $\hat{g}(\boldsymbol{X}_n)$ constructed by the observed data $\boldsymbol{X}_n = \{x_1, ..., x_n\}$. Once we specify the parameter values $\boldsymbol{\theta}$ of the specified model $f(x|\boldsymbol{\theta})$, the model will be fixed at a point on the surface. Therefore, the problem reduces to the parameter estimation problem.

The process determining the parameter values is called the estimation process. This estimation process adjusts the parameter values so that the specified model matches to the observed dataset. Figures 1.2–1.4 show a general image of the modeling process.

First we observe a set of data $\boldsymbol{X}_n = \{x_1, \ldots, x_n\}$. Depending on the problem, we might assume that the "true" model is contained within a set of models under our consideration. This is the M-closed framework (Bernardo and Smith (1994)). If the true model $g(x)$ belongs to the specified parametric family of distributions $f(x|\theta)$, then the true model $g(x)$ is on the surface of the specified model. In contrast to this assumption, one might follow Box (1976) in believing that "all models are wrong, but some are useful," or, none of the models under consideration is the true model. This view may be often realistic and called the M-open framework. Therefore, we usually assume that the data $\boldsymbol{X}_n$ are generated from unknown true distribution. Figure 1.2 gives an image of the relationship between the true model $g(x)$ and the empirical distribution $\hat{g}(\boldsymbol{X}_n)$ constructed by the observed data $\boldsymbol{X}_n$. Notice that we usually don't know the true model $g(x)$, but just observe the data $\boldsymbol{X}_n$. Once we specify the parameter values $\boldsymbol{\theta}$ of the specified model, the model will be fixed at a point on the model surface.

Of course, as shown in Figure 1.3, we can employ various types of probability models. In the above example, we specified the mode $f(x|\boldsymbol{\theta})$ as a normal distribution. However, we can also use Student-t distribution, Cauchy distribution and other distributions. In the model selection process, we usually

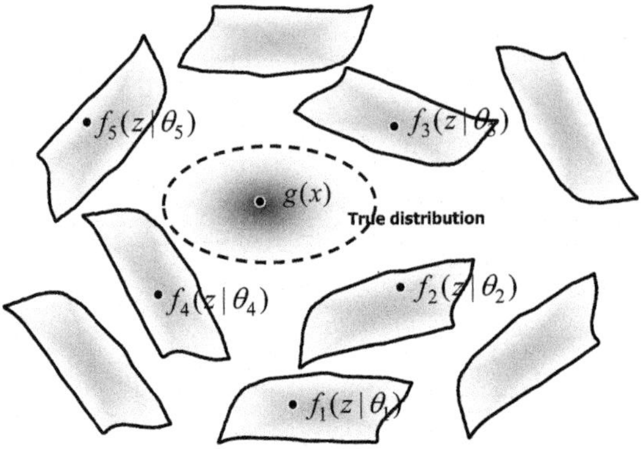

FIGURE 1.3: We can prepare various types of statistical models. For example, we can specify the probability distribution as $f_1(z|\boldsymbol{\theta}_1)$ normal distribution, $f_2(z|\boldsymbol{\theta}_2)$ Student-t distribution, $f_3(z|\boldsymbol{\theta}_3)$ Cauchy distribution, and so on.

pick one of the best models among these candidate models, or average over the set of probability models. In the context of Bayesian model selection, we usually select the model with the highest probability in the sense of the posterior model probabilities. Or, we use the Bayesian model averaging method, which averages over the set of probability models by using the posterior model probabilities.

Once we specify the probability model, the problem reduces to the parameter estimation problem. In other words, we want to specify the parameter values $\boldsymbol{\theta}$ of the specified model $f(x|\boldsymbol{\theta})$ so that the parameter value matches the true model $g(x)$. There are many approaches to estimate the unknown parameter values; the maximum likelihood method, the penalized maximum likelihood method, the generalized method of moments, robust estimation method, Bayesian estimation and so on. Figure 1.4 shows an image of the maximum likelihood method. The parameter values are determined to minimize the distance between the empirical distribution

$$\hat{g}\left(x; \boldsymbol{X}_n\right) = \frac{1}{n}\sum_{\alpha=1}^{n} I(x = x_\alpha),$$

and the specified model $f(x|\boldsymbol{\theta})$. Here $I(\cdot)$ is the indicator function that takes 1 if the relational expression in the blanket is true and 0 otherwise. The distance can be measured by using the minus of the likelihood function. Thus, the maximization of the likelihood function equals the minimized distance between the empirical distribution and the specified model $f(x|\boldsymbol{\theta})$.

Note, however, that we have to assess the closeness of the constructed statistical model $f(x|\hat{\boldsymbol{\theta}})$ not only to the true model $g(x)$, but also to the empirical

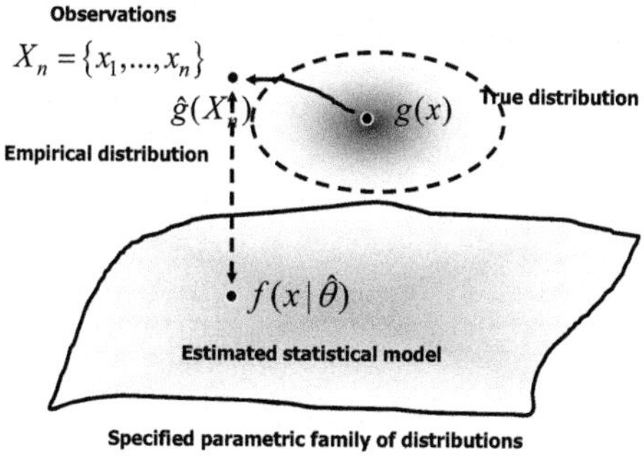

FIGURE 1.4: An image of the maximum likelihood method. The parameter values are determined so as to minimize the distance between the empirical distribution and the specified model $f(x|\boldsymbol{\theta})$. The distance can be measured by using the minus of the likelihood function. Thus, the maximization of the likelihood function is equal to minimizing the distance between the empirical distribution and the specified model $f(x|\boldsymbol{\theta})$.

distribution constructed by the observed data $\boldsymbol{X}_n = \{x_1, ..., x_n\}$. This is natural because the future observations are generated from the true model $g(x)$, but the empirical distribution $\hat{g}(\boldsymbol{X}_n)$. One of the most well-known methods is provided by Akaike (1973, 1974), who pionnered an information theoretic approach. One of the most well known methods is provided by Akaike (1973, 1974), who pionnered an information theoretic approach. In this framework, the model parameters are estimated by the maximum likelihood method and then the deviation of the estimated model $f(x|\hat{\boldsymbol{\theta}})$ from the true model $g(x)$ is measured by Kullback-Leibler information (Kullback and Leibler (1951)). The best model is chosen by minimizing Kullback-Leibler information among different statistical models.

Consider the situations that the specified parametric model $f(x|\boldsymbol{\theta})$ contains the true distribution $g(x)$, that is $g(x) = f(x|\boldsymbol{\theta}_0)$ for some $\boldsymbol{\theta}_0 \in \Theta$, and that the model is estimated by the maximum likelihood method. Together with certain mild regularity conditions, Akaike (1973, 1974) proposed Akaike's information criterion, known as AIC, for evaluating the constructed models. The following studies have been done by Takeuchi (1976), Konishi and Kitagawa (1996), where the specified parametric model does not necessarily contain the true distribution.

Once we select the best model that describes the true structures well, this model can be used for various purposes. In economic problems, for example, the constructed model can be used for quantifying the price and demand relationship, forecasting economic activity, developing a new economic policy and so on.

In contrast to an information theoretic approach, this book considers the statistical models estimated by the Bayesian approach. To obtain an intuition of the Bayesian approach, the next section provides a formal Bayesian statistical modeling framework.

1.2 Bayesian statistical modeling

As discussed in the previous section, the development of a statistical model requires specifications of a probability distribution function and a parameter value. One of the most common approaches is the maximum likelihood method, which simply maximizes the likelihood function to estimate model parameters. A frequentist thinks that the parameters of the specified model are unchanging.

In contrast to the frequentist approach, the Bayesian framework regards the parameters as random variables. As a consequence, the Bayesian analysis requires an additional component, a prior distribution on the parameter $\boldsymbol{\theta}$, $\pi(\boldsymbol{\theta})$. Incorporating prior information through the prior distribution, along with a given set of observations, the Bayesian inference on the specified model is performed. The prior information could come from related knowledge, past experiences, intuition, belief, or from operational data from previous comparable experiments, or from engineering knowledge.

Bayesian inference is therefore very useful when there is a limited observation but there is plenty prior information. Generally, Bayesian inference on the model parameter is done based on the posterior distribution $\pi(\boldsymbol{\theta}|\boldsymbol{X}_n)$, the conditional probability distribution given the data information $\boldsymbol{X}_n = \{\boldsymbol{x}_1, ..., \boldsymbol{x}_n\}$ and prior density. The results of the posterior inference are used for the decision making, forecasting and stochastic structure exploration problems. In principle, the Bayesian statistical modeling proceeds as follows (see also Figure 1.5, in which the corresponding chapters are also indicated):

Bayesian statistical modeling process:

1. Step 1. Model formulation: Express your unknown quantities using prior distribution $\pi(\boldsymbol{\theta})$ and specify the likelihood, or equivalently, the sampling density function $f(\boldsymbol{X}_n|\boldsymbol{\theta})$, which describes the process giving rise to a set of data $\boldsymbol{X}_n = \{\boldsymbol{x}_1, ..., \boldsymbol{x}_n\}$.

   To specify the prior distribution $\pi(\boldsymbol{\theta})$, various approaches have been proposed, including diffuse prior, Jeffreys' prior, conjugate prior, informative prior and so on.

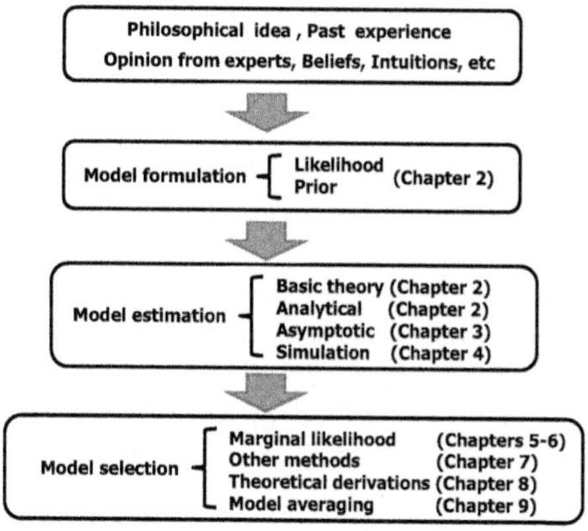

FIGURE 1.5: Bayesian statistical modeling process.

2. Step 2. Model estimation: Make inference on the unknown quantities based on the posterior distribution $\pi(\boldsymbol{\theta}|\boldsymbol{X}_n)$, the conditional probability distribution of the unknowns given the data information.

   Under a certain model specification, we can easily make an inference about the posterior distribution of parameters $\pi(\boldsymbol{\theta}|\boldsymbol{X}_n)$. In practice, however, we merely obtain an analytical expression of the posterior distribution of parameters. To make practical inference on the posterior distribution of parameters, there are mainly two approaches: the asymptotic approximation and simulation-based approach.

3. Step 3. Model selection: Evaluate the goodness of the estimated Bayesian models. If the developed Bayesian model does not satisfy you, then go back to Step 1.

   Once we specify the prior distribution and likelihood function, under a certain condition, the two approaches above automatically allows us to estimate the posterior distribution of the parameters. We then use the posterior inference results for the decision making, forecasting and stochastic structure exploration problems, etc. However, the quality of these solutions usually depends on the goodness of the constructed model. This crucial issue had been realized by researchers and practitioners, and thus Bayesian model selection criteria has been extensively investigated.

1.3 Book organization

Although a big picture is provided in Figure 1.5, we shall provide more details on each chapter. Chapter 2 first provides an introduction of the Bayesian analysis. Starting from the basic idea of Bayesian analysis, we provide the sampling density specifications in various studies, including quantification of a price elasticity of demand (Econometrics), describing a stock market behavior (Financial econometrics), tumor classification with gene expression data (Bioinformatics), factor analysis model (Psychometrics), survival analysis model for quantifying customer lifetime value (Marketing), logistic regression model (Medical science) and so on. In the Bayesian context, we also need to specify the prior density on unknown model parameters. Chapter 2 thus covers the prior specification methods (including diffuse prior, Jeffreys' prior, conjugate prior, informative prior, and so on).

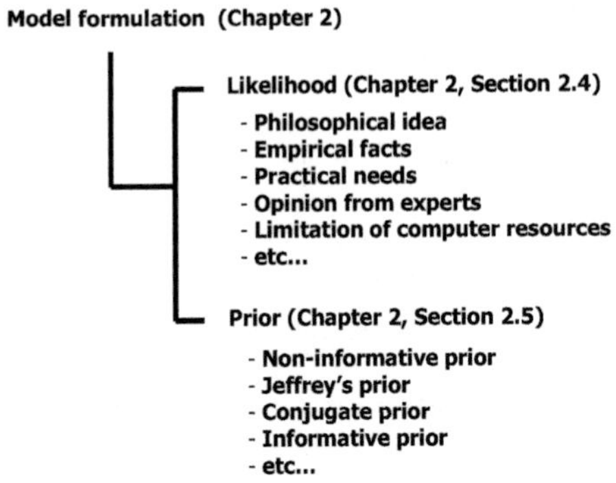

FIGURE 1.6: Model formulation.

Once we get a posterior distribution of parameter, we summarize the characteristics of the posterior distribution. We also review how to investigate the characteristics of the posterior distribution. As an example of Bayesian inference procedure, the Bayesian linear regression example is provided. Finally, we discuss the importance of model selection.

Chapters 3 and 4 cover the Bayesian estimation methods. As shown in Figure 1.7, Chapter 3 covers the asymptotic approximation approach for the Bayesian inference. Topics covered are consistency of posterior parameter estimate, the asymptotic normality of posterior parameter estimate, the asymptotic approximation of the posterior distribution based on the normal distribution, Laplace methods for approximating integrals, including posterior expectation of a function of model parameter and the predictive distribution.

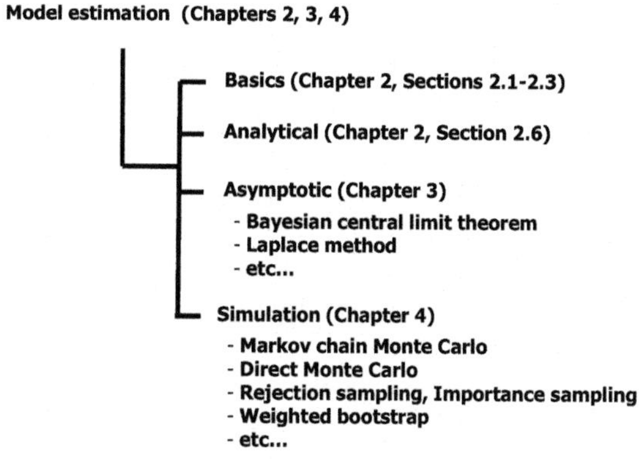

FIGURE 1.7: Model estimation.

Chapter 4 provides the simulation oriented approaches, including Markov chain Monte Carlo (MCMC) Gibbs sampling and Metropolis-Hastings approaches. In the Bayesian inference framework, we often need to implement the integrals. Reviewing the concept of Monte Carlo integration, we provide practical application of MCMC to the seemingly unrelated regression models. We also cover the Bayesian analysis of various models, including volatility time series, simultaneous equations, quantile regression, graphical models, multinomial probit models, and Markov switching models. The use of other simulation approaches such as importance sampling, rejection sampling, weighted bootstrap, direct Monte Carlo approaches is also discussed.

One of main purposes of this book is to cover various Bayesian model selection criteria for evaluating the goodness of the Bayesian models. From Chapter 5–9, we provide comprehensive explanations of the concepts and derivations of various Bayesian model selection criteria together with a wide range of practical examples. Details are given in Figure 1.8.

Chapter 5 provides a general framework of the Bayesian approach for model selection. This section provides definitions of the Bayes factor (Kass and Raftery (1995)), the Bayesian information criteria (BIC; Schwarz (1978)) and the generalized Bayesian information criteria (GBIC; Konishi et al. (2004)). In the Bayesian approach for model selection, the calculation of the posterior probabilities for a set of competing models is essential. One of the difficulties in the use of the Bayes factor is its sensitivity to prior distributions. Unfortunately, it is generally known that the use of improper priors for the parameters in alternative models results in Bayes factors that are not well defined. Therefore, Chapter 5 covers many attempts that have been made to define a convenient Bayes factor in the case of noninformative prior to the intrinsic Bayes factor (Berger and Pericchi (1996)), the fractional Bayes

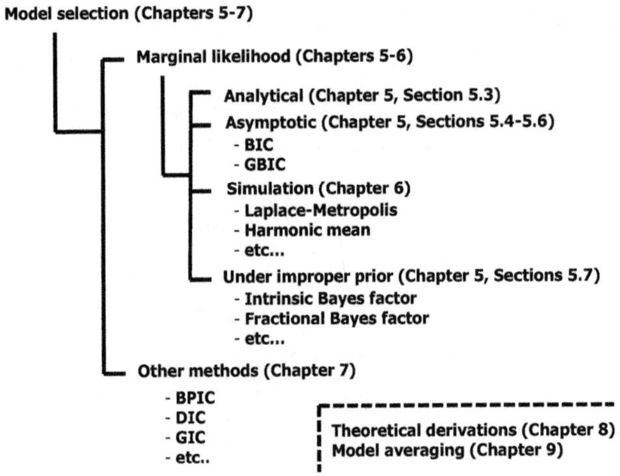

FIGURE 1.8: Model selection.

factor (O'Hagan (1995)), the pseudo Bayes factors based on cross validation (Gelfand et al. (1992)), the posterior Bayes factor (Aitkin (1991)), and the predictive likelihood approach (Ando and Tsay (2009)). Other related topics such as Bayesian p-values and sensitivity analysis are also covered.

In contrast to BIC and GBIC that are constructed by an asymptotic approach, we can construct the marginal likelihood based on the posterior simulations. Chapter 6 reviews many studies that take advantage of modern Bayesian computing methods, including the Laplace-Metropolis estimator (Lewis and Raftery (1997)), the so-called candidate formula (Chib (1995)), the harmonic mean estimator (Newton and Raftery (1994)), Gelfand and Dey's estimator (Gelfand and Dey (1994)), Bridge sampling estimator (Meng and Wong (1996)), and the Savage-Dickey density ratio (Verdinelalni and Wasserman (1995)) and Kernel density approach (Kim et al. (1998)). We also describe the reversible jump MCMC algorithm (Green (1995)), the product space search (Carlin and Chib (1995)), and the "Metropolizing" product space search (Dellaportas et al. (2002)) for computing the posterior model probabilities of each model. Bayesian variable selection studies (Geroge and McCulloch (1993)) are also covered.

Chapter 7 covers Bayesian predictive information criterion (BPIC, Ando (2007)). the deviance information criteria (Spiegelhalter et al. (2002)), the modified BIC (Eilers and Marx (1998)), the generalized information criteria (Konishi and Kitagawa (1996)) and a minimum posterior predictive loss approach (Gelfand and Ghosh (1998)). In contrast to the traditional Bayesian approach for model selection, these methods use other utility functions to evaluate the goodness of the Bayesian models. For example, BPIC is proposed as an estimator of the posterior mean of the expected log-likelihood of the predictive distribution when the specified family of probability distributions does not contain the true distribution.

Chapter 8 explains the theoretical developments of BIC, GBIC, BPIC and GIC. Comparisons of various Bayesian model selection criteria are also provided. We compare the properties of various Bayesian model selection criteria from several aspects, including the use of improper prior, computational amount, etc.

Chapter 9 covers the Bayesian model averaging, which provides a general probabilistic modeling framework that simultaneously treats both model and parameter uncertainties.

Chapter 2

Introduction to Bayesian analysis

Bayesian methods provide a useful tool for statistical inference under an uncertainty environment. Incorporating a scientific hypothesis, Bayesian methods introduced a new interpretation of "probability" as a rational, conditional measure of uncertainty. This chapter covers basic notions of Bayesian analysis, including Bayes' theorem, prior, posterior, predictive, densities, etc. In the next section, we discuss the concept of probability.

2.1 Probability and Bayes' theorem

Probability is a useful tool to express the likelihood of an occurrence of an event. As of now, the notion of probability has been used extensively in such areas of study as social sciences, natural sciences, medical sciences for the decision making, forecasting and stochastic structure exploration problems. In the field of probability theory, we have broadly two kinds of interpretations for the probability of an event. One regards a probability, defined as the relative frequency in the long run of outcomes, is called "objective" or "frequency" probability. The other, "Bayesian" probability represents a conditional measure of uncertainty associated with the occurrence of a particular event, given the available information and prior beliefs. In this framework, one might assign a probability to a nonrandom event if there is an uncertainty associated with the occurrence of the event.

To interpret the definitions of probability, consider tossing a fair coin. Thus, the probabilities that a head occurs and a tail occurs are equal and each are 0.5. This probability is an objective probability. It is defined based on frequency to which the event occurs, or defined by logic. We next consider the following question. What is the probability that the 77th digit of $\pi = 3.141\cdots$ is 3? There is no uncertainty in the 77th digit of π upon careful consideration. Therefore, the concept of objective probability does not fit in this statement and the answer would be either as 0 or 1 in this framework. On the other hand, the probability in the Bayes analysis depends on the prior knowledge. If the assayer has prior information that $0 \sim 9$ of the figure becomes the 77th digit at probability of $1/10$ digit, the probability that the statement is true becomes

1/10. This probability measure is a Bayesian probability. It is a probability as the numerical value shows the degree of trust to which a certain event occurs. Therefore, the Bayesian probability of each subject can have a different value.

In the Bayes analysis, the Bayes' theorem is a fundamental tool. We next review the Bayes' theorem. Let A and B be certain events. We denote $P(A|B)$ as a probability of event A occurs after event B occurs. Under a condition that a probability of event B occurs is positive, $P(B) > 0$, a conditional probability of event A when event B is given as

$$P(A|B) = \frac{P(A \cap B)}{P(B)}.$$

Here $P(A \cap B)$ is a probability that both events A and B happen. A conditional probability of event A given event B is adjusted by probability $P(B)$, because we consider a situation in which event B is given. Transforming above the expression of the conditional probability, we have

$$P(A \cap B) = P(A|B)P(B),$$

which is called the product rule for probabilities.

Bayes' theorem is derived according to the Law of Total Probability. Let A_1, A_2,...,A_m be disjoint events such that $P(A_i \cap A_j) = 0$, $i \neq j$ and $P(A_1 \cup \cdots \cup A_m) = 1$ for sure event Ω ($P(\Omega) = 1$). Then we have

$$P(B) = P(B|\Omega) = \sum_{j=1}^{m} P(B|A_j)P(A_j).$$

Thus sure event Ω is divided into a disjoint event of m pieces. Then a conditional probability of B given each divided event A_m is added together. A conditional probability of event A_k given event B, ($P(B) > 0$) is then given as

$$\begin{aligned} P(A_k|B) &= \frac{P(A_k \cap B)}{P(B)} \\ &= \frac{P(A_k \cap B)}{\sum_{j=1}^{m} P(B|A_j)P(A_j)} \\ &= \frac{P(B|A_k)P(A_k)}{\sum_{j=1}^{m} P(B|A_j)P(A_j)}, \quad k = 1...m. \end{aligned} \tag{2.1}$$

This expression is the so-called Bayes's theorem.

In the next section, the basic elements of the Bayesian inferential approach are illustrated through the basic inference problem.

2.2 Introduction to Bayesian analysis

As we discussed in the previous section, Bayesian methods regard the probability as a conditional measure of uncertainty associated with the occurrence of a particular event, given the available information and prior beliefs. The following example illustrates the subjective probability as a conditional measure of uncertainty along with Bayes' theorem.

Suppose that 3% of people from the human population are infected by a particular virus. We randomly select a person from this population. This person is subject to an initial test X which is known to yield positive results in 98% of infected people and in 40% of those not infected. Let V_+ denote the event that a person carries the virus and V_- denote the event that a person is not a carrier.

$$P(V_+) = 0.03, \quad \text{and} \quad P(V_-) = 0.97,$$

which come from the available knowledge of the infection. In the same way, let X_+ denote a positive result and X_- denote a negative result. Thus,

$$P(X_+|V_+) = 0.98, \quad \text{and} \quad P(X_+|V_-) = 0.40,$$

which is from the probability mechanism generating the test results. More generally, Bayesian methods interpret the conditional probability, e.g., $P(X_+|V_+)$ as a measure of belief in the occurrence of the event X_+ given assumptions V_+ and available knowledge. In this case, an available knowledge for $P(X_+|V_+)$ is the accuracy rate of test X. Thus we assigned a rational probability to $P(X_+|V_+)$.

Suppose that the result of the test turns out to be positive. Clearly, we are now interested in the probability that the person carries the virus, given the positive result. From the Bayes' theorem, we yield

$$\begin{aligned}
P(V_+|X_+) &= \frac{P(X_+|V_+)P(V_+)}{P(X_+)} \\
&= \frac{P(X_+|V_+)P(V_+)}{P(X_+|V_+)P(V_+) + P(X_+|V_-)P(V_-)} \\
&= \frac{(0.98) \times (0.03)}{(0.98) \times (0.03) + (0.40) \times (0.97)} \\
&\approx 0.07.
\end{aligned}$$

As a consequence, the information X_+ increases the probability from 3% to 7% that a person carries the virus.

A person takes the more accurate test Y, which relates to the probability that the person carries the virus as follows

$$P(Y_+|V_+) = 0.99, \quad \text{and} \quad P(Y_+|V_-) = 0.04.$$

Before we execute test Y, we can expect that test Y results in the positive as

$$\begin{aligned} P(Y_+|X_+) &= P(Y_+|X_+, V_+) + P(Y_+|X_+, V_-) \\ &= P(Y_+|V_+) \times P(V_+|X_+) + P(Y_+|V_-) \times P(V_-|X_+) \\ &= (0.99) \times (0.07) + (0.04) \times (0.93) \\ &\approx 0.11. \end{aligned}$$

Also, $P(Y_-|X_+) = 1 - P(Y_+|X_+) \approx 0.89$.

The more accurate test Y results in negative. We are next interested in the probability that the person carries the virus, given the positive result X_+ and the negative result Y_-. From the Bayes' theorem, we then yield

$$\begin{aligned} P(V_+|X_+, Y_-) &= \frac{P(Y_-|V_+)P(V_+|X_+)}{P(Y_-|X_+)} \\ &= \frac{P(Y_-|V_+)P(V_+|X_+)}{P(Y_-|V_+)P(V_+|X_+) + P(Y_-|V_-)P(V_-|X_+)} \\ &= \frac{(0.01) \times (0.07)}{(0.01) \times (0.07) + (1 - 0.04) \times (1 - 0.07)} \\ &\approx 0.00079. \end{aligned}$$

Therefore, test results of Y decreases the probability that a person carries the virus from 7% to 0.079%.

In summary, the probability that a person carries the virus has been changing as follows:

$$P(V_+|\text{Information}) \begin{cases} 3\% & \text{before } X \text{ and } Y \\ 7\% & \text{after } X_+ \text{ and before } Y \\ 0.079\% & \text{after } X_+ \text{ and } Y_- \end{cases}$$

Before making observations, one has prior information about the probability that the person carries the virus. After data with respect to test X have been observed, one updates the probability by computating the posterior probability. We also predicted the likely outcome of the following test Y results as positive. Finally, we again updated the probability that the person carries the virus by incorporating the test result of Y. Therefore, we can update the probability that a person carries the virus by properly taking into account all available information.

Also, notice that, given the information regarding positive result X_+ and the negative result Y_-, the same probability $P(V_+|X_+, Y_-)$ is obtained by considering both of the tests (X, Y) jointly. Noting that tests X and Y are independent, we have

$$P(X_+, Y_-|V_+) = P(X_+|V_+)P(Y_-|V_+) = 0.098,$$

and

$$P(X_+, Y_-|V_-) = P(X_+|V_-)P(Y_-|V_-) = 0.384.$$

From Bayes' theorem, we can also evaluate the probability $P(V_+|X_+, Y_-)$ as follows:

$$
\begin{aligned}
P(V_+|X_+, Y_-) &= \frac{P(Y_-, X_+|V_+)P(V_+)}{P(Y_-, X_+|V_+)P(V_+) + P(Y_-, X_+|V_-)P(V_-)} \\
&= \frac{(0.098) \times (0.03)}{(0.098) \times (0.03) + (0.384) \times (0.97)} \\
&\approx 0.00079.
\end{aligned}
$$

It indicates that the inference result given the available information, is the same, implemented sequentially or simultaneously.

This section described how the Bayes' theorem works in the inference on the conditional probability. The extension to an inference statistical model is straightforward and will be discussed in the next section.

2.3 Bayesian inference on statistical models

Suppose that we have a set of n observations $\boldsymbol{X}_n = \{\boldsymbol{x}_1, ..., \boldsymbol{x}_n\}$. To summarize data information, we can easily compute the sample mean, variance, kurtosis, skewness and other statistic values. In practical situations, it is difficult to obtain more precise information on the structure of a system or a process from a finite number of observed data. Therefore, one uses a parametric family of distributions with densities $\{f(\boldsymbol{x}|\boldsymbol{\theta}); \boldsymbol{\theta} \in \Theta \subset R^p\}$ to explore the nature of the system and to predict the system's future behavior. We derive statistical conclusions based on the assumed probability models.

In the frequentist approach, the unknown parameter vector $\boldsymbol{\theta}$ is a fixed value. One usually finds the maximum likelihood estimate by maximizing the likelihood function $f(\boldsymbol{X}_n|\boldsymbol{\theta})$. If the set of n observations $\boldsymbol{X}_n$ are independent, the likelihood function is given as the product of sampling density of each observation:

$$
f(\boldsymbol{X}_n|\boldsymbol{\theta}) = \prod_{\alpha=1}^{n} f(\boldsymbol{x}_\alpha|\boldsymbol{\theta}).
$$

The predictive density function $f(\boldsymbol{z}|\boldsymbol{\theta})$ for a future observation $\boldsymbol{z}$ can be constructed by replacing the unknown parameter vector $\boldsymbol{\theta}$ with the maximum likelihood estimate $\hat{\boldsymbol{\theta}}_{\rm MLE}$.

In the context of Bayesian approach, we use probability distributions to describe all relevant unknown quantities. In contrast to the frequentist approach, the unknown parameter vector $\boldsymbol{\theta}$ is regarded as a random variable. To describe the available knowledge, expert opinion, intuitions and beliefs, about the value of $\boldsymbol{\theta}$, usually, a prior probability distribution $\pi(\boldsymbol{\theta})$ over the parameter space Θ is prepared.

Let parameter $\boldsymbol{\theta}$ take only m values $\{\boldsymbol{\theta}_1, ..., \boldsymbol{\theta}_m\}$ with probabilities. $\{\pi(\boldsymbol{\theta}_1), ..., \pi(\boldsymbol{\theta}_m)\}$. Letting an event A_k be $\boldsymbol{\theta}$ takes $\boldsymbol{\theta} = \boldsymbol{\theta}_k$ and an event B be $\boldsymbol{X}_n$ is observed in the Bayes' theorem (2.1), it then follows from Bayes' theorem that all available information about the value of $\boldsymbol{\theta}$ is contained in the corresponding posterior distribution. We have

$$\pi(\boldsymbol{\theta} = \boldsymbol{\theta}_k|\boldsymbol{X}_n) = \frac{f(\boldsymbol{X}_n|\boldsymbol{\theta}_k)\pi(\boldsymbol{\theta}_k)}{\sum_{j=1}^{m} f(\boldsymbol{X}_n|\boldsymbol{\theta}_j)\pi(\boldsymbol{\theta}_j)},$$

where $f(\boldsymbol{X}_n|\boldsymbol{\theta})$ is the likelihood function. If the parameter $\boldsymbol{\theta}$ is a continuous random variable, then we have

$$\pi(\boldsymbol{\theta}|\boldsymbol{X}_n) = \frac{f(\boldsymbol{X}_n|\boldsymbol{\theta})\pi(\boldsymbol{\theta})}{\displaystyle\int f(\boldsymbol{X}_n|\boldsymbol{\theta})\pi(\boldsymbol{\theta})d\boldsymbol{\theta}} \propto f(\boldsymbol{X}_n|\boldsymbol{\theta})\pi(\boldsymbol{\theta}), \tag{2.2}$$

where the symbol $\propto$ means "proportional to," or equivalently, equal to a constant.

In the Bayesian framework, we often call the quantity $f(\boldsymbol{X}_n|\boldsymbol{\theta})\pi(\boldsymbol{\theta})$ a kernel. This is because multiplying some constant term Const. to the kernel, we again have the same posterior distribution.

$$\pi(\boldsymbol{\theta}|\boldsymbol{X}_n) = \frac{\text{Const.} \times f(\boldsymbol{X}_n|\boldsymbol{\theta})\pi(\boldsymbol{\theta})}{\displaystyle\int \text{Const.} \times f(\boldsymbol{X}_n|\boldsymbol{\theta})\pi(\boldsymbol{\theta})d\boldsymbol{\theta}} = \frac{f(\boldsymbol{X}_n|\boldsymbol{\theta})\pi(\boldsymbol{\theta})}{\displaystyle\int f(\boldsymbol{X}_n|\boldsymbol{\theta})\pi(\boldsymbol{\theta})d\boldsymbol{\theta}}.$$

Also, we call the term $f(\boldsymbol{X}_n|\boldsymbol{\theta})\pi(\boldsymbol{\theta})d\boldsymbol{\theta}$ in denominator, as a normalizing constant, or the marginal likelihood. As shown in a later chapter, this quantity plays an essential role in Bayesian model selection.

We can see that the posterior distribution depends on the prior distribution and the likelihood function. For example, it is obvious that any value of $\boldsymbol{\theta}$ with zero prior density has zero posterior density. Also, if the posterior is highly dependent on the likelihood function, then the posterior may not be unaffected by the prior distribution. Thus, a formulation of the sampling density for observations $\boldsymbol{X}_n$ and the prior density $\pi(\boldsymbol{\theta})$ is an important task in Bayesian statistical modeling.

To illustrate the Bayesian inference process, we provide the following example. Suppose that we have a set of n samples $\boldsymbol{X}_n = \{x_1, x_2, ..., x_n\}$ from the normal distribution with mean μ and known variance σ^2. The likelihood function is then

$$f(\boldsymbol{X}_n|\mu) = \prod_{\alpha=1}^{n} \frac{1}{(2\pi\sigma^2)^{1/2}} \exp\left[-\frac{(x_\alpha - \mu)^2}{2\sigma^2}\right].$$

For simplicity of explanation, let us specify the prior density for μ be just constant, i.e., $\pi(\mu) = \text{Const.}$. The posterior distribution, proportional to the likelihood time the prior, is then

$$\pi(\mu|\boldsymbol{X}_n) \propto f(\boldsymbol{X}_n|\mu)\pi(\mu) \propto \exp\left[\frac{n(x_\alpha - \bar{x}_n)}{2\sigma^2}\right],$$

with $\bar{x}_n = \sum_{\alpha=1}^n x_\alpha/n$ and we used the well known formulae,

$$\sum_{\alpha=1}^n (x_\alpha - \mu)^2 = \sum_{\alpha=1}^n (x_\alpha - \bar{x}_n + \bar{x}_n - \mu)^2 = \sum_{\alpha=1}^n (x_\alpha - \bar{x}_n)^2 + n(\bar{x}_n - \mu)^2.$$

Noting that the last term of $\pi(\mu|\boldsymbol{X}_n)$ is the kernel of normal distribution with the mean $\bar{x}_n$ and the variance σ^2/n, we know that the posterior density of μ is $N(\bar{x}_n, \sigma^2/n)$.

The following sections overview basic components of Bayesian inference on statistical models: a specification of sampling density and several prior density settings. We also describe how to summarize the posterior inference results, how to make prediction and so on.

2.4 Sampling density specification

In addition to the Non-Bayesian approach, a key component of Bayesian inference is an assumption on a certain class of statistical models, to which the data-generating mechanisms belong. One can specify the sampling density $f(\boldsymbol{x}|\boldsymbol{\theta})$ from various motivations and also from aspects. One of the most important principles for sampling density specification might be that your sampling density represents the real system generating the data $\boldsymbol{X}_n$.

If prior information about the sampling density is available, we can specify the sampling density based on such information. For example, experienced professionals related to a particular topic of our interest often have some intuitions and practical experiences. Usually, such information helps us to specify a sampling density. Also, we can employ a well known empirical phenomenon. One might specify the sampling density from a computational point of view. In this section, we provide several examples of sampling density specification.

2.4.1 Probability density specification

In Section 1.1, we fitted the normal distribution to the daily returns of Nikkei 225 index from August 28, 2001 to September 22, 2005. If we calculate the skewness and kurtosis, these statistics are given as 0.247 and 4.350, respectively. Noting that the kurtosis of the returns is above three, the true distribution of the transformed Nikkei 225 index data would be a fat-tailed distribution. In fact, there are some outliers in Figure 1.1. Note that, in addition to the normal distribution, we can also consider other sampling density specifications. Here, we consider to fit the Student-t distribution

$$f(x|\mu, \sigma^2, \nu) = \frac{\Gamma(\frac{\nu+1}{2})}{\Gamma(\frac{1}{2})\Gamma(\frac{\nu}{2})\sqrt{\nu\sigma^2}} \left\{1 + \frac{(x-\mu)^2}{\sigma^2\nu}\right\}^{-\frac{\nu+1}{2}}.$$

The Student-t distribution also contains the normal distribution by setting $\nu = \infty$. Thus Student-t model is more flexible than the normal distribution.

Figure 2.1 shows the fitted density function based on the maximum likelihood method. Histograms of data are also shown in the figure. In Figure 2.1, the dashed line represents the fitted normal density function. We can see that the fitted Student-t distribution has a fatter tail than that of normal density. This matches to the value of kurtosis. The maximum log-likelihood values are

$$\begin{aligned} \text{Normal distribution} &: \ -1746.61, \\ \text{Student}t \text{ distribution} &: \ -1739.23, \end{aligned}$$

respectively. Therefore, we can suspect that the true distribution of Nikkei 225 index return data is more close to the Student-t distribution than the normal.

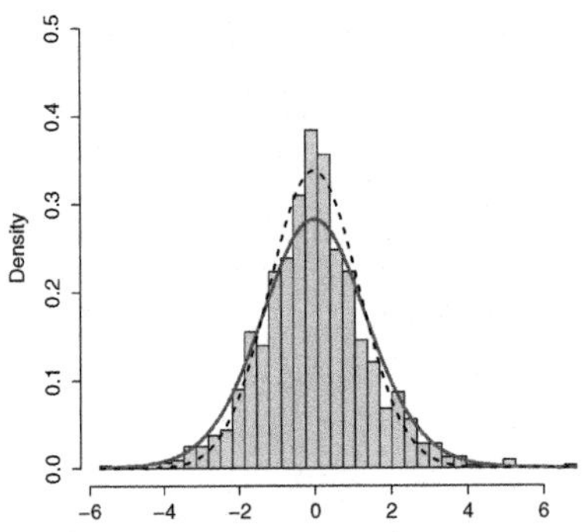

FIGURE 2.1: Histograms of Nikkei 225 index return data and the fitted Student-t density function based on the maximum likelihood method. The dashed line indicates the fitted normal density.

2.4.2 Econometrics: Quantifying price elasticity of demand

We provide an example of the sampling density specification based on the economic theory with a certain motivation. In business and economics studies, we often want to know the price elasticity of demand: the sensitivity measure of the quantity demanded (Q) for a product/service as a result of change in price (P) of the same product/service. Mathematically, the price elasticity (PE) of demand for a given product/service is defined as the percentage change in quantity demanded caused by a percent change in price:

$$\text{PE} = \frac{\Delta Q/Q}{\Delta P/P} = \frac{\%\Delta Q}{\%\Delta P}.$$

Intuitively, the price elasticity is generally negative due to the negative relationship between the price and quantity demanded.

To quantifying price elasticity of demand, one often can consider the log-linear demand curve

$$\log(Q) = \beta_0 + \beta_1 \log(P).$$

In a practical situation, note that one might also consider other factors, e.g., demand increase due to advertisement, effect from substitutes, effects from complements, income level, etc. Although one can put these factors into the log-linear demand curve model, we use the above model for simplicity. Noting that $\Delta \log(Q) \approx \%\Delta Q$ and $\Delta \log(P) \approx \%\Delta P$, one can quantify the price elasticity of demand as

$$\text{PE} = \frac{\%\Delta Q}{\%\Delta P} \approx \beta_1.$$

Thus, once one gets an estimator of β_1, the quantification of the price elasticity of demand can be done.

To estimate the model coefficients, we usually use a set of observations $\{(Q_\alpha, P_\alpha), \alpha = 1, ..., n$. Here Q_α is the observed quantity demanded and (P) is the observed price. Specifying the model as $\log(Q_\alpha) = \beta_0 + \beta_1 \log(P_\alpha) + \varepsilon_\alpha$ with normal error term $\varepsilon_\alpha \sim N(0, \sigma^2)$, the specified model can be estimated by using the Bayesian approach described in Section 2.7.

2.4.3 Financial econometrics: Describing a stock market behavior

This is an example that specifies the sampling density specification based on the empirical phenomenon and practical motivations. In a portfolio management, one often wants to explain the excess returns of a portfolio. One highly successful model is Fama-French's (1993) three-factor model. Observing empirically that the classes of market/value-weighted portfolios can explain the excess returns, they extended the Capital Asset Pricing Model (Sharpe (1964)) as

$$r_p - r_f = \alpha + \beta_1(r_m - r_f) + \beta_2\text{SMB} + \beta_3\text{HML} + \varepsilon,$$

where r_p is the portfolio's returns, r_f is the risk-free return rate, r_m is the return of the whole stock market, SMB (Small Minus Big) is the average return on the three small portfolios minus the average return on the three big portfolios, and HML (High Minus Low) is the average return on the two value portfolios minus the average return on the two growth portfolios. French's Data Library provides the details on these three factors as well as historical values of these factors. The model can also be estimated by the method using the Bayesian approach given in Section 2.7.

The three-factor model is currently used in various ways. For example, fund managers often employ this model to construct a portfolio. Also, based on the exposure to the factors, mutual funds are classified into several groups, which is useful information for investors. Moreover, this model can be used to evaluate the fund management skills.

A portfolio allocates wealth among various available assets. The standard mean-variance method of portfolio selection, pioneered by Markowitz (1952), has long attracted the attention of financial economics researchers and practitioners. In this context, an investor simply allocates wealth among m assets with weights $\boldsymbol{w} = (w_1, ..., w_m)^T$, over a one-period investment horizon. The optimal portfolio $\boldsymbol{w}$ is determined by solving the following problem:

$$\text{maximize} \ \ \boldsymbol{w}^T\boldsymbol{\mu} - \frac{\gamma}{2}\boldsymbol{w}^T\Sigma\boldsymbol{w}, \ \ \text{s.t.} \ \ \boldsymbol{w}^T\mathbf{1} = 1,$$

where $\boldsymbol{\mu}$ and Σ are the mean vector and covariance matrix of asset returns, respectively, γ is the investor's risk-aversion parameter, and $\mathbf{1}$ is a column vector of ones.

To quantify the covariance matrix of a set of m excess returns $\{r_{1p}, ..., r_{mp}\}$, one can consider the model

$$\begin{aligned}
&\begin{pmatrix} r_{1p} - r_f \\ \vdots \\ r_{mp} - r_f \end{pmatrix} \\
&= \begin{pmatrix} \alpha_1 \\ \vdots \\ \alpha_m \end{pmatrix} + \begin{pmatrix} \beta_{11} & \beta_{12} & \beta_{13} \\ \vdots & \ddots & \vdots \\ \beta_{m1} & \beta_{m2} & \beta_{m3} \end{pmatrix} \begin{pmatrix} r_m - r_f \\ \text{SMB} \\ \text{HML} \end{pmatrix} + \begin{pmatrix} \varepsilon_1 \\ \vdots \\ \varepsilon_m \end{pmatrix},
\end{aligned}$$

where $\boldsymbol{\varepsilon} = (\varepsilon_1, ..., \varepsilon_m)$ follows the multivariate normal with mean $\mathbf{0}$ and covariance matrix Σ. This model can be estimated by a Bayesian method described in Section 5.3.3.

Fund managers may want to use the different factors for different assets. As a more flexible model, the seemingly unrelated regression model (Zellner (1962)) is available. Bayesian analysis of this model is given in Section 4.2.4. See also Pastor (2000) and Pastor and Stambaugh (2000).

2.4.4 Bioinformatics: Tumor classification with gene expression data

With the recently developed microarray technology, we can measure thousands of genes' expression profiles simultaneously. In the bioinformatics field, a prediction of the tumor type of a new individual based on the gene expression profile is one of the most important research topics. Through the instrumentality of useful information included in gene expression profiles, a number of systematic methods to identify tumor types using gene expression data have been applied to tumor classification (see for example, Alon et al. (1999), Golub et al. (1999), Pollack et al. (1999), Veer and Jone (2002)).

When we consider the multiclass tumor classification, multinomial choice models provide a useful tool. In this model, a probability that an individual belongs to a tumor type k, $P(k|\boldsymbol{x})$, is expressed as

$$\Pr(k|\boldsymbol{x}) = \frac{\exp\left\{\boldsymbol{w}_k^T \boldsymbol{x}\right\}}{\sum_{j=1}^{J} \exp\left\{\boldsymbol{w}_j^T \boldsymbol{x}\right\}}, \qquad k = 1, \ldots, J,$$

where J is the number of tumor types, $\boldsymbol{x} = (x_1, ..., x_p)^T$ is p-dimensional gene expression profiles, $\boldsymbol{w}_k = (w_{k1}, \ldots, w_{kp})^T$ is a p-dimensional parameter vector. Since total of sum of probabilities $\{\Pr(1|\boldsymbol{x}), ..., \Pr(J|\boldsymbol{x})\}$ is one, only $G-1$ probabilities can be identified. For the identification of the model, normalization process is usually made, the restriction of one of the parameters is zero e.g., $\boldsymbol{w}_1 = \boldsymbol{0}$.

Given the gene expression profiles, we can predict a new individual's tumor type. Also, we can identify a set of genes that contribute the tumor classification. In Section 5.6.2, Bayesian inference procedure for multinomial logistic model with basis expansion predictors is provided. Section 7.5.2 applied the model to real gene expression data.

2.4.5 Psychometrics: Factor analysis model

Factor analysis is widely used in various research fields as a convenient tool for extracting information from multivariate data. The key idea behind factor analysis is that the observations are generated from some lower-dimensional structure, which is not directly observable. Factor analysis model is often used for intelligence studies, where the observables x_j $(j = 1, ..., p)$ are often examination scores (e.g., score of mathematics etc.), the factors are regarded as unobservable ability (e.g, verbal intelligence), and the factor loadings give a weight to each factor so that the model adjusts to the observed values x_j.

Suppose that there is a set of p observable random variables, $\boldsymbol{x} = (x_1, ..., x_p)^T$ with mean $\boldsymbol{\mu} = (\mu_1, ..., \mu_p)^T$. For any specified number of factors m, the factor analysis model assumes the following structure:

$$\begin{pmatrix} x_1 - \mu_1 \\ \vdots \\ x_p - \mu_p \end{pmatrix} = \begin{pmatrix} \lambda_{11} & \cdots & \lambda_{1m} \\ \vdots & \ddots & \vdots \\ \lambda_{pm} & \cdots & \lambda_{pm} \end{pmatrix} \begin{pmatrix} f_1 \\ \vdots \\ f_m \end{pmatrix} + \begin{pmatrix} \varepsilon_1 \\ \vdots \\ \varepsilon_p \end{pmatrix},$$

where $\boldsymbol{f} = (f_1, ..., f_m)$ is the m-dimensional common factor, λ_{ij} is the factor loadings (weight for each of the factors), and ε_i is the zero-mean noise.

Bayesian factor analysis methodology has received considerable attention in a wide variety of application areas such as economics, finance, psychology, and genome science. Bayesian factor analysis (Press and Sigemasu (1989, 1999)) is an useful tool to incorporate the prior information regarding parameter values into the model Press. Many studies have been conducted for the

Bayesian analysis of factor model (see for example, Aguilar and West (2000), Lopes and West (2004), Press and Shigemasu (1989), Ando (2009a)). Section 5.5.3 provides the Bayesian analysis of factor models.

2.4.6 Marketing: Survival analysis model for quantifying customer lifetime value

In marketing research, a quantification of customer lifetime value (CLV), the sum of the lifetime values of the company's customers, is one of the most important studies (see for example, Gupta and Donald (2003), Gupta, S. et al. (2006)). To illustrate an idea, consider a simple situation. Let $P(t)$ be a profit from a customer at time period t (price paid by the consumer at time t minus direct cost of servicing the customer at time t)C $S(t)$ be probability of customer repeat buying or being alive at time t, and $D(t)$ be the discount rate or cost of capital for a firm. Then CLV for a customer (without an acquisition cost) is given as

$$\mathrm{CLV} = \sum_{t=1}^{\infty} P(t) \times S(t) \times D(t),$$

where the discount rate $D(t)$ may be obtained from the financial market data, and $P(t)$ may be forecast by using the customer's past expenditures, economic variables and so on.

To quantify the term structure of $S(t)$, one may employ the survival models. Let T be a random variable that represents customer being alive at time T with probability density $f(t)$. Then the $S(t)$ can be expressed as

$$S(t) = \Pr(T > t) = \int_t^{\infty} f(x)dx.$$

This survival probability $S(t)$ can take into account customer's purchase history, demographic variables, and so on. Let us denote this information through the covariates $x_1, ..., x_p$. If the random variable T follows a Weibull distribution, then the $S(t)$ is

$$S\left(t|\boldsymbol{x}, \boldsymbol{\theta}\right) = \exp\left\{-t^{\alpha} \exp\left(\sum_{j=1}^{p} \beta_j x_j\right)\right\},$$

where $\boldsymbol{\theta} = (\alpha, \beta_1, ..., \beta_p)^T$ is the model parameter. Bayesian inference on this model is given in Section 5.5.4. There are many applications of the survival modeling methods in marketing research. For example, Chintagunta and Prasad (1998) jointly investigated the purchase timing and brand-choice decisions of households.

2.4.7 Medical science: Nonlinear logistic regression models

In various studies, we often see the nonlinear structure in the data. Hastie and Tibshirani (1990) investigated multiple level thoracic and lumbar laminectomy, in particular, a corrective spinal surgery performed in children for tumor and congenital/developmental abnormalities. The purpose of the study is to estimate the unknown incidence and nature of spinal deformities following the surgery and assess the importance of age at time of surgery, and the effect of the number and location of vertebrae levels decompressed. The data consist of 83 patients undergoing corrective spinal surgery. The response variable indicates kyphosis after the operation and was coded as either 1 (presence) or 0 (absence). The predictor variables are "age" in months at time of operation, the starting point of vertebra level involved in the operation "start" and the number of levels involved "number." Figure 2.2 shows a matrix of scatterplots.

There seems to be a nonlinear relationship between the kyphosis and the age. In order to investigate the nonlinear relationship between the kyphosis and the predictor age, one can consider nonlinear logistic regression with predictor variable. A probability is expressed as

$$\Pr(\text{kyphosis} = 1|\text{age}) = \frac{\exp\{h(\text{age})\}}{1 + \exp\{h(\text{age})\}},$$

where $h(\cdot)$ is some nonlinear function of the age. Several specification examples for $h(\cdot)$ can be found in Section 5.6.1. Inference on this model is given in Section 7.4.2.

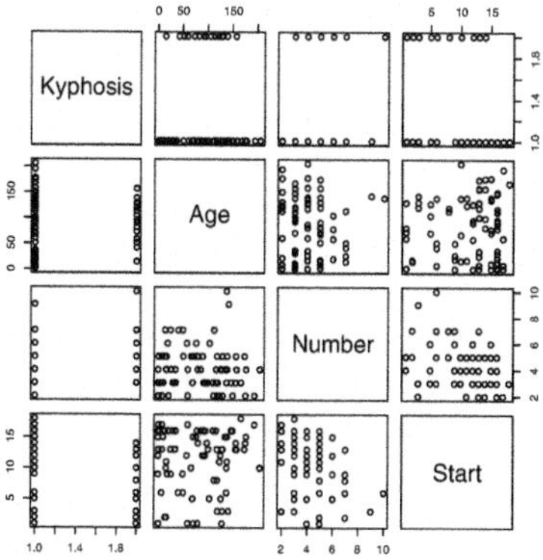

FIGURE 2.2: A matrix of scatterplots of kyphosis data.

2.4.8 Under the limited computer resources

If our computer resource is limited, one might specify the sampling density from a computational point of view. For example, it is well-known that the volatility of financial asset return changes randomly over time (Clark (1973)). When we focus on this information, the stochastic volatility (SV; Taylor (1982)) type models and autoregressive conditional heteroschedastisity (ARCH; Engle (1982)) type models would attract our attention.

It is known that a maximum likelihood estimation of SV-type models is very time consuming, because the SV-type volatility models generally specify the volatility as a stochastic process and the likelihood function depends upon high-dimensional integrals. On the other hand, ARCH-type volatility models specify the volatility of the current return as a nonstochastic function of past observations; it is easy to evaluate the likelihood function. Therefore, one might employ ARCH-type volatility models for ease of computation. Section 5.9.3.1 implements the Bayesian inference on GARCH(1,1) models.

2.5 Prior distribution

One of critical features of Bayesian analysis is the prior specification. Prior distribution represents the value of the parameter, a random realization from the prior distribution. This section describes several prior specifications.

2.5.1 Diffuse priors

One of the most common priors is the diffuse (uninformative, or flat) prior where the prior is specified as

$$\pi(\boldsymbol{\theta}) = \text{Const.}, \quad \boldsymbol{\theta} \in A.$$

This indicates that there is no a priori reason to favor any particular parameter value, while we just know its range $\boldsymbol{\theta} \in A$. Under the diffuse prior, the posterior distribution (2.2) is proportional to a constant times the likelihood,

$$\pi(\boldsymbol{\theta}|\boldsymbol{X}_n) \propto f(\boldsymbol{X}_n|\boldsymbol{\theta}), \quad \boldsymbol{\theta} \in A,$$

where $A \subset \Theta$ is the distributional range of $\boldsymbol{\theta}$. Therefore, the distributional range of $\boldsymbol{\theta}$ cannot go outside of range A.

If we specify region A as the parameter space $A = \Theta$, Bayesian analysis assuming a flat prior reduces to frequentist statistical analysis in many cases. Note also that, the normalizing constant term of the diffuse prior sometimes does not exist, i.e.,

$$\int \pi(\boldsymbol{\theta})d\boldsymbol{\theta} = \infty.$$

In such cases, the diffuse prior is referred to as an improper prior. Consider a diffuse prior for one dimensional parameter $\pi(\theta)$, $\theta \in A$. If the parameter of interest θ ranges over $A \in (-\infty, a)$, $A \in (b, \infty)$ or $A \in (-\infty, \infty)$ with constant values a and b, then the integral of the diffuse prior does not exist. Bayesian analysis of a Bernoulli distribution with a uniform prior is provided in Section 3.3.3.

2.5.2 The Jeffreys' prior

Jeffreys (1961) proposed a general rule for the choice of a noninformative prior. It is proportional to the square root of the determinant of the Fisher information matrix:

$$\pi(\boldsymbol{\theta}) \propto |J(\boldsymbol{\theta})|^{1/2}.$$

The Fisher information is given as

$$J(\boldsymbol{\theta}) = -\int \left[\frac{\partial^2 \log f(\boldsymbol{x}|\boldsymbol{\theta})}{\partial\boldsymbol{\theta}\partial\boldsymbol{\theta}^T}\right] f(\boldsymbol{x}|\boldsymbol{\theta})d\boldsymbol{x},$$

where the expactation is taken with respect to the sampling distribution of $\boldsymbol{x}$.

The Jeffreys' prior gives an automated method for finding a noninformative prior for any parametric model. Also, it is known that the Jeffreys' prior is invariant to transformation. For an alternate parameterization $\boldsymbol{\psi} = (r_1(\boldsymbol{\theta}), ..., r_p(\boldsymbol{\theta}))^T$, we again have $\pi(\boldsymbol{\psi}) \propto |J(\boldsymbol{\psi})|^{1/2}$. As an example, Section 4.2.4 provides Bayesian analysis of a seemingly unrelated regression model with the Jeffreys' prior.

2.5.3 Conjugate priors

A prior is conjugate for a family of distributions if the prior and the posterior are of the same family. The conjugate priors are useful because we always obtain the posterior distributions in analytical form. Thus there is a mathematical convenience. The exponential family includes many common distributions (normal, gamma, Poisson, binomial, etc.). A conjugate prior for the exponential family has the form of an exponential family. See for example, Section 2.7, where Bayesian inference on linear regression model with conjugate prior is illustrated.

2.5.4 Informative priors

An informative prior is a prior family that is not dominated by the likelihood, and thus has an impact on the posterior distribution. Gathering the available knowledge, expert opinion, intuitions and beliefs, an informative prior is designed. We can therefore interpret an informative prior as state of prior knowledge, because it expresses our prior knowledge. Since the elicitation process for a prior specification might be biased, one has to design an informative prior carefully. Section 2.8.2 illustrates the effect of informative priors on the inference results.

2.5.5 Other priors

We can also use other prior specifications from various perspectives. For example, we can use a truncated normal distribution for a particular parameter so that we can restrict the range of parameter regions but have some belief about its mean and variance. Zellner (1977) developed the maximal data information prior, which provides a general mathematical approach for deriving and justifying prior densities. In this framework, the prior is chosen to maximize the average information in the data density relative to that in the prior. For more details, we refer to Zellner (1996). Under the no (weak) prior knowledge, reference prior is also available (Bernardo (1979)).

2.6 Summarizing the posterior inference

Once we get a posterior distribution of parameter, we usually use some statistics that summarize the characteristics of the posterior distribution.

2.6.1 Point estimates

Point estimates are quantities of interest in the Bayesian analysis. There are mainly three point estimates, the posterior mean, the posterior mode and the posterior median. Although these estimators cannot obtain the full form of the posterior distribution, they provide useful information for us. For simplicity, let us consider the one dimensional parameter θ.

1. **Posterior mean**:

$$\bar{\theta}_n = \int \theta \pi(\theta|\boldsymbol{X}_n)d\theta.$$

2. **Posterior mode**:

$$\hat{\theta}_n = \text{argmax}_\theta \pi(\theta|\boldsymbol{X}_n).$$

3. **Posterior median**:

$$\tilde{\theta}_n \text{ such that } \int_{-\infty}^{\tilde{\theta}_n} \pi(\theta|\boldsymbol{X}_n)d\theta = 0.5.$$

Even if the analytical expression of the posterior distribution is not available, we can generally obtain the posterior samples by simulation approach.

2.6.2 Interval estimates

Generally, the point estimates are contrasted with interval estimates. In the Bayesian analysis, a $100(1-\alpha)\%$ posterior credible interval is given by any region R satisfying

$$\int_R \pi(\theta|\boldsymbol{X}_n)d\theta = 1-\alpha.$$

There are mainly two posterior credible intervals. The highest posterior density region and the equal-tailed posterior credible intervals.

1. **Highest posterior density region**: Mathematically, the highest posterior density region of context α for θ, R, is defined as follows:

$$\pi(\theta_a|\boldsymbol{X}_n) \geq \pi(\theta_b|\boldsymbol{X}_n), \quad \theta_a \in R, \ \theta_b \notin R,$$

with $\int_R \pi(\theta|\boldsymbol{X}_n)d\theta = 1-\alpha$. Intuitively, it is the region of values containing $100(1-\alpha)\%$ of the posterior probability and the density within the region is never lower than outside. For multiple parameters, the highest posterior density region is those with the smallest volume in the parameter space.

2. **Equal-tailed posterior credible intervals**: It is the range of values above and below which lies exactly $100(1-\alpha)\%$ of the posterior probability $[L_{\alpha/2}, R_{\alpha/2}]$ such that

$$\int_{-\infty}^{L_{\alpha/2}} \pi(\theta|\boldsymbol{X}_n)d\theta = \alpha/2 \quad \text{and} \quad \int_{R_{\alpha/2}}^{\infty} \pi(\theta|\boldsymbol{X}_n)d\theta = \alpha/2.$$

2.6.3 Densities

It often happens that we are concerned only on a subset of the unknown parameters. In such cases, Bayesian analysis removes the effects of the unconcerned parameters by simply integrating them out of the posterior distribution. This generates a marginal posterior distribution for the parameters of interest. Let the parameter vector $\boldsymbol{\theta}$ can be divided into two groups $\boldsymbol{\theta}_1$ and $\boldsymbol{\theta}_2$, $\boldsymbol{\theta} = (\boldsymbol{\theta}_1^T, \boldsymbol{\theta}_2^T)^T$. We are interested in just $\boldsymbol{\theta}_1$, but $\boldsymbol{\theta}_2$. Then the marginal posterior distribution for $\boldsymbol{\theta}_1$ is given as follows.

1. **Marginal posterior distribution**:

$$\pi(\boldsymbol{\theta}_1|\boldsymbol{X}_n) = \int \pi(\boldsymbol{\theta}_1, \boldsymbol{\theta}_2|\boldsymbol{X}_n)d\boldsymbol{\theta}_2.$$

Also, we often know the posterior distribution of $\boldsymbol{\theta}_1$ conditional on $\boldsymbol{\theta}_2$. In such cases, the conditional posterior distribution is used.

2. **Conditional posterior distribution**: We often can obtain the conditional posterior distribution in analytical form, while the analytical expression of the joint posterior distribution is not available. For example, the conditional posterior distribution of $\boldsymbol{\theta}_1$ given $\boldsymbol{\theta}_2$ is

$$\pi(\boldsymbol{\theta}_1|\boldsymbol{X}_n, \boldsymbol{\theta}_2^*) = \frac{\pi(\boldsymbol{\theta}_1, \boldsymbol{\theta}_2 = \boldsymbol{\theta}_2^*|\boldsymbol{X}_n)}{\int \pi(\boldsymbol{\theta}_1, \boldsymbol{\theta}_2 = \boldsymbol{\theta}_2^*|\boldsymbol{X}_n)d\boldsymbol{\theta}_1},$$

where the value of $\boldsymbol{\theta}_2^*$ is a fixed value.

2.6.4 Predictive distributions

Bayesian analysis is often interested in making inference about observables $\boldsymbol{X}_n$. There are mainly two important predictive distributions, the marginal likelihood and the predictive distribution.

1. **Marginal likelihood**:

$$P(\boldsymbol{X}_n) = \int f(\boldsymbol{X}_n|\boldsymbol{\theta})\pi(\boldsymbol{\theta})d\boldsymbol{\theta}.$$

This is the normalizing constant of the posterior distribution $\pi(\boldsymbol{\theta}|\boldsymbol{X}_n)$. As described in Chapter 5, the marginal likelihood plays a main role in Bayesian statistical analysis.

Once we obtain the data, we immediately obtain the posterior distribution $\pi(\boldsymbol{\theta}|\boldsymbol{X}_n)$. To predict future value of $\boldsymbol{z}$, we can use the predictive distribution.

2. **Predictive distribution**:

$$f(\boldsymbol{z}|\boldsymbol{X}_n) = \int f(\boldsymbol{z}|\boldsymbol{\theta})\pi(\boldsymbol{\theta}|\boldsymbol{X}_n)d\boldsymbol{\theta}.$$

This density is usually used for prediction.

2.7 Bayesian inference on linear regression models

Suppose we have n independent observations $\{(y_\alpha, \boldsymbol{x}_\alpha); \alpha = 1, 2, ..., n\}$, where y_α are random response variables and $\boldsymbol{x}_\alpha$ are vectors of p-dimensional explanatory variables. The problem to be considered is how to estimate the relationship between the response variable and the explanatory variables from the observed data. Generally, assuming the linear combination of the explanatory variables, one uses the Gaussian linear regression model

$$y_\alpha = \sum_{j=1}^{p} \beta_j x_{j\alpha} + \varepsilon_\alpha, \qquad \alpha = 1, ..., n, \tag{2.3}$$

where errors ε_α are independently, normally distributed with mean zero and variance σ^2. This model can be expressed in a matrix form

$$\boldsymbol{y}_n = X_n \boldsymbol{\beta} + \boldsymbol{\varepsilon}_n, \quad \boldsymbol{\varepsilon}_n \sim N(0, \sigma^2 I),$$

or equivalently, in a density form

$$f\left(\boldsymbol{y}_n | X_n, \boldsymbol{\beta}, \sigma^2\right) = \frac{1}{(2\pi\sigma^2)^{n/2}} \exp\left[-\frac{(\boldsymbol{y}_n - X_n\boldsymbol{\beta})^T(\boldsymbol{y}_n - X_n\boldsymbol{\beta})}{2\sigma^2}\right],$$

with

$$X_n = \begin{pmatrix} \boldsymbol{x}_1^T \\ \vdots \\ \boldsymbol{x}_n^T \end{pmatrix} = \begin{pmatrix} x_{11} & \cdots & x_{1p} \\ \vdots & \ddots & \vdots \\ x_{n1} & \cdots & x_{np} \end{pmatrix} \quad \text{and} \quad \boldsymbol{y}_n = \begin{pmatrix} y_1 \\ \vdots \\ y_n \end{pmatrix}.$$

Use a conjugate normal inverse-gamma prior $\pi(\boldsymbol{\beta}, \sigma^2) = \pi(\boldsymbol{\beta}|\sigma^2)\pi(\sigma^2)$ with

$$\begin{aligned} \pi(\boldsymbol{\beta}|\sigma^2) &= N\left(\boldsymbol{\beta}_0, \sigma^2 A^{-1}\right) = \frac{1}{(2\pi\sigma^2)^{p/2}} |A|^{1/2} \exp\left[-\frac{(\boldsymbol{\beta} - \boldsymbol{\beta}_0)^T A (\boldsymbol{\beta} - \boldsymbol{\beta}_0)}{2\sigma^2}\right], \\ \pi(\sigma^2) &= IG\left(\frac{\nu_0}{2}, \frac{\lambda_0}{2}\right) = \frac{\left(\frac{\lambda_0}{2}\right)^{\nu_0/2}}{\Gamma\left(\frac{\nu_0}{2}\right)} (\sigma^2)^{-\left(\frac{\nu_0}{2}+1\right)} \exp\left[-\frac{\lambda_0}{2\sigma^2}\right], \end{aligned}$$

which leads to the posterior distribution

$$\pi\left(\boldsymbol{\beta}, \sigma^2 \middle| \boldsymbol{y}_n, X_n\right) = \pi\left(\boldsymbol{\beta} \middle| \sigma^2, \boldsymbol{y}_n, X_n\right) \pi\left(\sigma^2 \middle| \boldsymbol{y}_n, X_n\right),$$

with

$$\pi\left(\boldsymbol{\beta} \middle| \sigma^2, \boldsymbol{y}_n, X_n\right) = N\left(\hat{\boldsymbol{\beta}}_n, \sigma^2 \hat{A}_n\right), \quad \pi\left(\sigma^2 \middle| \boldsymbol{y}_n, X_n\right) = IG\left(\frac{\hat{\nu}_n}{2}, \frac{\hat{\lambda}_n}{2}\right).$$

The conditional posterior distribution of $\boldsymbol{\beta}$ is normal, and the marginal posterior distribution of σ^2 is inverse-gamma distribution. Here

$$\begin{aligned} &\hat{\boldsymbol{\beta}}_n = \left(X_n^T X_n + A\right)^{-1} \left(X_n^T X_n \hat{\boldsymbol{\beta}}_{\text{MLE}} + A\boldsymbol{\beta}_0\right), \\ &\hat{\boldsymbol{\beta}}_{\text{MLE}} = \left(X_n^T X_n\right)^{-1} X_n^T \boldsymbol{y}_n, \quad \hat{A}_n = (X_n^T X_n + A)^{-1}, \\ &\hat{\nu}_n = \nu_0 + n, \\ &\hat{\lambda}_n = \lambda_0 + \left(\boldsymbol{y}_n - X_n \hat{\boldsymbol{\beta}}_{\text{MLE}}\right)^T \left(\boldsymbol{y}_n - X_n \hat{\boldsymbol{\beta}}_{\text{MLE}}\right) \\ &\qquad + \left(\boldsymbol{\beta}_0 - \hat{\boldsymbol{\beta}}_{\text{MLE}}\right)^T \left((X_n^T X_n)^{-1} + A^{-1}\right)^{-1} \left(\boldsymbol{\beta}_0 - \hat{\boldsymbol{\beta}}_{\text{MLE}}\right). \end{aligned}$$

Since we obtain a posterior distribution of parameter, we can summarize the characteristics of the posterior distribution: posterior mean, posterior mode, posterior median, $100(1-\alpha)$ posterior credible intervals, and so on.

Often, there is concern about a subset of the unknown parameters $\boldsymbol{\beta}$, while the variance parameter σ^2 is really of no concern to us. Simply integrating the variance parameter σ^2 out of the posterior distribution, we obtain the marginal posterior distribution for the parameter of $\boldsymbol{\beta}$. The marginal posterior distribution of $\boldsymbol{\beta}$ is

$$\begin{aligned}\pi\left(\boldsymbol{\beta}\middle|\boldsymbol{y}_n, X_n\right) &= \int \pi\left(\boldsymbol{\beta}, \sigma^2\middle|\boldsymbol{y}_n, X_n\right) d\sigma^2 \\ &\propto \left[1+\frac{1}{\hat{\nu}_n}\left(\boldsymbol{\beta}-\hat{\boldsymbol{\beta}}_n\right)^T\left(\frac{\hat{\lambda}_n}{\hat{\nu}_n}\hat{A}_n\right)^{-1}\left(\boldsymbol{\beta}-\hat{\boldsymbol{\beta}}_n\right)\right]^{-\frac{\hat{\nu}_n+p}{2}},\end{aligned}$$

which is the density of a p-dimensional Student-t distribution. Therefore, the marginal posterior distribution of $\boldsymbol{\beta}$ is Student-t distribution with

$$\begin{aligned}\text{Mean} &: \hat{\boldsymbol{\beta}}_n, \\ \text{Covariance} &: \frac{\hat{\lambda}_n}{\hat{\nu}_n-2}\hat{A}_n.\end{aligned}$$

We next obtain the predictive distribution for the future observation $\boldsymbol{z}_n$, given X_n. Integrating the probability density function $f\left(\boldsymbol{z}_n|X_n, \boldsymbol{\beta}, \sigma^2\right)$ with respect to the posterior distributions of $\boldsymbol{\beta}$ and σ^2, we obtain the predictive distribution for the future observation $\boldsymbol{z}_n$, given X_n. This turns out to be the multivariate Student-t distribution

$$\begin{aligned}f(\boldsymbol{z}_n|\boldsymbol{y}_n, X_n) &= \int f\left(\boldsymbol{z}_n|X_n, \boldsymbol{\beta}, \sigma^2\right)\pi\left(\boldsymbol{\beta}, \sigma^2|\boldsymbol{y}_n, X_n\right) d\boldsymbol{\beta} d\sigma^2 \\ &= \frac{\Gamma\left(\frac{\hat{\nu}_n+n}{2}\right)}{\Gamma\left(\frac{\hat{\nu}_n}{2}\right)(\pi\hat{\nu}_n)^{\frac{n}{2}}}|\Sigma_n^*|^{-\frac{1}{2}}\left\{1+\frac{1}{\hat{\nu}_n}(\boldsymbol{z}_n-\hat{\boldsymbol{\mu}}_n)^T\Sigma^{*-1}(\boldsymbol{z}_n-\hat{\boldsymbol{\mu}}_n)\right\}^{-\frac{\hat{\nu}_n+n}{2}}, \quad (2.4)\end{aligned}$$

where

$$\hat{\boldsymbol{\mu}}_n = X_n\hat{\boldsymbol{\beta}}_n, \quad \text{and} \quad \Sigma^* = \frac{\hat{\lambda}_n}{\hat{\nu}_n}\left(I+X_n\hat{A}_n^{-1}X_n^T\right).$$

The predictive mean and variance matrix of $\boldsymbol{z}_n$ given X_n is

$$\begin{aligned}\text{Mean} &: X_n\hat{\boldsymbol{\beta}}_n, \\ \text{Covariance} &: \frac{\hat{\lambda}_n}{\hat{\nu}_n-2}\left(I+X_n\hat{A}_n^{-1}X_n^T\right).\end{aligned}$$

See Zellner and Chetty (1965), who discussed how to incorporate parameter uncertainty into the predictive distribution.

Lastly, we obtain the marginal likelihood. Noting that the three terms (the prior, the likelihood and the posterior) in the posterior distribution are analytically available, we obtain the marginal likelihood as follows:

$$
\begin{aligned}
P\left(\boldsymbol{y}_n \middle| X_n, M\right) &= \frac{f\left(\boldsymbol{y}_n|X_n, \boldsymbol{\beta}, \sigma^2\right)\pi(\boldsymbol{\beta}, \sigma^2)}{\pi\left(\boldsymbol{\beta}, \sigma^2|\boldsymbol{y}_n, X_n\right)} \\
&= \frac{\left|\hat{A}_n\right|^{1/2}|A|^{1/2}\left(\frac{\lambda_0}{2}\right)^{\frac{\nu_0}{2}}\Gamma\left(\frac{\hat{\nu}_n}{2}\right)}{\pi^{\frac{n}{2}}\Gamma\left(\frac{\nu_0}{2}\right)}\left(\frac{\hat{\lambda}_n}{2}\right)^{-\frac{\hat{\nu}_n}{2}}. \qquad (2.5)
\end{aligned}
$$

Remark: We obtained the posterior distribution of parameters, the predictive distribution, and the marginal likelihood analytically. In practice, however, often the analytical expression of the posterior distribution as well as other related densities are not available. There are mainly two approaches to make practical inference on the posterior distribution of parameters. One is based on the asymptotic approximation, which relies on the asymptotic statistical theory. The other is simulation based approach, including Markov chain Monte Carlo (MCMC) methods. Several standard MCMC approaches, including Gibbs sampling, Metropolis-Hastings and reversible jump algorithms, are applicable to the Bayesian inference problems. Using these algorithms it is possible to generate posterior samples using computer simulation. Once we specify the prior distribution and likelihood function, under a certain condition, these two approaches estimate the posterior distribution of the parameters, the predictive distribution, and the marginal likelihood with other related quantities.

In the next section, we discuss the importance of model selection using this normal regression as an example.

2.8 Bayesian model selection problems

The task of selecting a statistical model from a set of competing models is a crucial issue in various fields of study. A main goal of this section is to clarify the essential points in constructing statistical models. This section, the concepts of over-fitting and under-fitting will also be described throgh the Bayesian linear regression model with a conjugate prior.

2.8.1 Example: Subset variable selection problem

Regression analysis is one of the most popular and widely used statistical techniques. The main purpose of regression analysis is to explore the relationship between the explanatory variable and response variable.

A set of random samples $\{(x_{1\alpha}, ..., x_{5\alpha}, y_\alpha); \alpha = 1, ..., 30\}$ are generated from the true model

$$y_\alpha = -0.25x_{1\alpha} + 0.5x_{2\alpha} + \varepsilon_\alpha, \quad \alpha = 1, ..., 30,$$

where the noises ε_α are generated from the normal with mean 0, and the standard deviation $s = 0.3$ and $x_{j\alpha}$ $j = 1, ..., 5$ are uniformly distributed within $[-2, 2]$.

We estimate the true function $h(\boldsymbol{x}) = -0.25x_1 + 0.5x_2$ by using the Bayesian linear regression model with a conjugate prior. Especially, we shall consider the following three models:

$$\begin{aligned}
M_1 &: \ y_\alpha = \beta_1 x_{1\alpha} + \varepsilon_\alpha, \\
M_2 &: \ y_\alpha = \beta_1 x_{1\alpha} + \beta_2 x_{2\alpha} + \varepsilon_\alpha, \\
M_3 &: \ y_\alpha = \beta_1 x_{1\alpha} + \beta_2 x_{2\alpha} + \cdots + \beta_5 x_{5\alpha} + \varepsilon_\alpha,
\end{aligned}$$

where ε_α are generated from the normal with mean 0 and the variance σ^2. Note that model M_1 lacks the true predictor x_2, while model M_2 is the correct specification. Moreover, model M_3 contains superfluous predictors $x_3 \sim x_5$. This type of identification problem is called Subset variable selection problem. To implement the Bayesian inference, the normal inverse-gamma prior $\pi(\boldsymbol{\beta}, \sigma^2) = \pi(\boldsymbol{\beta}|\sigma^2)\pi(\sigma^2) = N(\mathbf{0}, \sigma^2 A)IG(a, b)$ in example 5 is used. In this experiment, we set $A = 10^{-5} \times I_p$ and $a = b = 10^{-10}$, which makes the prior diffused.

Figure 2.3 compares the true surface $h(\boldsymbol{x})$ and predicts future value of $\boldsymbol{z}_n$ given X_n based on the models $M_1 \sim M_3$. To predict the future value of $\boldsymbol{z}_n$ given X_n, we used the predictive mean $\hat{\boldsymbol{\mu}}_n$ of the predictive distribution (2.4). The fitted surface based on model M_1 cannot capture the true surface because model M_1 doesn't have enough model flexibility to capture the true surface. We call this phenomenon "under-fitting." On the other hand, as shown in the predict future value based on model M_3, a model with unnecessary predictors results in the complicated surface. It should be reemphasized that the fitting surface with a suitable predictor, the model M_2, can capture the true structure very well.

Table 2.1 summarizes the marginal likelihood value, the training error (TE)

$$\mathrm{TE} = \frac{1}{30} \sum_{\alpha=1}^{30} \left\{ y_\alpha - \hat{\mu}_\alpha \right\}^2$$

and the prediction error (PE)

$$\mathrm{PE} = \frac{1}{30} \sum_{\alpha=1}^{30} \left\{ h(\boldsymbol{x}_\alpha) - \hat{\mu}_\alpha \right\}^2$$

for various models. In addition to $M_1 \sim M_3$, we also fit the following models.

$$\begin{aligned}
M_4 &: \ y_\alpha = \beta_1 x_{1\alpha} + \beta_3 x_{3\alpha} + \varepsilon_\alpha, \\
M_5 &: \ y_\alpha = \beta_2 x_{2\alpha} + \beta_4 x_{4\alpha} + \beta_5 x_{5\alpha} + \varepsilon_\alpha, \\
M_6 &: \ y_\alpha = \beta_3 x_{3\alpha} + \beta_4 x_{4\alpha} + \beta_5 x_{5\alpha} + \varepsilon_\alpha.
\end{aligned}$$

Here $\hat{\mu}_\alpha$ is the predictive mean value for the α-th observations. We can see from Table 2.1 that the training error becomes smaller as the number of predictors becomes larger. On the other hand, the prediction error for the model M_3 is not the minimum value, while that of the training error is minimum. It is obvious that the choice of an optimal set of predictors is an essential issue. Therefore, we need some model evaluation criteria from a certain point of view, e.g., from a predictive point of view.

Note also that the marginal likelihood value of the model M_2 is minimum among the fitted models. As shown in the next section, the marginal likelihood provides a useful tool for evaluating statistical models. It has many attractive properties in practical applications. In the next chapter, we discuss more details of the marginal likelihood and related concepts.

TABLE 2.1: The number of predictors, the training error (TE) TE $= \sum_{\alpha=1}^{n} \{y_\alpha - \hat{\mu}_\alpha\}^2 / n$, the prediction error (PE) PE $= \sum_{\alpha=1}^{n} \{h(\boldsymbol{x}_\alpha) - \hat{\mu}_\alpha\}^2 / n$ and the log-marginal likelihood value (Log-ML), for the models $M_1 \sim M_6$.

Model	Predictors	TE	PE	Log-ML
True	$(\boldsymbol{x}_1, \boldsymbol{x}_2)$	–	–	–
M_1	$(\boldsymbol{x}_1)$	0.5644	0.4523	-54.4016
M_2	$(\boldsymbol{x}_1, \boldsymbol{x}_2)$	0.0822	0.0013	-33.2596
M_3	$(\boldsymbol{x}_1, \boldsymbol{x}_2, x_3, x_4, x_5)$	0.0665	0.0171	-52.5751
M_4	$(\boldsymbol{x}_1, x_3)$	0.4974	0.3533	-60.0235
M_5	$(\boldsymbol{x}_2, x_4, x_5)$	0.1078	0.0635	-45.1262
M_6	(x_3, x_4, x_5)	0.4812	0.3614	-67.3170

2.8.2 Example: Smoothing parameter selection problem

Although the linear regression model would be a useful and convenient technique, in practical situations, the assumption that the response variables depend linearly on the explanatory variable is not always guaranteed. In social and natural sciences, it is often the case that a theory describes a functional relationship between a response variable and explanatory variables is more complicated rather than that expressed by the linear regression models.

Many researchers therefore have been moving away from linear functions and model the dependence of y on $\boldsymbol{x}$ in a more nonlinear fashion. For example, let us believe that the relationship between the response and the one quantitative predictor follows a particular functional relationship; say the 15-th polynomial function

$$y_\alpha = \beta_1 x_\alpha + \beta_2 x_\alpha^2 + \cdots + \beta_{15} x_\alpha^{15} + \varepsilon_\alpha, \quad \alpha = 1, ..., n.$$

Assuming that the error terms ε_α follow normal with mean 0 and variance σ^2,

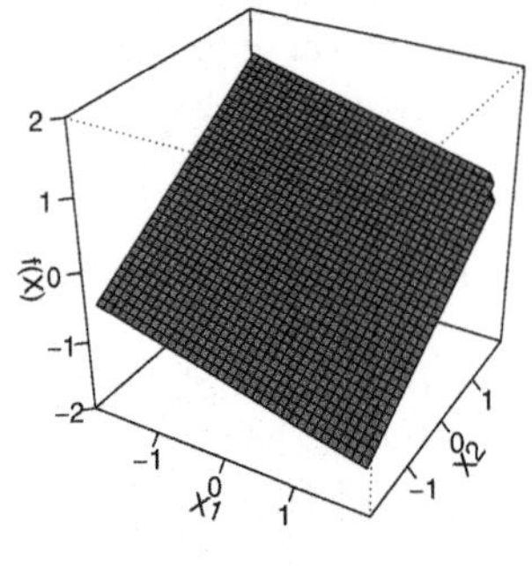

(a): True surface $h(\boldsymbol{x})$.

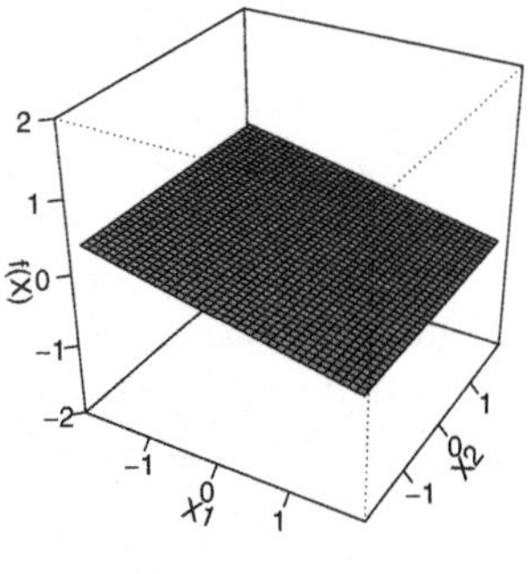

(b): M_1.

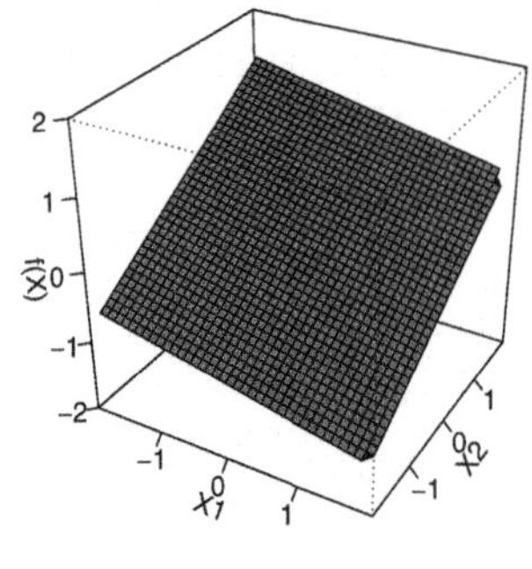

(c): M_2.

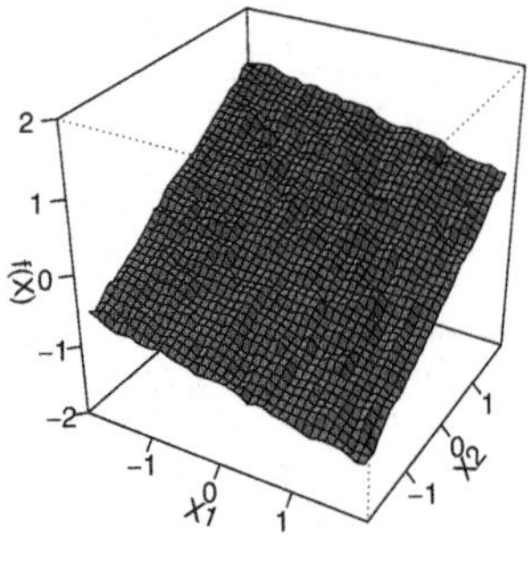

(d): M_3.

FIGURE 2.3: Comparison of the true surface $f(\boldsymbol{x}) = -0.25x_1 + 0.5x_2$ and the predicted surfaces based on the models $M_1 \sim M_3$.

this model can be expressed as a density form

$$f\left(\boldsymbol{y}_n|X_n, \boldsymbol{\beta}, \sigma^2\right) = \frac{1}{(2\pi\sigma^2)^{n/2}} \exp\left[-\frac{(\boldsymbol{y}_n - B_n\boldsymbol{\beta})^T(\boldsymbol{y}_n - B_n\boldsymbol{\beta})}{2\sigma^2}\right],$$

with

$$B_n = \begin{pmatrix} x_1 & x_1^2 & \cdots & x_1^{15} \\ x_2 & x_2^2 & \cdots & x_2^{15} \\ \vdots & \vdots & \ddots & \vdots \\ x_n & x_n^2 & \cdots & x_n^{15} \end{pmatrix} \quad \text{and} \quad \boldsymbol{y}_n = \begin{pmatrix} y_1 \\ \vdots \\ y_n \end{pmatrix}.$$

This model is called the polynomial regression model. Replacing the design matrix X_n with B_n, we can use the analytical result of linear regression model in (2.3).

We generated a set of $n = 50$ data $\{y_\alpha, x_\alpha; \alpha = 1, ..., 50\}$ from the true model

$$y_\alpha = 0.3\cos(\pi x_\alpha) + 0.5\sin(2\pi x_\alpha) + \varepsilon_\alpha,$$

where the design points are uniformly distributed in $[-1, 1]$ and the error terms ε_α follows the normal with mean 0 and variance $\sigma^2 = 0.2$.

We consider fitting the 15-th polynomial regression model. To implement the Bayesian inference, the normal inverse-gamma prior $\pi(\boldsymbol{\beta}, \sigma^2) = \pi(\boldsymbol{\beta}|\sigma^2)\pi(\sigma^2) = N(\mathbf{0}, \sigma^2 A)IG(a, b)$ in example 5 is used. In this experiment, we set $a = b = 10^{-10}$, which make the prior on σ^2 to be diffuse. Setting $A = \lambda I_{15}$, we considered the following prior settings on the coefficients

$$\begin{aligned} M_1 &: \ \boldsymbol{\beta} \sim N\left(\mathbf{0}, \sigma^2 \times 100,000 I_{15}\right), \\ M_2 &: \ \boldsymbol{\beta} \sim N\left(\mathbf{0}, \sigma^2 \times 1,000 I_{15}\right), \\ M_3 &: \ \boldsymbol{\beta} \sim N\left(\mathbf{0}, \sigma^2 \times 10 I_{15}\right), \\ M_4 &: \ \boldsymbol{\beta} \sim N\left(\mathbf{0}, \sigma^2 \times 0.1 I_{15}\right), \end{aligned}$$

where λ is a smoothing parameter, which has an effect on the posterior variances of the regression coefficients and also the predictive distribution of y. Also, I_{15} is the 15 dimensional unit diagonal matrix. When we set the value of smoothing parameter to be $\lambda = 100,000$, the prior variance of $\boldsymbol{\beta}$ is very large. It indicates that the resulting prior distribution is diffuse. On the other hand, if we set $\lambda = 0.1$, we have strong confidence that the regression coefficients are close to zero.

Figure 2.4 shows the predicted future value of $\boldsymbol{z}_n$ given X_n based on the models $M_1 \sim M_4$. The fitting curve corresponding to a relatively large value of smoothing parameter $\lambda = 100,000$ is obviously under smoothed. The fitting curve is capturing the noise. On the other hand, a too small smoothing parameter value $\lambda = 0.1$ gives nearly a linear curve. It should be reemphasized that the fitting curve with a suitable smoothing parameter can capture the true structure very well.

2.8.3 Summary

As we have seen through these two examples, the predictive performance of the estimated Bayesian model depends on our model speciation. Therefore, there has recently been substantial interest in the Bayesian model selection problem. In the Bayesian approach for model selection, model evaluation procedures have relied on Bayes factors and its extensions. The Bayes factor (Kass and Raftery (1995)) has played a major role in the evaluation of the goodness of the Bayesian models. Despite its popularity, the Bayes factor has come under increasing criticism. The main weakness of the Bayes factor is its sensitivity to the improper prior distribution. In particular, a noninformative prior may lead to the severe situation in which the Bayes factor is not well-defined. These topics are covered in Chapter 5.

Finally, more information about the Bayesian analysis can be found in many books, including Albert (2007), Bauwens et al. (1999), Berger (1985), Bernardo and Smith (1994), Box and Tiao (1973), Carlin and Louis (2000), Chen et al. (2000), Congdon (2001, 2007), Gelman et al. (1995), Geweke

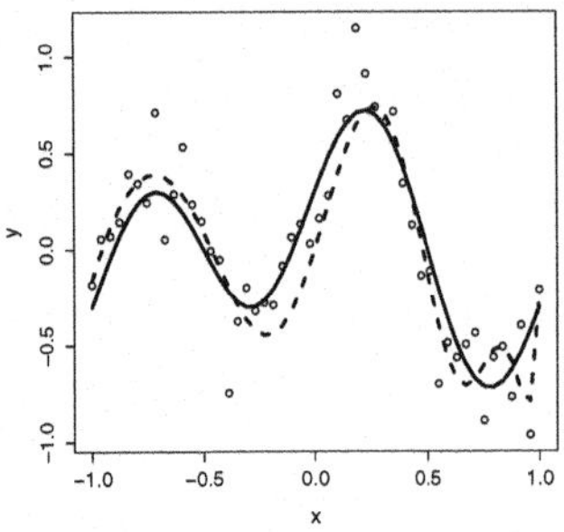

(a): M_1.

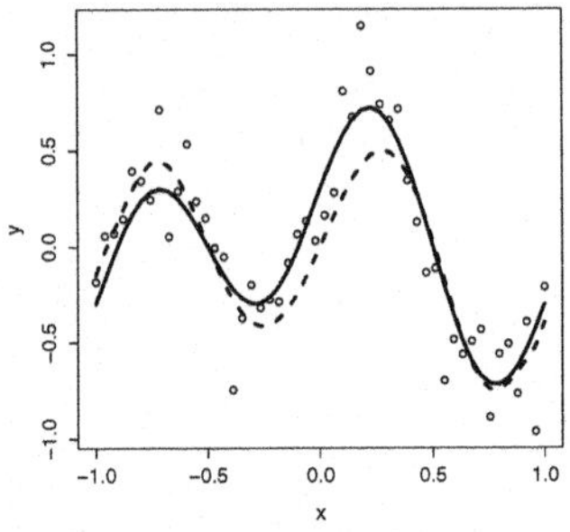

(b): M_2.

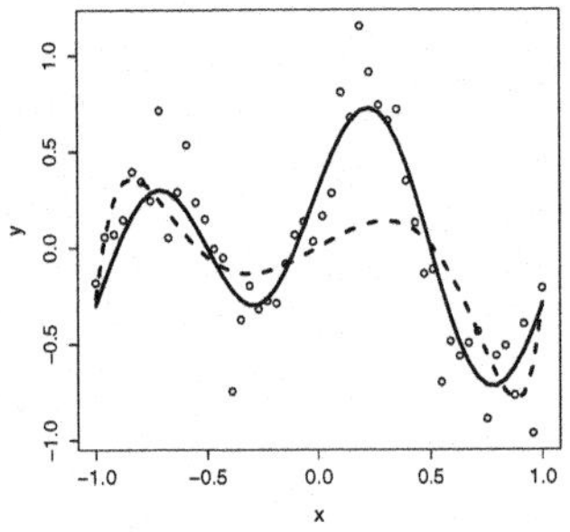

(c): M_3.

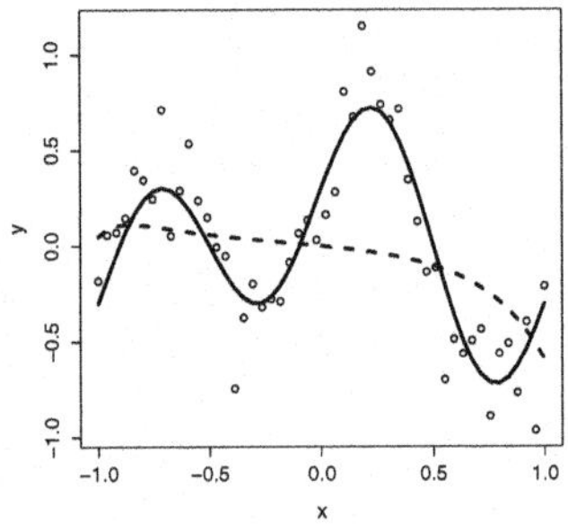

(d): M_4.

FIGURE 2.4: Comparison of the true curve $f(x) = 0.3\cos(\pi x) + 0.5\sin(2\pi x)$ and the predicted curves based on the models $M_1 \sim M_4$. To predict future value of $\boldsymbol{z}$ given X_n, we used the predictive mean $\hat{\boldsymbol{\mu}}_n$ of the predictive distribution.

(2005), Ibrahim et al. (2007), Koop (2003), Koop et al. (2007), Lancaster (2004), Lee (2004), Lee (2007), Liu (1994), Pole et al. (2007), Press (2003), Robert (2001), Sivia (1996) and Zellner (1971).

Exercises

1. *In Section 2.2 we illustrated the subjective probability as a conditional measure of uncertainty along with Bayes' theorem. Change the settings given below and reanalyze the problem.*

a. 10% of people from human population are infected by a particular virus.

b. A person is subject to the test X which is known to yield positive results in 95% of infected people and in 20% of noninfected.

c. A person takes the test Y, which relates to the probability that the person carries the virus, yields positive results in 95% of infected people and in 20% of noninfected.

2. *Suppose that $\boldsymbol{X}_n = \{x_1, x_2, ..., x_n\}$ are a set of independent random draws from the same Bernoulli distribution with parameter p. Thus, $y_n = \sum_{\alpha=1}^{n} x_\alpha$ follows draws from a binomial distribution with parameter n and p. If there is no a priori reason to favor any particular parameter value, a reasonable prior distribution for p must be bounded between zero and one $\pi(p) = \text{Const.}$, $p \in (0,1)$. Show that the posterior distribution of p is the beta density*

$$\pi(p|\boldsymbol{X}_n) = \frac{\Gamma(n+2)}{\Gamma(n-y_n+1)\Gamma(y_n+1)} p^{y_n}(1-p)^{n-y_n},$$

with parameter $y_n + 1$ and $n - y_n + 1$.

3. *Consider again that $\boldsymbol{X}_n = \{x_1, x_2, ..., x_n\}$ are a set of independent random draws from the same Bernoulli distribution with parameter p, and thus $y_n = \sum_{\alpha=1}^{n} x_\alpha$ follows draws from a binomial distribution with parameter n and p. Show that the Jeffreys's prior becomes*

$$\pi(p) \propto \left[\frac{n}{p(1-p)}\right]^{1/2} \propto p^{-1/2}(1-p)^{-1/2},$$

which is a Beta distribution with parameter 0.5 and 0.5.

4. *We have a set of n independent samples $\boldsymbol{X}_n = \{x_1, x_2, ..., x_n\}$ from the Bernoulli distribution with parameter p. Show that a conjugate prior for p, a binomial distribution, is a Beta distribution.*

5. *We have a set of n independent samples $\boldsymbol{X}_n = \{x_1, x_2, ..., x_n\}$ from the Poisson distribution with parameter λ. Show that a conjugate prior for λ is a Gamma distribution with parameter (α, β), $\text{Ga}(\alpha, \beta)$. Also, show that the posterior is $\text{Ga}(\alpha + \sum_{\alpha=1}^{n} x_\alpha, \beta + n)$.*

6. *Suppose that we have a set of n independent samples x_1, x_2,...,x_n from the normal distribution with mean μ and known variance $\sigma^2 = 5$, so that the likelihood function is*

$$f(\boldsymbol{X}_n|\mu) = \prod_{\alpha=1}^{n} \frac{1}{\sqrt{2\pi 5}} \exp\left\{-\frac{1}{2}\frac{(x_\alpha - \mu)^2}{5}\right\}.$$

Assuming the normal prior $\pi(\mu)$ with mean zero and variance s^2, then the posterior distribution is

$$\pi(\mu|\boldsymbol{X}_n) \propto \exp\left\{-\frac{1}{2}\frac{n}{5}\left(\bar{x}_n - \mu\right)^2\right\} \times \exp\left\{-\frac{1}{2}\frac{1}{s^2}\mu^2\right\},$$

with $\bar{x}_n = n^{-1}\sum_{\alpha=1}^{n} x_\alpha$ the sample mean.

Changing the values of the sample size n and the prior variance s^2, investigate the behavior of the posterior density.

7. *Section 2.7 considered the Bayesian analysis of the linear regression models with a conjugate normal inverse-gamma prior. Show the following identities*

$$\begin{aligned}
&(\boldsymbol{y}_n - X_n\boldsymbol{\beta})^T(\boldsymbol{y}_n - X_n\boldsymbol{\beta}) \\
&= \left(\boldsymbol{y}_n - X_n\hat{\boldsymbol{\beta}}_{\text{MLE}}\right)^T\left(\boldsymbol{y}_n - X_n\hat{\boldsymbol{\beta}}_{\text{MLE}}\right) \\
&\quad + \left(\boldsymbol{\beta} - \hat{\boldsymbol{\beta}}_{\text{MLE}}\right)^T X_n^T X_n \left(\boldsymbol{\beta} - \hat{\boldsymbol{\beta}}_{\text{MLE}}\right)
\end{aligned}$$

and

$$\begin{aligned}
&\left(\boldsymbol{\beta} - \hat{\boldsymbol{\beta}}_{\text{MLE}}\right)^T X_n^T X_n \left(\boldsymbol{\beta} - \hat{\boldsymbol{\beta}}_{\text{MLE}}\right) + (\boldsymbol{\beta} - \boldsymbol{\beta}_0)^T A(\boldsymbol{\beta} - \boldsymbol{\beta}_0) \\
&= \left(\boldsymbol{\beta}_0 - \hat{\boldsymbol{\beta}}_{\text{MLE}}\right)^T \left((X_n^T X_n)^{-1} + A^{-1}\right)^{-1}\left(\boldsymbol{\beta}_0 - \hat{\boldsymbol{\beta}}_{\text{MLE}}\right) \\
&\quad + \left(\boldsymbol{\beta} - \hat{\boldsymbol{\beta}}_n\right)^T \hat{A}_n^{-1}\left(\boldsymbol{\beta} - \hat{\boldsymbol{\beta}}_n\right),
\end{aligned}$$

where $\hat{\boldsymbol{\beta}}_n$, $\hat{\boldsymbol{\beta}}_{\text{MLE}}$, and $\hat{A}_n$ are given in Section 2.7.

8. *(Continued.) Joint posterior density of $\boldsymbol{\beta}$ and σ^2 is expressed as*

$$\begin{aligned}
&\pi\left(\boldsymbol{\beta}, \sigma^2 \middle| \boldsymbol{y}_n, X_n\right) \propto f\left(\boldsymbol{y}_n | X_n, \boldsymbol{\beta}, \sigma^2\right)\pi(\boldsymbol{\beta}, \sigma^2) \\
&\propto \frac{1}{(\sigma^2)^{n/2}}\exp\left[-\frac{(\boldsymbol{y}_n - X_n\boldsymbol{\beta})^T(\boldsymbol{y}_n - X_n\boldsymbol{\beta})}{2\sigma^2}\right] \\
&\quad \times \frac{1}{(\sigma^2)^{(p+\nu_0)/2+1}}\exp\left[-\frac{\lambda_0 + (\boldsymbol{\beta} - \boldsymbol{\beta}_0)^T A(\boldsymbol{\beta} - \boldsymbol{\beta}_0)}{2\sigma^2}\right].
\end{aligned}$$

Using the identities in the above problem, show that

$$\begin{aligned}
\pi\left(\boldsymbol{\beta}, \sigma^2 \middle| \boldsymbol{y}_n, X_n\right) &\propto \left|\sigma^2 \hat{A}_n\right|^{-1/2} \exp\left[-\frac{\left(\boldsymbol{\beta} - \hat{\boldsymbol{\beta}}_n\right)^T \hat{A}_n^{-1}\left(\boldsymbol{\beta} - \hat{\boldsymbol{\beta}}_n\right)}{2\sigma^2}\right] \\
&\quad \times \frac{1}{(\sigma^2)^{\hat{\nu}_n/2+1}}\exp\left[-\frac{\hat{\lambda}_n}{2\sigma^2}\right],
\end{aligned}$$

where $\hat{\nu}_n$ and $\hat{\lambda}_n$ are given in Section 2.7.

9. *Generate a set of* $n = 100$ *random samples* $\{(x_{1\alpha}, ..., x_{8\alpha}, y_\alpha); \alpha = 1, ..., 100\}$ *from the true model*

$$y_\alpha = -1.25x_{1\alpha} + 4.5x_{4\alpha} + 4.5x_{7\alpha} + \varepsilon_\alpha, \quad \alpha = 1, ..., 100,$$

where the noises ε_α *are generated from the normal with mean* 0 *and the standard deviation* $s = 0.9$ *and* $x_{j\alpha}$ $j = 1, ..., 8$ *are uniformly distributed within* $[-5, 5]$.

Implement the Bayesian linear regression model with a conjugate prior (the normal inverse-gamma prior $\pi(\boldsymbol{\beta}, \sigma^2) = \pi(\boldsymbol{\beta}|\sigma^2)\pi(\sigma^2) = N(\mathbf{0}, \sigma^2 A)IG(a, b)$*) through the following six models:*

$$\begin{aligned}
M_1 &: \; y_\alpha = \beta_1 x_{1\alpha} + \varepsilon_\alpha, \\
M_2 &: \; y_\alpha = \beta_1 x_{1\alpha} + \beta_4 x_{4\alpha} + \varepsilon_\alpha, \\
M_3 &: \; y_\alpha = \beta_1 x_{1\alpha} + \beta_4 x_{4\alpha} + \beta_5 x_{5\alpha} + \varepsilon_\alpha, \\
M_4 &: \; y_\alpha = \beta_1 x_{1\alpha} + \beta_4 x_{4\alpha} + \beta_7 x_{7\alpha} + \varepsilon_\alpha, \\
M_5 &: \; y_\alpha = \beta_1 x_{1\alpha} + \beta_4 x_{4\alpha} + \beta_7 x_{7\alpha} + \beta_8 x_{8\alpha} + \varepsilon_\alpha, \\
M_6 &: \; y_\alpha = \beta_1 x_{1\alpha} + \beta_2 x_{2\alpha} + \cdots + \beta_8 x_{8\alpha} + \varepsilon_\alpha,
\end{aligned}$$

where ε_α *are generated from the normal with mean* 0 *and the variance* σ^2. *When implementing the analysis, set* $A = 10^5 \times I_p$ *and* $a = b = 10^{-10}$, *which make the prior to be diffused.*

Similar to Table 2.1, summarize the marginal likelihood value, the training error and the prediction error for various models.

10. *Generate a set of* $n = 100$ *observations* $\{y_\alpha, x_\alpha; \alpha = 1, ..., 100\}$ *from the true model*

$$y_\alpha = \cos(\pi x_\alpha) + 0.5x_\alpha^2 + \varepsilon_\alpha,$$

where the design points are uniformly distributed in $[-2, 1]$ *and the error terms* ε_α *follow the normal with mean 0 and variance* $\sigma^2 = 0.5$.

Implement the Bayesian inference on the $p = 10$*-th order polynomial regression model*

$$y_\alpha = \beta_0 + \beta_1 x_\alpha + \cdots + \beta_p x_\alpha^p + \varepsilon_\alpha, \quad \alpha = 1, ..., n,$$

with the normal inverse-gamma prior $\pi(\boldsymbol{\beta}, \sigma^2) = \pi(\boldsymbol{\beta}|\sigma^2)\pi(\sigma^2) = N(\mathbf{0}, \sigma^2 A)IG(a, b)$.

Setting $a = b = 10^{-10}$ *and* $A = \lambda I$, *consider the following prior settings on the coefficients*

$$\begin{aligned}
M_1 &: \; \boldsymbol{\beta} \sim N\left(\mathbf{0}, \sigma^2 \times 10{,}000I\right), \\
M_2 &: \; \boldsymbol{\beta} \sim N\left(\mathbf{0}, \sigma^2 \times 100I\right), \\
M_3 &: \; \boldsymbol{\beta} \sim N\left(\mathbf{0}, \sigma^2 \times 1I\right), \\
M_4 &: \; \boldsymbol{\beta} \sim N\left(\mathbf{0}, \sigma^2 \times 0.01I\right).
\end{aligned}$$

where λ is a smoothing parameter, which have an effect on the posterior variances of the regression coefficients and also the predictive distribution of y. Similar to Table 2.1, summarize the marginal likelihood value, the training error and the prediction error for various models.

Chapter 3

Asymptotic approach for Bayesian inference

3.1 Asymptotic properties of the posterior distribution

We study consistency and asymptotic normality of posterior distributions. We also provide the Bayesian central limit theorem, which shows that the posterior distributions concentrate in neighborhoods of a certain parameter point and can be approximated by an appropriate normal distribution.

3.1.1 Consistency

In this section, we describe the consistency of the Bayesian parameter estimators. Let $\boldsymbol{\theta}_0$ be the mode of the expected penalized log-likelihood function

$$\int \{\log f(\boldsymbol{x}|\boldsymbol{\theta}) + \log \pi_0(\boldsymbol{\theta})\} \, g(\boldsymbol{x}) d\boldsymbol{x},$$

with $\log \pi_0(\boldsymbol{\theta}) = \lim_{n\to\infty} n^{-1} \log \pi(\boldsymbol{\theta})$. Similarly, let $\hat{\boldsymbol{\theta}}_n$ be the posterior mode, the mode of the empirical penalized log-likelihood function

$$n^{-1} \log\{f(\boldsymbol{X}_n|\boldsymbol{\theta})\pi(\boldsymbol{\theta})\},$$

which is obtained by replacing the unknown distribution $G(\boldsymbol{x})$ by the empirical distribution based on the observed data $\boldsymbol{X}_n = \{\boldsymbol{x}_1, ..., \boldsymbol{x}_n\}$. It follows from the law of large numbers that

$$n^{-1} \log\{f(\boldsymbol{X}_n|\boldsymbol{\theta})\pi(\boldsymbol{\theta})\} \to \int \log\{f(\boldsymbol{x}|\boldsymbol{\theta})\pi_0(\boldsymbol{\theta})\} dG(\boldsymbol{x})$$

as n tends to infinity. Then

$$\hat{\boldsymbol{\theta}}_n \to \boldsymbol{\theta}_0$$

in probability as n tends to infinity. Thus, the posterior mode is consistent for $\boldsymbol{\theta}_0$.

Consider the case where $\log \pi(\boldsymbol{\theta}) = O_p(1)$. Then $n^{-1} \log \pi(\boldsymbol{\theta}) \to 0$ as $n \to \infty$ and the prior information can be ignored for a sufficiently large n. In this case, the mode $\boldsymbol{\theta}_0$ is the pseudo parameter value, which minimizes the Kullback-Leibler (Kullback and Leibler (1951)) distance between the specified model $f(\boldsymbol{x}|\boldsymbol{\theta})$ and the true model $g(\boldsymbol{x})$. Under the model specified situation, i.e., $g(\boldsymbol{x}) = f(\boldsymbol{x}|\boldsymbol{\theta}_t)$ for some $\boldsymbol{\theta}$, the mode $\boldsymbol{\theta}_0$ reduces to the true parameter value $\boldsymbol{\theta}_0 = \boldsymbol{\theta}_t$.

Next consider the case $\log \pi(\boldsymbol{\theta}) = O_p(n)$, i.e., the prior information grows with the sample size. Then $\log \pi_0(\boldsymbol{\theta}) = O_p(1)$, and the prior information cannot be ignored even when the sample size n is large. In this case, $\boldsymbol{\theta}_0$ is neither the pseudo parameter value, nor the true parameter value. However, even in this case, the posterior mode $\hat{\boldsymbol{\theta}}_n$ converges to $\boldsymbol{\theta}_0$.

3.1.2 Asymptotic normality of the posterior mode

Under certain regularity conditions, the distribution of the posterior mode approaches the normal distribution as the sample size increases. In this section, we give the asymptotic normality of the posterior mode.

Asymptotic normality of the posterior mode

Suppose that $\boldsymbol{X}_n$ *are independent observations from* $g(\boldsymbol{x})$. *Let* $f(\boldsymbol{x}|\boldsymbol{\theta})$ *be the specified parametric model, which does not necessarily contain the true model generating the data* $g(x)$. *Let* $\pi(\boldsymbol{\theta})$ *be the prior density which may be improper.* $\sqrt{n}(\hat{\boldsymbol{\theta}}_n - \boldsymbol{\theta}_0)$ *is asymptotically normally distributed as* $N\{\boldsymbol{0}, S^{-1}(\boldsymbol{\theta}_0)Q(\boldsymbol{\theta}_0)S^{-1}(\boldsymbol{\theta}_0)\}$, *where* $Q(\boldsymbol{\theta})$ *and* $S(\boldsymbol{\theta})$ *are the* $p \times p$ *matrices, respectively, defined by*

$$Q(\boldsymbol{\theta}) = \int \frac{\partial \log\{f(\boldsymbol{x}|\boldsymbol{\theta})\pi_0(\boldsymbol{\theta})\}}{\partial \boldsymbol{\theta}} \frac{\partial \log\{f(\boldsymbol{x}|\boldsymbol{\theta})\pi_0(\boldsymbol{\theta})\}}{\partial \boldsymbol{\theta}^T} dG(\boldsymbol{x}),$$

$$S(\boldsymbol{\theta}) = -\int \frac{\partial^2 \log\{f(\boldsymbol{x}|\boldsymbol{\theta})\pi_0(\boldsymbol{\theta})\}}{\partial \boldsymbol{\theta} \partial \boldsymbol{\theta}^T} dG(\boldsymbol{x}).$$

Outline of the Proof. Since $\hat{\boldsymbol{\theta}}_n$ is the mode of $f(\boldsymbol{X}_n|\boldsymbol{\theta})\pi(\boldsymbol{\theta})$, it satisfies the score equation

$$\left. \frac{\partial [\log\{f(\boldsymbol{X}_n|\boldsymbol{\theta})\pi(\boldsymbol{\theta})\}]}{\partial \boldsymbol{\theta}} \right|_{\boldsymbol{\theta}=\hat{\boldsymbol{\theta}}_n} = \boldsymbol{0}.$$

Taylor expansion leads to

$$-\frac{1}{n} \left. \frac{\partial^2 \log\{f(\boldsymbol{X}_n|\boldsymbol{\theta})\pi(\boldsymbol{\theta})\}}{\partial \boldsymbol{\theta} \partial \boldsymbol{\theta}^T} \right|_{\boldsymbol{\theta}=\boldsymbol{\theta}_0} \sqrt{n}(\hat{\boldsymbol{\theta}}_n - \boldsymbol{\theta}_0)$$
$$= \frac{1}{\sqrt{n}} \left. \frac{\partial \log\{f(\boldsymbol{X}_n|\boldsymbol{\theta})\pi(\boldsymbol{\theta})\}}{\partial \boldsymbol{\theta}} \right|_{\boldsymbol{\theta}=\boldsymbol{\theta}_0} + O_p\left(\frac{1}{\sqrt{n}}\right).$$

It follows from the central limit theorem that the right-hand side is asymptotically distributed as $N\{\mathbf{0}, Q(\boldsymbol{\theta}_0)\}$, i.e.,

$$\sqrt{n} \times \frac{1}{n} \frac{\partial \log\{f(\boldsymbol{X}_n|\boldsymbol{\theta})\pi(\boldsymbol{\theta})\}}{\partial \boldsymbol{\theta}}\bigg|_{\boldsymbol{\theta}=\boldsymbol{\theta}_0} \rightarrow N(\mathbf{0}, Q(\boldsymbol{\theta}_0)).$$

By the law of large numbers, when $n \rightarrow \infty$, it can be shown that

$$-\frac{1}{n} \frac{\partial^2 \log\{f(\boldsymbol{X}_n|\boldsymbol{\theta})\pi(\boldsymbol{\theta})\}}{\partial \boldsymbol{\theta} \partial \boldsymbol{\theta}^T}\bigg|_{\boldsymbol{\theta}=\boldsymbol{\theta}_0} \rightarrow S(\boldsymbol{\theta}_0).$$

We obtain

$$S(\boldsymbol{\theta}_0)\sqrt{n}(\hat{\boldsymbol{\theta}}_n - \boldsymbol{\theta}_0) \rightarrow N(\mathbf{0}, Q(\boldsymbol{\theta}_0)).$$

Therefore, the convergence in law

$$\sqrt{n}(\hat{\boldsymbol{\theta}}_n - \boldsymbol{\theta}_0) \rightarrow N\left(\mathbf{0}, S^{-1}(\boldsymbol{\theta}_0)Q(\boldsymbol{\theta}_0)S^{-1}(\boldsymbol{\theta}_0)\right)$$

holds as $n \rightarrow \infty$. Thus we obtained the desired result.

3.1.3 Example: Asymptotic normality of the posterior mode of logistic regression

Let $y_1, ..., y_n$ be binary random variables taking values of 0 and 1 with conditional probabilities

$$\Pr(y_\alpha = 1|x_\alpha) := \pi(x_\alpha) = \frac{\exp\{\beta_0 + \beta_1 x_\alpha\}}{1 + \exp\{\beta_0 + \beta_1 x_\alpha\}}, \tag{3.1}$$

where x_α are explanatory variables. The likelihood function for $\boldsymbol{\beta} = (\beta_0, \beta_1)^T$ is

$$\begin{aligned} f(\boldsymbol{y}_n|X_n; \boldsymbol{\beta}) &= \prod_{\alpha=1}^n \left[\frac{\exp\{\beta_0 + \beta_1 x_\alpha\}}{1 + \exp\{\beta_0 + \beta_1 x_\alpha\}}\right]^{y_\alpha} \left[\frac{1}{1 + \exp\{\beta_0 + \beta_1 x_\alpha\}}\right]^{1-y_\alpha} \\ &= \exp\left[\sum_{\alpha=1}^n [y_\alpha(\beta_0 + \beta_1 x_\alpha) - \log\{1 + \exp(\beta_0 + \beta_1 x_\alpha)\}]\right], \end{aligned}$$

where $\boldsymbol{y}_n = \{y_1, ..., y_n\}$ and $X_n = \{x_1, ..., x_n\}$.

We specify a normal prior with mean zero and diagonal covariance λI for $\boldsymbol{\beta}$:

$$\pi(\boldsymbol{\beta}) \propto \exp\left[-\frac{\beta_0^2 + \beta_1^2}{2\lambda}\right].$$

The posterior mode $\hat{\boldsymbol{\beta}}_n$ can be found by maximizing the penalized log-likelihood function

$$\begin{aligned}&\log\{f(\boldsymbol{y}_n|X_n,\boldsymbol{\beta})\pi(\boldsymbol{\beta})\}\\&=\left[\sum_{\alpha=1}^{n}y_\alpha(\beta_0+\beta_1x_\alpha)-\log\{1+\exp(\beta_0+\beta_1x_\alpha)\}\right]-\frac{\beta_0^2+\beta_1^2}{2\lambda}.\end{aligned}$$

Although this optimization process with respect to unknown parameter $\boldsymbol{\beta}$ is nonlinear, the mode $\hat{\boldsymbol{\beta}}_n$ is obtained by the Fisher scoring iterations (see for example, Green and Silverman (1994)). Starting the initial avlue $\boldsymbol{\beta}^{(0)}$, we update parameter vector

$$\boldsymbol{\beta}^{new}=\left(X_n^TWX_n+\lambda^{-1}I_2\right)^{-1}X_n^TW\boldsymbol{\zeta},$$

until a suitable convergence criterion is satisfied. Here W is an $n\times n$ diagonal matrix and $\boldsymbol{\zeta}$ is an n-dimensional vector, respectively, given as

$$X_n=\begin{pmatrix}1 & x_1\\ \vdots & \vdots\\ 1 & x_n\end{pmatrix},\quad \boldsymbol{\zeta}=\begin{pmatrix}\frac{y_1-\pi(x_1)}{\pi(x_1)(1-\pi(x_1))}+\beta_0+\beta_1x_1\\ \vdots\\ \frac{y_n-\pi(x_n)}{\pi(x_n)(1-\pi(x_n))}+\beta_0+\beta_1x_n\end{pmatrix},$$

$$W=\begin{pmatrix}\pi(x_1)(1-\pi(x_1)) & & \\ & \ddots & \\ & & \pi(x_n)(1-\pi(x_n))\end{pmatrix},$$

Setting the true parameter value $\boldsymbol{\beta}=(0.5,0.35)^T$, we repeatedly generated a set of n independent binary observations $\{y_1,...,y_n\}$ with true conditional probabilities $\pi(x_\alpha)$ given in (3.1). The predictors are eually palced on the space between $[0,1]$. Under various settings of n, we estimated the posterior mode over 200 repeated Monte Carlo trials $\{\hat{\boldsymbol{\beta}}_n^{(1)},...,\hat{\boldsymbol{\beta}}_n^{(200)}\}$. We set the prior variance matrix to be diffuse by setting $\lambda=1,000$. Figure 3.1 shows the empirical density function of the estimated posterior mode. Here we used the **R** functions **persp**, **KernSmooth** and **MASS**. When the sample size is small $n=5$, the variance of the estimated posterior mode is very large. However, we can see that the constructed density for the estimated posterior mode is shrinking to the normal density with mean true value as the sample becomes larger.

3.2 Bayesian central limit theorem

Under a large sample situation, we can approximate the Bayesian posterior distribution of the model parameters by multivariate normal distribution.

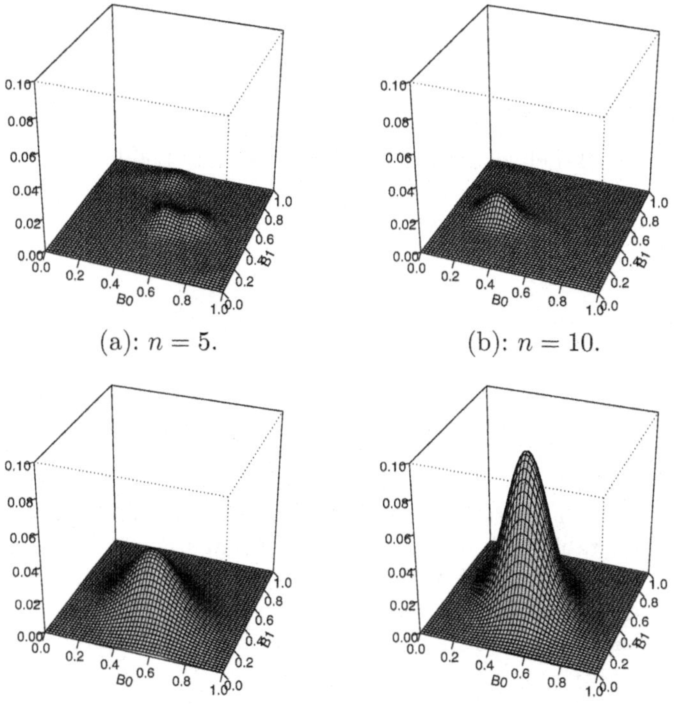

(a): $n = 5$. (b): $n = 10$.

(c): $n = 25$. (d): $n = 50$.

FIGURE 3.1: Comparison of the constructed density function of the posterior mode $\hat{\boldsymbol{\beta}}_n$ under the various sample sizes n. Under a diffuse prior setting $\lambda = 0.00001$, the posterior mode was repeatedly estimated.

This large-sample approximation is guaranteed by the Bayesian central limit theorem. The Bayesian central limit theorem is the simplest, convenient and workable approximation of the posterior distribution. In this section, we discuss the details of this theorem.

3.2.1 Bayesian central limit theorem

The following assumptions (A1–A9) similar to those of White (1982) are used to proof the asymptotic normality.

A1: The independent observations in $\boldsymbol{X}_n$ have common joint distribution function G on a measurable Euclidean space, and G has a measurable Radon-Nikodym density $g = dG/d\nu$. Note that G is unknown a priori, therefore, we use a family of distribution functions $F(\boldsymbol{x}|\boldsymbol{\theta})$ as an approximation of the true model G.

A2: The family of distribution functions $F(\boldsymbol{x}|\boldsymbol{\theta})$ has Radon-Nikodym densities

$f(\boldsymbol{x}|\boldsymbol{\theta}) = dF(\boldsymbol{x}|\boldsymbol{\theta})/d\nu$ which are measurable in $\boldsymbol{x}$ and continuous for every $\boldsymbol{\theta}$ in a compact subset of a p-dimensional Euclidean space.

A3: $E[\log g(\boldsymbol{x})] = \int \log g(\boldsymbol{x}) g(\boldsymbol{x}) d\boldsymbol{x}$ exists and $|\log f(\boldsymbol{x}|\boldsymbol{\theta})| \leq w(\boldsymbol{x})$ for all $\boldsymbol{\theta}$ in a compact subset of a p-dimensional Euclidean space, where $w(\boldsymbol{x})$ is integrable with respect to true distribution G.

A4: The Kullback-Leibler information (Kullback and Leibler (1951)) between $g(\boldsymbol{x})$ and $f(\boldsymbol{x}|\boldsymbol{\theta})$ has a unique minimum in a compact subset of a p-dimensional Euclidean space.

A5: The quantities

$$\left| \frac{\partial^2 \{\log f(\boldsymbol{x}|\boldsymbol{\theta}) + \log \pi_0(\boldsymbol{\theta})\}}{\partial \theta_i \partial \theta_j} \right|, \quad \left| \frac{\partial^3 \{\log f(\boldsymbol{x}|\boldsymbol{\theta}) + \log \pi_0(\boldsymbol{\theta})\}}{\partial \theta_i \partial \theta_j \partial \theta_k} \right|,$$

$$\left| \frac{\partial \{\log f(\boldsymbol{x}|\boldsymbol{\theta}) + \log \pi_0(\boldsymbol{\theta})\}}{\partial \theta_i} \cdot \frac{\partial \{\log f(\boldsymbol{x}|\boldsymbol{\theta}) + \log \pi_0(\boldsymbol{\theta})\}}{\partial \theta_j} \right|, \quad \text{and}$$

$i, j, k = 1, ..., p$, exist and these quantities are dominated by some functions integrable with respect to G for all $\boldsymbol{x}$ and $\boldsymbol{\theta}$.

A6: $\partial\{\log f(\boldsymbol{x}|\boldsymbol{\theta}) + \log \pi_0(\boldsymbol{\theta})\}/\partial \theta_i$, $i = 1, \ldots, p$, are measurable functions of $\boldsymbol{x}$ for each $\boldsymbol{\theta}$ and continuously differentiable functions of $\boldsymbol{\theta}$ for each $\boldsymbol{x}$.

A7: The value of $\boldsymbol{\theta}_0$ is interior to a compact subset of a p-dimensional Euclidean space.

A8: The $p \times p$ expected Hessian matrix

$$S(\boldsymbol{\theta}) = -\int \left[\frac{\partial^2 \{\log f(\boldsymbol{x}|\boldsymbol{\theta}) + \log \pi_0(\boldsymbol{\theta})\}}{\partial \boldsymbol{\theta} \partial \boldsymbol{\theta}^T} \right] g(\boldsymbol{x}) d\boldsymbol{x}$$

is nonsingular.

A9: $\boldsymbol{\theta}_0$ is a regular point of the expected Hessian matrix. In other words, S has constant rank in some open neighborhood of $\boldsymbol{\theta}$ (White (1982)).

Under these assumptions 1–9, we have the Bayesian central limit theorem.

Bayesian central limit theorem

Suppose that $\boldsymbol{X}_n$ are independent observations from $g(\boldsymbol{x})$. Let $f(\boldsymbol{x}|\boldsymbol{\theta})$ be the specified parametric model, which does not necessarily contain the true model generating the data $g(\boldsymbol{x})$. Let $\pi(\boldsymbol{\theta})$ be the prior density which may be improper. Further assume that the posterior distribution is proper.

Then, under a large sample situation, we can approximate the Bayesian posterior distribution of the model parameters $\pi(\boldsymbol{\theta}|\boldsymbol{X}_n)$ by multivariate normal distribution with mean the posterior mode $\hat{\boldsymbol{\theta}}_n$ and covariance matrix $n^{-1} S_n^{-1}(\hat{\boldsymbol{\theta}}_n)$ with

$$S_n(\hat{\boldsymbol{\theta}}_n) = -\frac{1}{n} \frac{\partial^2 \log\{f(\boldsymbol{X}_n|\boldsymbol{\theta})\pi(\boldsymbol{\theta})\}}{\partial \boldsymbol{\theta} \partial \boldsymbol{\theta}^T} \bigg|_{\boldsymbol{\theta} = \hat{\boldsymbol{\theta}}_n}.$$

Outline of the Proof. Noting that the first derivative of the empirical penalized log-likelihood function $n^{-1}\log\{f(\boldsymbol{X}_n|\boldsymbol{\theta})\pi(\boldsymbol{\theta})\}$ evaluated at the posterior mode $\hat{\boldsymbol{\theta}}_n$ equals zero, we have the following Taylor expansion of the posterior distribution around the posterior mode:

$$\pi(\boldsymbol{\theta}|\boldsymbol{X}_n) = \exp\left\{\log\pi(\hat{\boldsymbol{\theta}}_n|\boldsymbol{X}_n) - \frac{n}{2}(\boldsymbol{\theta}-\hat{\boldsymbol{\theta}}_n)^T S_n(\hat{\boldsymbol{\theta}}_n)(\boldsymbol{\theta}-\hat{\boldsymbol{\theta}}_n) + O_p\left(\frac{1}{\sqrt{n}}\right)\right\}.$$

Note that, the first term $\log\pi(\hat{\boldsymbol{\theta}}_n|\boldsymbol{X}_n)$ doesn't involve $\boldsymbol{\theta}$, so it is absorbed into the normalizing constant. The third term $O_p(n^{-1/2})$ becomes negligible as $n \to \infty$. Therefore, we obtain

$$\pi(\boldsymbol{\theta}|\boldsymbol{X}_n) \approx \exp\left\{-\frac{n}{2}(\boldsymbol{\theta}-\hat{\boldsymbol{\theta}}_n)^T S_n(\hat{\boldsymbol{\theta}}_n)(\boldsymbol{\theta}-\hat{\boldsymbol{\theta}}_n)\right\} \quad \text{as} \quad n \to \infty.$$

Note that this is the kernel of multivariate normal density centered at $\hat{\boldsymbol{\theta}}_n$ and covariance matrix $n^{-1}S_n^{-1}(\hat{\boldsymbol{\theta}}_n)$. This gives rise to the approximation. $\pi(\boldsymbol{\theta}|\boldsymbol{X}_n) \approx N(\hat{\boldsymbol{\theta}}_n, n^{-1}S_n^{-1}(\hat{\boldsymbol{\theta}}_n))$.

Some Remarks. The Bayesian central limit theorem says that the posterior samples $\boldsymbol{\theta}$ from the posterior distribution $\pi(\boldsymbol{\theta}|\boldsymbol{X}_n)$ are in the neighborhood of the posterior mode $\hat{\boldsymbol{\theta}}_n$ as $n \to \infty$. Since the posterior mode $\hat{\boldsymbol{\theta}}_n$ is consistent for $\boldsymbol{\theta}_0$, the posterior samples become close to $\boldsymbol{\theta}_0$. The value of $\boldsymbol{\theta}_0$ depends on the order of $\log\pi(\boldsymbol{\theta})$.

Although the large sample results are not necessary for making inference, they are useful as tools for understanding and as approximations to the posterior distribution. When sample size is large (relative to the number of parameters), this approximation is a reasonable way to proceed. However, note that this approximation is accurate when the number of data points is large relative to the number of parameters.

If the true posterior distribution is a multivariate Student-t distribution, the accuracy of this approximation on the tails may not be inaccurate. This is because a multivariate Student-t distribution can be much heavier than those of a multivariate normal distribution.

3.2.2 Example: Poisson distribution with conjugate prior

Assume that a set of n independent observations $\boldsymbol{X}_n = \{x_1, ..., x_n\}$ are generated from a Poisson distribution with parameter λ. We already know that the conjugate prior of the Poisson distribution is the gamma distribution with parameter α and β. As a result, we again obtain a gamma posterior

$$\pi\left(\lambda\middle|\boldsymbol{X}_n\right) \propto \lambda^{n\bar{x}_n+\alpha-1}\exp\{-\lambda(n+\beta)\}$$

with parameter $n\bar{x}_n + \alpha$ and $n + \beta$. The posterior mode is found by solving

$$\begin{aligned}\frac{\partial \log \pi\left(\lambda \middle| \boldsymbol{X}_n\right)}{\partial \lambda} &= \frac{\partial}{\partial \lambda}\left[(n\bar{x}_n + \alpha - 1)\log \lambda - \lambda(n+\beta)\right] \\ &= \frac{n\bar{x}_n + \alpha - 1}{\lambda} - (n+\beta) = 0.\end{aligned}$$

We then yield

$$\hat{\lambda}_n = \frac{n\bar{x}_n + \alpha - 1}{(n+\beta)}.$$

Calculating the negative Hessian of $\log\{f(\boldsymbol{X}_n|\lambda)\pi(\lambda)\}$ evaluating at the posterior mode

$$\begin{aligned}S_n(\hat{\lambda}_n) &= -\frac{1}{n}\frac{\partial^2 \log\{f(\boldsymbol{X}_n|\lambda)\pi(\lambda)\}}{\partial \lambda^2}\Bigg|_{\lambda=\hat{\lambda}_n} \\ &= \frac{1}{n}\frac{n\bar{x}_n + \alpha - 1}{\lambda^2}\Bigg|_{\lambda=\hat{\lambda}_n} \\ &= \frac{1}{n}\frac{(n+\beta)^2}{n\bar{x}_n + \alpha - 1},\end{aligned}$$

we can approximate the posterior distribution $\pi\left(\lambda \middle| \boldsymbol{X}_n\right)$ by the normal with mean $\hat{\lambda}_n$ and variance $n^{-1}S_n^{-1}(\hat{\lambda}_n)$.

We generated a set of n random samples from the Poisson distribution with parameter $\lambda = 4$. Under a diffuse prior setting $\alpha = \beta = 0.1$, we approximated the posterior distribution by using the Bayesian central limit theorem. Figure 3.2 shows the approximated posterior and the true posterior densities. We can see that approximation is accurate when the number of observations n is large. Even when $n = 10$, the approximated posterior is close to the true posterior distribution. Also, the posterior density is shrinking to the true value $\lambda = 4$ as the sample becomes larger.

3.2.3 Example: Confidence intervals

The Bayesian central limit theorem is also useful for providing approximate posterior confidence intervals for elements of $\boldsymbol{\theta}$, or for an arbitrary linear transformation of $\boldsymbol{\theta}$, $\boldsymbol{c}^T\boldsymbol{\theta}$. Noting that we can approximate the Bayesian posterior distribution of the model parameters $\pi(\boldsymbol{\theta}|\boldsymbol{X}_n)$ by multivariate normal distribution with mean the posterior mode $\hat{\boldsymbol{\theta}}_n$ and covariance matrix $n^{-1}S_n^{-1}(\hat{\boldsymbol{\theta}}_n)$, an approximate 95% posterior confidence interval for an arbitrary linear transformation of $\boldsymbol{\theta}$, $\boldsymbol{c}^T\boldsymbol{\theta}$, is given by

$$\left[\boldsymbol{c}^T\hat{\boldsymbol{\theta}}_n - 1.96\left(\boldsymbol{c}^T n^{-1}S_n^{-1}(\hat{\boldsymbol{\theta}}_n)\boldsymbol{c}\right), \boldsymbol{c}^T\hat{\boldsymbol{\theta}}_n + 1.96\left(\boldsymbol{c}^T n^{-1}S_n^{-1}(\hat{\boldsymbol{\theta}}_n)\boldsymbol{c}\right)\right]$$

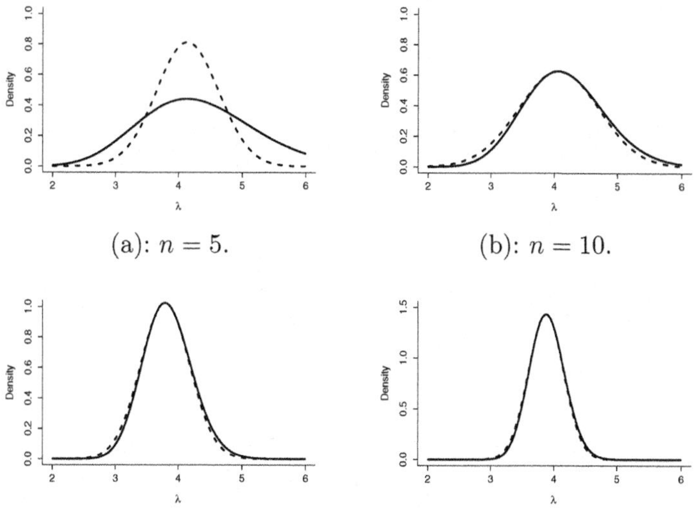

(a): $n = 5$. (b): $n = 10$.

(c): $n = 25$. (d): $n = 50$.

FIGURE 3.2: Comparison of the approximated posterior and the true posterior densities under the various sample sizes n. We generated a set of n random samples from the Poisson distribution with parameter $\lambda = 4$. Under a diffuse prior setting $\alpha = \beta = 0.1$, we approximated the posterior distribution by using the Bayesian central limit theorem.

3.3 Laplace method

In the previous section, we showed that the posterior distribution can be approximated by the normal distribution. However, if the true posterior distribution is far away from the normal distribution, this approximation may not be accurate. One of the treatments is to consider higher order derivatives in the Taylor expansion of the posterior distribution. As another approach, we can apply Laplace method to obtain an analytical approximation to integrals. Tierney and Kadane (1986) provided an early use of Laplace's method in the Bayesian context. We first overview a general version of Laplace's method for approximating integrals. Then we apply the Laplace method to Bayesian inference problems.

3.3.1 Laplace method for integral

Let $h(\boldsymbol{\theta})$ be a smooth, positive function of the p-dimensional vector $\boldsymbol{\theta} = (\theta_1, ..., \theta_p)^T$, and $q(\boldsymbol{\theta}, n)$ be a smooth function of n and $\boldsymbol{\theta}$. We are interested in the following integral:

$$U = \int h(\boldsymbol{\theta}) \exp\{s(\boldsymbol{\theta}, n)\} d\boldsymbol{\theta}. \tag{3.2}$$

First, we provide several key assumptions to use in Laplace's method.

B1: The function $s(\boldsymbol{\theta}, n)$ has a unique global maximum $\hat{\boldsymbol{\theta}}_n$ in the interior of the parameter space Θ and also not be too close to the boundary of Θ.

B2: The function $s(\boldsymbol{\theta}, n)$ is thrice continuously differentiable as a function of $\boldsymbol{\theta}$ on Θ.

B3: The function $h(\boldsymbol{\theta})$ is sufficiently smooth, i.e., it is continuously differentiable, bounded and positive on Θ. Also, the first-order partial derivatives of $h(\boldsymbol{\theta})$ are bounded on Θ.

B4: The negative of the Hessian matrix of $n^{-1}s(\boldsymbol{\theta}, n)$,

$$S\left(\hat{\boldsymbol{\theta}}_n, n\right) = -\frac{1}{n} \frac{\partial^2 \{s(\boldsymbol{\theta}, n)\}}{\partial \boldsymbol{\theta} \partial \boldsymbol{\theta}^T}\bigg|_{\boldsymbol{\theta}=\hat{\boldsymbol{\theta}}_n}$$

is positive definite. Also, the smallest eigenvalue of $n \times S\left(\hat{\boldsymbol{\theta}}_n, n\right)$ should tend to infinity as $n \to \infty$ so that the quadratic approximation of $s(\boldsymbol{\theta}, n)$ is accurate.

Although these are key components of the assumption of the Laplace's method for integrals, these are additional assumptions. For more details on the remaining assumptions, we refer to Barndorff-Nielsen and Cox (1989).

Under the regularity conditions, we can obtain the Laplace approximation of the integral in (3.2) as follows. Noting that the first derivative of the function $s(\boldsymbol{\theta}, n)$ evaluated at the mode $\hat{\boldsymbol{\theta}}_n$ equals to zero, we obtain the following expression of the integral in (3.2) using the Taylor expansion of both $h(\boldsymbol{\theta})$ and $s(\boldsymbol{\theta}, n)$ about $\hat{\boldsymbol{\theta}}_n$

$$\begin{aligned} U &\approx \int \Bigg[\left\{ h\left(\hat{\boldsymbol{\theta}}_n\right) + (\boldsymbol{\theta} - \hat{\boldsymbol{\theta}}_n)^T \frac{\partial h(\hat{\boldsymbol{\theta}}_n)}{\partial \boldsymbol{\theta}} + \cdots \right\} \\ &\quad \times \exp\left\{ s\left(\hat{\boldsymbol{\theta}}_n, n\right) - \frac{n}{2}(\boldsymbol{\theta} - \hat{\boldsymbol{\theta}}_n)^T S\left(\hat{\boldsymbol{\theta}}_n, n\right)(\boldsymbol{\theta} - \hat{\boldsymbol{\theta}}_n)^T \right\} \Bigg] d\boldsymbol{\theta} \\ &= \exp\left\{ s\left(\hat{\boldsymbol{\theta}}_n, n\right) \right\} \int \left\{ h\left(\hat{\boldsymbol{\theta}}_n\right) + (\boldsymbol{\theta} - \hat{\boldsymbol{\theta}}_n)^T \frac{\partial h(\hat{\boldsymbol{\theta}}_n)}{\partial \boldsymbol{\theta}} + \cdots \right\} \\ &\quad \times \exp\left\{ -\frac{n}{2}(\boldsymbol{\theta} - \hat{\boldsymbol{\theta}}_n)^T S\left(\hat{\boldsymbol{\theta}}_n, n\right)(\boldsymbol{\theta} - \hat{\boldsymbol{\theta}}_n)^T \right\} d\boldsymbol{\theta}, \end{aligned}$$

where $S(\hat{\boldsymbol{\theta}}_n, n)$ is the negative of the Hessian matrix of $s(\boldsymbol{\theta}, n)$.

Noting that

$$\exp\left\{ -\frac{n}{2}(\boldsymbol{\theta} - \hat{\boldsymbol{\theta}}_n)^T S\left(\hat{\boldsymbol{\theta}}_n, n\right)(\boldsymbol{\theta} - \hat{\boldsymbol{\theta}}_n) \right\}$$

is the kernel of the normal distribution with mean $\hat{\boldsymbol{\theta}}_n$ and variance matrix $n^{-1}S\left(\hat{\boldsymbol{\theta}}_n, n\right)^{-1}$, the second term in the brackets vanishes

$$\begin{aligned}
&\int \left\{ (\boldsymbol{\theta} - \hat{\boldsymbol{\theta}}_n)^T \frac{\partial h(\hat{\boldsymbol{\theta}}_n)}{\partial \boldsymbol{\theta}} \right\} \exp\left\{ -\frac{n}{2} (\boldsymbol{\theta} - \hat{\boldsymbol{\theta}}_n)^T S\left(\hat{\boldsymbol{\theta}}_n, n\right) (\boldsymbol{\theta} - \hat{\boldsymbol{\theta}}_n) \right\} d\boldsymbol{\theta} \\
&= \left[\frac{\partial h(\hat{\boldsymbol{\theta}}_n)}{\partial \boldsymbol{\theta}} \right]^T \int (\boldsymbol{\theta} - \hat{\boldsymbol{\theta}}_n) \exp\left\{ -\frac{n}{2} (\boldsymbol{\theta} - \hat{\boldsymbol{\theta}}_n)^T S\left(\hat{\boldsymbol{\theta}}_n, n\right) (\boldsymbol{\theta} - \hat{\boldsymbol{\theta}}_n) \right\} d\boldsymbol{\theta} \\
&= \mathbf{0}.
\end{aligned}$$

Therefore, the integral (3.2) can be approximated as

$$U \approx \exp\left\{ s(\hat{\boldsymbol{\theta}}_n, n) \right\} h(\hat{\boldsymbol{\theta}}_n) \frac{(2\pi)^{\frac{p}{2}}}{n^{\frac{p}{2}} \left| S\left(\hat{\boldsymbol{\theta}}_n, n\right) \right|^{1/2}} \times (1 + o(1)). \tag{3.3}$$

The order of the second term in the brackets is $o(1)$, which comes from the regularity condition of the Laplace method. Therefore, one can show that $U \approx \hat{U}\{1 + o(1)\}$ with

$$\hat{U} = \exp\left\{ s(\hat{\boldsymbol{\theta}}_n, n) \right\} h\left(\hat{\boldsymbol{\theta}}_n\right) \frac{(2\pi)^{\frac{p}{2}}}{n^{\frac{p}{2}} \left| S\left(\hat{\boldsymbol{\theta}}_n, n\right) \right|^{1/2}}.$$

The next section applies the Laplace method to the Bayesian inference problems.

3.3.2 Posterior expectation of a function of parameter

Suppose that we want to calculate the posterior expectation of a function of parameter $r(\boldsymbol{\theta})$.

$$\int r(\boldsymbol{\theta}) \pi\left(\boldsymbol{\theta} | \boldsymbol{X}_n\right) d\boldsymbol{\theta} = \frac{\int r(\boldsymbol{\theta}) f(\boldsymbol{X}_n|\boldsymbol{\theta}) \pi(\boldsymbol{\theta}) d\boldsymbol{\theta}}{\int f(\boldsymbol{X}_n|\boldsymbol{\theta}) \pi(\boldsymbol{\theta}) d\boldsymbol{\theta}}.$$

To apply the Laplace method, let us link the function $s(\boldsymbol{\theta}, n)$ in (3.2) as the unnormalized log-posterior density $\log\{f(\boldsymbol{X}_n|\boldsymbol{\theta})\pi(\boldsymbol{\theta})\}$. In this case, the mode of $\log\{f(\boldsymbol{X}_n|\boldsymbol{\theta})\pi(\boldsymbol{\theta})\}$ in (3.3) is the posterior mode, $\hat{\boldsymbol{\theta}}_n$, and the approximation error $o(1)$ in the bracket of Equation (3.3) becomes $O(n^{-1})$. This is from the Bayesian central limit theorem.

From the Laplace method to the numerator (by specifying $h(\boldsymbol{\theta}) = r(\boldsymbol{\theta})$)

and the denominator (by specifying $h(\boldsymbol{\theta}) = 1$), we obtain

$$\begin{aligned}
&\int r(\boldsymbol{\theta})\pi(\boldsymbol{\theta}|\boldsymbol{X}_n)\,d\boldsymbol{\theta} \\
&= \frac{f\left(\boldsymbol{X}_n|\hat{\boldsymbol{\theta}}_n\right)\pi\left(\hat{\boldsymbol{\theta}}_n\right)r\left(\hat{\boldsymbol{\theta}}_n\right)\dfrac{(2\pi)^{\frac{p}{2}}}{n^{\frac{p}{2}}\left|S_n\left(\hat{\boldsymbol{\theta}}_n\right)\right|^{1/2}}\left\{1+O\left(\frac{1}{n}\right)\right\}}{f\left(\boldsymbol{X}_n|\hat{\boldsymbol{\theta}}_n\right)\pi\left(\hat{\boldsymbol{\theta}}_n\right)\dfrac{(2\pi)^{\frac{p}{2}}}{n^{\frac{p}{2}}\left|S_n\left(\hat{\boldsymbol{\theta}}_n\right)\right|^{1/2}}\left\{1+O\left(\frac{1}{n}\right)\right\}} \\
&= r\left(\hat{\boldsymbol{\theta}}_n\right)\left\{1+O\left(\frac{1}{n}\right)\right\}, \qquad (3.4)
\end{aligned}$$

where

$$S_n\left(\hat{\boldsymbol{\theta}}_n\right) = -\frac{1}{n}\frac{\partial^2 \log\{f(\boldsymbol{X}_n|\boldsymbol{\theta})\pi(\boldsymbol{\theta})\}}{\partial\boldsymbol{\theta}\partial\boldsymbol{\theta}^T}\Bigg|_{\boldsymbol{\theta}=\hat{\boldsymbol{\theta}}_n}$$

is the negative of the Hessian matrix of $n^{-1}\log\{f(\boldsymbol{X}_n|\boldsymbol{\theta})\pi(\boldsymbol{\theta})\}$ evaluated at $\hat{\boldsymbol{\theta}}_n$. It indicates that we can approximate the posterior expectation of a function of parameter $r(\boldsymbol{\theta})$ by $r\left(\hat{\boldsymbol{\theta}}_n\right)$.

Although the first order approximation is convenient, the remaining error term might be relatively large, when we seriously consider the accuracy of the posterior expectation of a function of parameter. To solve this problem, Tierney and Kanade (1986) proposed the following idea to get more accurate approximation based on the Laplace method.

Let $r(\boldsymbol{\theta}) > 0$ be a positive valued function. Rewrite the numerator integrand in the posterior expectation of a function of parameter as

$$\int r(\boldsymbol{\theta})f(\boldsymbol{X}_n|\boldsymbol{\theta})\pi(\boldsymbol{\theta})d\boldsymbol{\theta} = \int \exp\left[\log r(\boldsymbol{\theta}) + \log\{f(\boldsymbol{X}_n|\boldsymbol{\theta})\pi(\boldsymbol{\theta})\}\right]d\boldsymbol{\theta},$$

and then link the function $s(\boldsymbol{\theta}, n)$ in (3.2) as $\log r(\boldsymbol{\theta}) + \log\{f(\boldsymbol{X}_n|\boldsymbol{\theta})\pi(\boldsymbol{\theta})\}$. From the Laplce method to the integrad (by specifying $h(\boldsymbol{\theta}) = 1$), we obtain

$$\begin{aligned}
&\int r(\boldsymbol{\theta})f(\boldsymbol{X}_n|\boldsymbol{\theta})\pi(\boldsymbol{\theta})d\boldsymbol{\theta} \\
&= f\left(\boldsymbol{X}_n|\hat{\boldsymbol{\theta}}_n^*\right)\pi\left(\hat{\boldsymbol{\theta}}_n^*\right)r\left(\hat{\boldsymbol{\theta}}_n^*\right)\frac{(2\pi)^{\frac{p}{2}}}{n^{\frac{p}{2}}\left|S_n^*\left(\hat{\boldsymbol{\theta}}_n^*\right)\right|^{1/2}}\left\{1+O\left(\frac{1}{n}\right)\right\},
\end{aligned}$$

where $\hat{\boldsymbol{\theta}}_n^*$ is the mode of $\log r(\boldsymbol{\theta}) + \log\{f(\boldsymbol{X}_n|\boldsymbol{\theta})\pi(\boldsymbol{\theta})\}$, and $S_n^*\left(\hat{\boldsymbol{\theta}}_n^*\right)$ is the negative of the Hessian matrix of $n^{-1}[\log r(\boldsymbol{\theta})+\log\{f(\boldsymbol{X}_n|\boldsymbol{\theta})\pi(\boldsymbol{\theta})\}]$ evaluated at $\hat{\boldsymbol{\theta}}_n^*$,

$$S_n^*\left(\hat{\boldsymbol{\theta}}_n^*\right) = -\frac{1}{n}\left.\frac{\partial^2\left[\log r(\boldsymbol{\theta})+\log\{f(\boldsymbol{X}_n|\boldsymbol{\theta})\pi(\boldsymbol{\theta})\}\right]}{\partial\boldsymbol{\theta}\partial\boldsymbol{\theta}^T}\right|_{\boldsymbol{\theta}=\hat{\boldsymbol{\theta}}_n^*}.$$

Putting this result into the numerator, we have

$$\begin{aligned}
&\int r(\boldsymbol{\theta})\pi\left(\boldsymbol{\theta}|\boldsymbol{X}_n\right)d\boldsymbol{\theta}\\
&= \frac{f\left(\boldsymbol{X}_n|\hat{\boldsymbol{\theta}}_n^*\right)\pi\left(\hat{\boldsymbol{\theta}}_n^*\right)r\left(\hat{\boldsymbol{\theta}}_n^*\right)\dfrac{(2\pi)^{\frac{p}{2}}}{n^{\frac{p}{2}}\left|S_n^*\left(\hat{\boldsymbol{\theta}}_n^*\right)\right|^{1/2}}\left\{1+O\left(\dfrac{1}{n}\right)\right\}}{f\left(\boldsymbol{X}_n|\hat{\boldsymbol{\theta}}_n\right)\pi\left(\hat{\boldsymbol{\theta}}_n\right)\dfrac{(2\pi)^{\frac{p}{2}}}{n^{\frac{p}{2}}\left|S_n\left(\hat{\boldsymbol{\theta}}_n\right)\right|^{1/2}}\left\{1+O\left(\dfrac{1}{n}\right)\right\}}\\
&= r\left(\hat{\boldsymbol{\theta}}_n^*\right)\frac{f\left(\boldsymbol{X}_n|\hat{\boldsymbol{\theta}}_n^*\right)\pi\left(\hat{\boldsymbol{\theta}}_n^*\right)\left|S_n\left(\hat{\boldsymbol{\theta}}_n\right)\right|^{1/2}}{f\left(\boldsymbol{X}_n|\hat{\boldsymbol{\theta}}_n\right)\pi\left(\hat{\boldsymbol{\theta}}_n\right)\left|S_n^*\left(\hat{\boldsymbol{\theta}}_n^*\right)\right|^{1/2}}\left\{1+O\left(\frac{1}{n^2}\right)\right\}.
\end{aligned}$$

The improvement of accuracy is from the identical leading terms $O\left(n^{-1}\right)$ both in the numerator and denominator. If the function is not strictly positive, Carlin and Louis (2000) recommended adding a large constant C to $r(\boldsymbol{\theta})$. After we apply the Laplace method by using $r(\boldsymbol{\theta})+C$, we can subtract the constant C from the computation results.

The Laplace method cannot be used when the posterior has multimodal, because the method is valid as long as the posterior is unimodal. Also, Laplace's method may not be accurate when numerical computation of the associated Hessian matrices is difficult. However, it is often easier and more stable numerically, because the Laplace method replaces a numerical integration problem with numerical differentiation and optimization problems.

3.3.3 Example: Bernoulli distribution with a uniform prior

Let $\boldsymbol{X}_n = \{x_1, ..., x_n\}$ be a set of n independent samples from the Bernoulli distribution with parameter p. The use of a uniform prior for $\pi(p) = \text{Const.}$, leads to a Beta posterior density with parameter $\sum_{\alpha=1}^n x_\alpha + \alpha$ and $n - \sum_{\alpha=1}^n x_\alpha + \beta$.

Suppose that we want to calculate the posterior mean. Although we know the exact expression of the posterior mean, we shall apply the Laplace method to this problem:

$$\begin{aligned}
&\int p\pi\left(p|\boldsymbol{X}_n\right)dp \\
&= \frac{\displaystyle\int pf\left(\boldsymbol{X}_n|p\right)\pi\left(p\right)dp}{\displaystyle\int f\left(\boldsymbol{X}_n|p\right)\pi\left(p\right)dp} \\
&= \frac{\displaystyle\int p^{\sum_{\alpha=1}^n x_\alpha+1}(1-p)^{n-\sum_{\alpha=1}^n x_\alpha}dp}{\displaystyle\int p^{\sum_{\alpha=1}^n x_\alpha}(1-p)^{n-\sum_{\alpha=1}^n x_\alpha}dp} \\
&= \frac{\displaystyle\int \exp\left\{\left(\sum_{\alpha=1}^n x_\alpha+1\right)\log p+\left(n-\sum_{\alpha=1}^n x_\alpha\right)\log(1-p)\right\}dp}{\displaystyle\int \exp\left\{\left(\sum_{\alpha=1}^n x_\alpha\right)\log p+\left(n-\sum_{\alpha=1}^n x_\alpha\right)\log(1-p)\right\}dp}.
\end{aligned}$$

The modes of the functions in the bracket in the numerator $\hat{p}_n^*$ and denominator $\hat{p}_n$ can be found as

$$\hat{p}_n^* = \frac{\sum_{\alpha=1}^n x_\alpha+1}{n+1} \quad \text{and} \quad \hat{p}_n = \frac{\sum_{\alpha=1}^n x_\alpha}{n}.$$

Also, the negative of the second derivatives of the functions in the bracket in the numerator evaluated at the mode $\hat{p}_n^*$ is

$$\begin{aligned}
S_n^*\left(\hat{p}_n^*\right) &= -\frac{1}{n}\frac{d^2}{dp^2}\left\{\left(\sum_{\alpha=1}^n x_\alpha+1\right)\log p+\left(n-\sum_{\alpha=1}^n x_\alpha\right)\log(1-p)\right\}\Bigg|_{p=\hat{p}_n^*} \\
&= \frac{1}{n}\left[\frac{(n+1)^2}{\sum_{\alpha=1}^n x_\alpha+1}+\frac{(n+1)^2}{n-\sum_{\alpha=1}^n x_\alpha}\right].
\end{aligned}$$

Similarly, we have

$$\begin{aligned}
S_n\left(\hat{p}_n\right) &= -\frac{1}{n}\frac{d^2}{dp^2}\left\{\left(\sum_{\alpha=1}^n x_\alpha\right)\log p+\left(n-\sum_{\alpha=1}^n x_\alpha\right)\log(1-p)\right\}\Bigg|_{p=\hat{p}_n} \\
&= \frac{1}{n}\left[\frac{n^2}{\sum_{\alpha=1}^n x_\alpha}+\frac{n^2}{n-\sum_{\alpha=1}^n x_\alpha}\right].
\end{aligned}$$

Thus we obtain

$$\int p\pi\left(p|\boldsymbol{X}_n\right)dp \approx \hat{p}_n^*\frac{f\left(\boldsymbol{X}_n|\hat{p}_n^*\right)S_n\left(\hat{p}_n\right)^{1/2}}{f\left(\boldsymbol{X}_n|\hat{p}_n\right)S_n^*\left(\hat{p}_n^*\right)^{1/2}}. \tag{3.5}$$

Note that we know the true posterior mean

$$\int p\pi\left(p|\boldsymbol{X}_n\right)dp = \frac{\sum_{\alpha=1}^{n} x_\alpha + 1}{n+2}.$$

Thus we can compare the accuracy of the Laplace approximation. We generated a set of $n = 10$ independent samples from the Bernoulli distribution with parameter $p = 0.3$. The generated samples are $\boldsymbol{X}_n = \{0,1,0,1,0,0,0,0,1,0\}$. We then obtain the following values

$$\begin{aligned} \hat{p}_n^* &= 0.3326, \quad \text{and} \quad \hat{p}_n = 0.3000, \\ S_n^*\left(\hat{p}_n^*\right) &= 4.75357, \quad \text{and} \quad S_n\left(\hat{p}_n\right) = 4.76190. \end{aligned}$$

Putting these numbers into Equation (3.5), we obtain the approximated posterior mean value 0.3326, which is close to the true posterior mean 0.3333. The first order Laplace approximation (3.4) is 0.3000. This is a natural result because the second order approximation provides more accurate results.

3.3.4 Asymptotic approximation of the Bayesian predictive distribution

One generic problem of the calculation of the Bayesian predictive density

$$f(\boldsymbol{z}|\boldsymbol{X}_n) = \int f(\boldsymbol{z}|\boldsymbol{\theta})\pi(\boldsymbol{\theta}|\boldsymbol{X}_n)d\boldsymbol{\theta}$$

is that because the Bayesian predictive density is obtained by integrating the sampling density $f(\boldsymbol{z}|\boldsymbol{\theta})$ with respect to the posterior distribution of the parameters $\pi(\boldsymbol{\theta}|\boldsymbol{X}_n)$. Even we know the analytical form of the posterior distribution of the parameters $\pi(\boldsymbol{\theta}|\boldsymbol{X}_n)$, analytic evaluation of the integral is often impossible.

We can approximate the Bayesian predictive density for future observation by using the Laplace method. Using the basic Laplace approximation and a ratio of integrals (Gelfand and Day (1994)), we can express the Bayesian predictive density as

$$\begin{aligned} f(\boldsymbol{z}|\boldsymbol{X}_n) &= \frac{\displaystyle\int f(\boldsymbol{z}|\boldsymbol{\theta})f(\boldsymbol{X}_n|\boldsymbol{\theta})\pi(\boldsymbol{\theta})d\boldsymbol{\theta}}{\displaystyle\int f(\boldsymbol{X}_n|\boldsymbol{\theta})\pi(\boldsymbol{\theta})d\boldsymbol{\theta}} \\ &= \frac{f(\boldsymbol{z}|\hat{\boldsymbol{\theta}}_n(\boldsymbol{z}))f\{\boldsymbol{X}_n|\hat{\boldsymbol{\theta}}_n(\boldsymbol{z})\}\pi\{\hat{\boldsymbol{\theta}}_n(\boldsymbol{z})\}}{f(\boldsymbol{X}_n|\hat{\boldsymbol{\theta}}_n\}\pi\{\hat{\boldsymbol{\theta}}_n\}} \\ &\quad \times \left[\frac{\left|R_n^{-1}\{\boldsymbol{z},\hat{\boldsymbol{\theta}}_n(\boldsymbol{z})\}\right|}{\left|R_n^{-1}(\hat{\boldsymbol{\theta}}_n)\right|}\right]^{1/2}\left\{1+O_p\left(\frac{1}{n^2}\right)\right\}, \end{aligned}$$

where $\hat{\boldsymbol{\theta}}_n(z)$ and $\hat{\boldsymbol{\theta}}_n$ are defined as

$$\hat{\boldsymbol{\theta}}_n(\boldsymbol{z}) = \text{argmax}_{\boldsymbol{\theta}} f(\boldsymbol{z}|\boldsymbol{\theta}) f(\boldsymbol{X}_n|\boldsymbol{\theta})\pi(\boldsymbol{\theta}),$$
$$\hat{\boldsymbol{\theta}}_n = \text{argmax}_{\boldsymbol{\theta}} f(\boldsymbol{X}_n|\boldsymbol{\theta})\pi(\boldsymbol{\theta}).$$

The $p \times p$ matrices $R_n\{\boldsymbol{z}, \hat{\boldsymbol{\theta}}_n(\boldsymbol{z})\}$ and $R_n(\hat{\boldsymbol{\theta}}_n)$ are given as

$$R_n\{\boldsymbol{z}, \hat{\boldsymbol{\theta}}_n(\boldsymbol{z})\} = -\frac{1}{n} \frac{\partial^2 \{\log f(\boldsymbol{z}|\boldsymbol{\theta}) + \log f(\boldsymbol{X}_n|\boldsymbol{\theta}) + \log \pi(\boldsymbol{\theta})\}}{\partial \boldsymbol{\theta} \partial \boldsymbol{\theta}^T}\Bigg|_{\boldsymbol{\theta}=\hat{\boldsymbol{\theta}}_n(\boldsymbol{z})},$$

$$R_n(\hat{\boldsymbol{\theta}}_n) = -\frac{1}{n} \frac{\partial^2 \{\log f(\boldsymbol{X}_n|\boldsymbol{\theta}) + \log \pi(\boldsymbol{\theta})\}}{\partial \boldsymbol{\theta} \partial \boldsymbol{\theta}^T}\Bigg|_{\boldsymbol{\theta}=\hat{\boldsymbol{\theta}}_n}.$$

The improvement of accuracy is from the identical leading terms $O\left(n^{-1}\right)$ both in the numerator and denominator.

Noting that $\hat{\boldsymbol{\theta}}_n - \hat{\boldsymbol{\theta}}_{\text{MLE}} = O_p(n^{-1})$ and $\hat{\boldsymbol{\theta}}_n(\boldsymbol{z}) - \hat{\boldsymbol{\theta}}_{\text{MLE}} = O_p(n^{-1})$ for the maximum likelihood estimator $\hat{\boldsymbol{\theta}}_{\text{MLE}}$ and using the result

$$\left|R_n^{-1}\{\boldsymbol{z}, \hat{\boldsymbol{\theta}}_n(\boldsymbol{z})\}\right| \Big/ \left|R_n^{-1}(\hat{\boldsymbol{\theta}}_n)\right| = 1 + O_p(n^{-1}),$$

the Bayesian predictive density can be further approximated as

$$f(\boldsymbol{z}|\boldsymbol{X}_n) = f(\boldsymbol{z}|\hat{\boldsymbol{\theta}}_{\text{MLE}}) + O_p(n^{-1}).$$

3.3.5 Laplace method for approximating marginal posterior distribution

Tierney and Kadane (1986) showed that the Laplace's method can be used to approximate marginal posterior densities of parameters. Suppose that the parameter vector $\boldsymbol{\theta}$ is divided into two parts, $\boldsymbol{\theta} = (\theta_1, \boldsymbol{\theta}_2^T)^T$. Applying Laplace's method to the integrals in the numerator and denominator of the marginal density of θ_1

$$\pi(\theta_1|\boldsymbol{X}_n) = \frac{\displaystyle\int f(\boldsymbol{X}_n|\theta_1, \boldsymbol{\theta}_2)\pi(\theta_1, \boldsymbol{\theta}_2) d\boldsymbol{\theta}_2}{\displaystyle\int f(\boldsymbol{X}_n|\boldsymbol{\theta})\pi(\boldsymbol{\theta}) d\boldsymbol{\theta}}$$

we obtain the approximation as

$$\hat{\pi}(\theta_1|\boldsymbol{X}_n) = \frac{f\left(\boldsymbol{X}_n|\theta_1, \hat{\boldsymbol{\theta}}_{2,n}(\theta_1)\right) \pi\left(\theta_1, \hat{\boldsymbol{\theta}}_{2,n}(\theta_1)\right)}{f\left(\boldsymbol{X}_n|\hat{\boldsymbol{\theta}}_n\right) \pi\left(\hat{\boldsymbol{\theta}}_n\right)} \left[\frac{n\left|S_n\left(\hat{\boldsymbol{\theta}}_n\right)\right|}{2\pi \left|S_n\left\{\hat{\boldsymbol{\theta}}_{2,n}(\theta_1)\right\}\right|}\right]^{1/2},$$

where $\hat{\boldsymbol{\theta}}_n$ and $\hat{\boldsymbol{\theta}}_{2,n}(\theta_1)$ are defined as

$$\hat{\boldsymbol{\theta}}_n = \text{argmax}_{\boldsymbol{\theta}}\{f(\boldsymbol{X}_n|\boldsymbol{\theta})\pi(\boldsymbol{\theta})\}$$

and

$$\hat{\boldsymbol{\theta}}_{2,n}(\theta_1) = \text{argmax}_{\boldsymbol{\theta}_2}\{f(\boldsymbol{X}_n|\theta_1, \boldsymbol{\theta}_2)\pi(\theta_1, \boldsymbol{\theta}_2)\},$$

with given value of θ_1. The $p \times p$ matrix $S_n\{\hat{\boldsymbol{\theta}}_n\}$ and the $(p-1) \times (p-1)$ matrix $S_n(\hat{\boldsymbol{\theta}}_{2,n}(\theta_1))$ are given as

$$S_n\left(\hat{\boldsymbol{\theta}}_n\right) = -\frac{1}{n}\frac{\partial^2\{\log f(\boldsymbol{X}_n|\boldsymbol{\theta}) + \log\pi(\boldsymbol{\theta})\}}{\partial\boldsymbol{\theta}\partial\boldsymbol{\theta}^T}\Bigg|_{\boldsymbol{\theta}=\hat{\boldsymbol{\theta}}_n(z)},$$

$$S_n\left(\hat{\boldsymbol{\theta}}_{2,n}(\theta_1)\right) = -\frac{1}{n}\frac{\partial^2\{\log f(\boldsymbol{X}_n|\theta_1, \boldsymbol{\theta}_2) + \log\pi(\theta_1, \boldsymbol{\theta}_2)\}}{\partial\boldsymbol{\theta}_2\partial\boldsymbol{\theta}_2^T}\Bigg|_{\boldsymbol{\theta}=\hat{\boldsymbol{\theta}}_{2,n}(\theta_1)}.$$

Exercise

1. *In Example 3.1.3, we illustrated asymptotic normality of the posterior mode of logistic regression. Generate a set of n independent observations $\boldsymbol{X}_n = \{x_1, ..., x_n\}$ from a Poisson distribution with parameter $\lambda = 5$. We already know that the conjugate prior of the Poisson distribution is the gamma distribution with parameter α and β, which results in gamma posterior with parameter $n\bar{x}_n + \alpha$ and $n + \beta$. The posterior mode is $\hat{\lambda}_n = (n\bar{x}_n + \alpha - 1)/(n + \beta)$.*

   Under various setting n, estimate the posterior mode over 1,000 repeated Monte Carlo trials. Also plot the empirical density function of the estimated posterior mode. The **R** *function* **density** *may be useful.*

2. *Recall the Bayesian analysis logistic regression models used in Example 3.1.3. Show that the asymptotic covariance matrix of $\boldsymbol{\beta}$ from Bayesian central limit theorem can be estimated by*

$$n^{-1}S_n^{-1}(\hat{\boldsymbol{\beta}}_n) = \left(X_n^T W X_n + \lambda^{-1} I_p\right)^{-1},$$

   where X_n is the design matrix, and W is a $n \times n$ diagonal matrix given in Example 3.1.3.

3. *In Example 3.3.3, we analyzed a Bernoulli distribution with a uniform prior. Show that the modes of*

$$h_1(p) = \left(\sum_{\alpha=1}^n x_\alpha + 1\right)\log p + \left(n - \sum_{\alpha=1}^n x_\alpha\right)\log(1-p)$$

$$h_2(p) = \left(\sum_{\alpha=1}^n x_\alpha\right)\log p + \left(n - \sum_{\alpha=1}^n x_\alpha\right)\log(1-p)$$

are given by $\hat{p}_n^*$ *and* $\hat{p}_n$*, respectively. Here* $\hat{p}_n^*$ *and* $\hat{p}_n$ *are given in Example 3.3.3.*

4. *A regression model generally consists of a random component, which specifies the distribution of the response* y_α*, and a systematic component, which specifies the structure of the conditional expectation* $\mu_\alpha = E[y_\alpha|\boldsymbol{x}_\alpha]$*. In generalized linear models (McCullagh and Nelder (1989), Nelder and Wedderburn (1972)), the* y_α *are drawn from the exponential family of distributions*

$$f(y_\alpha|\boldsymbol{x}_\alpha;\xi_\alpha,\phi) = \exp\left\{\frac{y_\alpha\xi_\alpha - u(\xi_\alpha)}{\phi} + v(y_\alpha,\phi)\right\},$$

where $u(\cdot)$ *and* $v(\cdot,\cdot)$ *are functions specific to each distribution and* ϕ *is an unknown scale parameter. The conditional expectation* μ_α*, first derivative of* $u(\xi_\alpha)$ *such that* $\mu_\alpha = E[Y_\alpha|\boldsymbol{x}_\alpha] = u'(\xi_\alpha)$*, is related by a linking function* $h(\mu_\alpha) = \eta_\alpha$*. One often uses the linear predictor:*

$$\eta_\alpha = \sum_{k=1}^{p} \beta_k x_{k\alpha}, \qquad \alpha = 1,2,...,n.$$

Show that Gaussian regression model can be derived by taking

$$u(\xi_\alpha) = \xi_\alpha^2/2,\ \phi = \sigma^2,\ v(y_\alpha,\phi) = -\frac{y_\alpha^2}{2\sigma^2} - \log\left(\sigma\sqrt{2\pi}\right),\ h(\mu_\alpha) = \mu_\alpha.$$

5. *(Continued). Show that logistic regression model can be derived by taking*

$$u(\xi_\alpha) = \log\{1+\exp(\xi_\alpha)\},\ v(y_\alpha,\phi) = 0,\ h(\mu_\alpha) = \log\frac{\mu_\alpha}{1-\mu_\alpha},\ \phi = 1.$$

6. *(Continued). Show that Poisson regression model can be derived by taking*

$$u(\xi_\alpha) = \exp(\xi_\alpha),\ \phi = 1,\ v(y_\alpha,\phi) = -\log(y_\alpha!),\ h(\mu_\alpha) = \log(\mu_\alpha).$$

7. *(Continued.) Consider the prior*

$$\pi(\boldsymbol{\beta},\phi) \propto \exp\left\{-n\lambda\frac{\boldsymbol{\beta}^T\boldsymbol{\beta}}{2}\right\},$$

with λ *is a smoothing parameter. Show that the asymptotic covariance matrix of* $\boldsymbol{\theta} = (\boldsymbol{\beta}^T\phi)^T$ *from Bayesian central limit theorem is given as* $n^{-1}S_n^{-1}(\hat{\boldsymbol{\theta}}_n)$*. Here*

$$S_n(\hat{\boldsymbol{\theta}}_n) = \frac{1}{n}\begin{pmatrix} X_n^T\Gamma X_n/\hat{\phi}_n + n\lambda I_p & X_n^T\Lambda\mathbf{1}_n/\hat{\phi}_n^2 \\ \mathbf{1}_n^T\Lambda X_n/\hat{\phi}_n^2 & -\boldsymbol{q}^T\mathbf{1}_n \end{pmatrix}.$$

where $\hat{\boldsymbol{\theta}}_n$ is the posterior mode. Here X_n is the design matrix, and $\mathbf{1}_n = (1, ..., 1)^T$. Λ and Γ are $n \times n$ diagonal matrices, and $\boldsymbol{p}$ and $\boldsymbol{q}$ are n-dimensional vectors. Their elements are

$$\Lambda_{\alpha\alpha} = \frac{y_\alpha - \hat{\mu}_\alpha}{u''(\hat{\xi_\alpha})h'(\hat{\mu}_\alpha)},$$

$$\Gamma_{\alpha\alpha} = \frac{(y_\alpha - \hat{\mu}_\alpha)\{u'''(\hat{\xi_\alpha})h'(\hat{\mu}_\alpha) + u''(\hat{\xi_\alpha})^2 h''(\hat{\mu}_\alpha)\}}{\{u''(\hat{\xi_\alpha})h'(\hat{\mu}_\alpha)\}^3} + \frac{1}{u''(\hat{\xi_\alpha})h'(\hat{\mu}_\alpha)^2},$$

$$p_\alpha = -\frac{y_\alpha r\{\hat{\boldsymbol{\beta}}_n^T \boldsymbol{x}_\alpha\} - s\{\hat{\boldsymbol{\beta}}_n^T \boldsymbol{x}_\alpha\}}{\hat{\phi}_n^2} + \left.\frac{\partial}{\partial \phi} v(y_\alpha, \phi)\right|_{\phi=\hat{\phi}_n},$$

$$q_\alpha = \left.\frac{\partial p_\alpha}{\partial \phi}\right|_{\phi=\hat{\phi}_n}.$$

Also, $r(\cdot)$ and $s(\cdot)$ are defined by $r(\cdot) = u'^{-1} \circ h^{-1}(\cdot)$ and $s(\cdot) = u \circ u'^{-1} \circ h^{-1}(\cdot)$, respectively.

8. *In Section 2.7, we considered the Bayesian inference on linear regression models. As a result, we found that the conditional posterior distribution of $\boldsymbol{\beta}$ is normal, and the marginal posterior distribution of σ^2 is inverse-gamma distribution:*

$$\pi\left(\boldsymbol{\beta}, \sigma^2 \middle| \boldsymbol{y}_n, X_n\right) = \pi\left(\boldsymbol{\beta} \middle| \sigma^2, \boldsymbol{y}_n, X_n\right) \pi\left(\sigma^2 \middle| \boldsymbol{y}_n, X_n\right)$$

with

$$\pi\left(\boldsymbol{\beta} \middle| \sigma^2, \boldsymbol{y}_n, X_n\right) = N\left(\hat{\boldsymbol{\beta}}_n, \sigma^2 \hat{A}_n\right), \quad \pi\left(\sigma^2 \middle| \boldsymbol{y}_n, X_n\right) = IG\left(\frac{\hat{\nu}_n}{2}, \frac{\hat{\lambda}_n}{2}\right).$$

Using the Laplace method for approximating marginal posterior distribution, obtain the marginal density of σ^2. Then compare the approximated result with the true marginal density $\pi\left(\sigma^2 \middle| \boldsymbol{y}_n, X_n\right)$.

9. *(Continued.) Consider the Bayesian inference on linear regression models in Section 2.7. Obtain the predictive distribution $f(z|\boldsymbol{y}_n, X_n)$ based on the asymptotic approximation of the Bayesian predictive distribution. Then compare the analytical result of the Bayesian predictive distribution.*

Chapter 4

Computational approach for Bayesian inference

In principal, Bayesian inference is easily implemented based on the posterior distribution of parameters $\pi(\boldsymbol{\theta}|\boldsymbol{X}_n)$ conditional on observed data $\boldsymbol{X}_n$ and prior distribution $\pi(\boldsymbol{\theta})$. However, in most of practical situations, we don't have the joint posterior distribution of $\boldsymbol{\theta}$ in analytical form. In such a case, we can employ a simulation based approach. This section first introduces the concept of Monte Carlo integration. Then computational approaches for Bayesian inference will be explained.

For more information about the computational approach for Bayesian inference, we refer to the books by Congdon (2001), Gamerman and Lopes (2006), Geweke (2005), Gilks et al. (1996), Rossi, Allenby and McCulloch (2005).

4.1 Monte Carlo integration

As we mentioned, the analytical evaluation of the posterior expectation of our interest is difficult in most situations. A solution is an approximation of integrals carried out by Monte Carlo integration. Let $\boldsymbol{\theta}$ have a density $s(\boldsymbol{\theta})$ and we seek

$$\gamma = \int h(\boldsymbol{\theta})s(\boldsymbol{\theta})d\boldsymbol{\theta}.$$

In Bayesian context, a density $s(\boldsymbol{\theta})$ will often be the posterior distribution $\pi(\boldsymbol{\theta}|\boldsymbol{X}_n)$ and $h(\boldsymbol{\theta})$ may be a sampling density $f(\boldsymbol{x}|\boldsymbol{\theta})$, or some functional form of parameter. Then if $\boldsymbol{\theta}^{(j)}$, $j = 1, ..., L$ are independent samples from $s(\boldsymbol{\theta})$, we have

$$\hat{\gamma} = \frac{1}{L}\sum_{j=1}^{L} h\left(\boldsymbol{\theta}^{(j)}\right),$$

which converges to γ almost surely as $L \to \infty$ by the Strong Law of Large Numbers. This is a definition of the Monte Carlo approximation. The quality of this approximation increases as the number of samples L increases, while it requires more computational time. With respect to the Monte Carlo approximation, we refer to many textbooks.

Using the Monte Carlo approximation, we can estimate various quantities based on the posterior samples $\boldsymbol{\theta}^{(j)} \sim \pi(\boldsymbol{\theta}|\boldsymbol{X}_n)$. Following are some examples.

1. **Posterior mean**:

$$\bar{\boldsymbol{\theta}}_n = \int \boldsymbol{\theta}\pi(\boldsymbol{\theta}|\boldsymbol{X}_n)d\boldsymbol{\theta} \quad \leftarrow \quad \frac{1}{L}\sum_{j=1}^{L}\boldsymbol{\theta}^{(j)}.$$

2. **Posterior mode**:

$$\hat{\boldsymbol{\theta}}_n = \text{argmax}_{\boldsymbol{\theta}}\pi(\boldsymbol{\theta}|\boldsymbol{X}_n) \quad \leftarrow \quad \text{argmax}_j\pi(\boldsymbol{\theta}^{(j)}|\boldsymbol{X}_n).$$

3. **Posterior probability in a particular region Q**:

$$\pi(\boldsymbol{\theta} \in Q|\boldsymbol{X}_n) \quad \leftarrow \quad \frac{1}{L}\sum_{j=1}^{L} I\left(\boldsymbol{\theta}^{(j)} \in Q\right),$$

where I is an indicator function.

4. **Marginal posterior distribution**:

$$\pi(\boldsymbol{\theta}_1|\boldsymbol{X}_n) = \int \pi(\boldsymbol{\theta}_1, \boldsymbol{\theta}_2|\boldsymbol{X}_n)d\boldsymbol{\theta}_2 \quad \leftarrow \quad \frac{1}{L}\sum_{j=1}^{L}\pi\left(\boldsymbol{\theta}_1, \boldsymbol{\theta}_2^{(j)}|\boldsymbol{X}_n\right).$$

5. **Predictive Distribution**:

$$f(\boldsymbol{z}|\boldsymbol{X}_n) = \int f(\boldsymbol{z}|\boldsymbol{\theta})\pi(\boldsymbol{\theta}|\boldsymbol{X}_n)d\boldsymbol{\theta} \quad \leftarrow \quad \frac{1}{L}\sum_{j=1}^{L} f(\boldsymbol{z}|\boldsymbol{\theta}^{(j)}).$$

Note that to compute the posterior expectation based on the Monte Carlo approximation requires a sample of size L from the posterior distribution. We therefore describe how to generate posterior samples from the posterior distribution in the next section.

4.2 Markov chain Monte Carlo methods for Bayesian inference

This section provides a description of the foundations of Markov chain Monte Carlo algorithms, which provide a useful tool for exploring the posterior distribution of parameters $\pi(\boldsymbol{\theta}|\boldsymbol{X}_n)$. We cover the Gibbs sampler (Geman and Geman (1984)), the Metropolis-Hastings algorithm (Metropolis et al.

(1953) and Hastings (1970)), and theoretical convergence properties of MCMC algorithms.

4.2.1 Gibbs sampler

We first consider the simplest situation, where the complete conditional posterior densities are available. Assume that the parameter vector $\boldsymbol{\theta}$ can be partitioned into B parameter blocks $\boldsymbol{\theta} = (\boldsymbol{\theta}_1^T, ..., \boldsymbol{\theta}_B^T)^T$ such that the complete conditional posterior densities for each block

$$\begin{aligned}
&\pi(\boldsymbol{\theta}_1|\boldsymbol{X}_n, \boldsymbol{\theta}_2, \boldsymbol{\theta}_3, ..., \boldsymbol{\theta}_B),\\
&\pi(\boldsymbol{\theta}_2|\boldsymbol{X}_n, \boldsymbol{\theta}_1, \boldsymbol{\theta}_3, ..., \boldsymbol{\theta}_B),\\
&\qquad\vdots\\
&\pi(\boldsymbol{\theta}_B|\boldsymbol{X}_n, \boldsymbol{\theta}_1, \boldsymbol{\theta}_2, ..., \boldsymbol{\theta}_{B-1}),
\end{aligned}$$

are available in closed form.

A Gibbs sampler directly samples iteratively from all of the complete conditional posterior distributions. Initializing the parameter value $\boldsymbol{\theta}^{(0)}$, the Gibbs sampler generates a sequence of random variables:

$$\begin{aligned}
&\text{Step 1. Draw } \boldsymbol{\theta}_1^{(1)} \sim \pi(\boldsymbol{\theta}_1|\boldsymbol{X}_n, \boldsymbol{\theta}_2^{(0)}, \boldsymbol{\theta}_3^{(0)}, ..., \boldsymbol{\theta}_B^{(0)}),\\
&\text{Step 2. Draw } \boldsymbol{\theta}_2^{(1)} \sim \pi(\boldsymbol{\theta}_2|\boldsymbol{X}_n, \boldsymbol{\theta}_1^{(1)}, \boldsymbol{\theta}_3^{(0)}, ..., \boldsymbol{\theta}_B^{(0)}),\\
&\qquad\qquad\vdots\\
&\text{Step B. Draw } \boldsymbol{\theta}_B^{(1)} \sim \pi(\boldsymbol{\theta}_B|\boldsymbol{X}_n, \boldsymbol{\theta}_1^{(1)}, \boldsymbol{\theta}_2^{(1)}, ..., \boldsymbol{\theta}_{B-1}^{(1)}).
\end{aligned}$$

Once we approach Step B, then go back to Step 1. Notice that we have to make a draw based on the current values of $\boldsymbol{\theta}$ in the next loop. For example, the seond repitition makes a draw $\boldsymbol{\theta}_1^{(2)}$ from $\pi(\boldsymbol{\theta}_1|\boldsymbol{X}_n, \boldsymbol{\theta}_2^{(1)}, \boldsymbol{\theta}_3^{(1)}, ..., \boldsymbol{\theta}_B^{(1)})$ and so on. Continuing the Step 1–Steps B, Gibbs sampler sequentially generates a set of random variables $\{\boldsymbol{\theta}^{(j)}\}$, $j = 1, 2, ...L$. After a certain step, the samples from Gibbs sampler can be regarded as a sample from the joint posterior distribution of $\boldsymbol{\theta}$. Thus, discarding the first part of Gibbs sampler, a set of samples drawn from $\pi(\boldsymbol{\theta}|\boldsymbol{X}_n)$ is used for Bayesian inference.

4.2.2 Metropolis-Hastings sampler

Though the Gibbs sampler is a simple MCMC algorithm, it requires the complete set of conditional posterior distributions. In most of cases, one or more of the conditional posterior distribution of parameters are not available and we cannot apply the Gibbs sampler. Consider the case where one of the conditional posterior distributions of $\boldsymbol{\theta}_k$, $\pi(\boldsymbol{\theta}_k|\boldsymbol{X}_n, \boldsymbol{\theta}_1, \ldots, \boldsymbol{\theta}_{k-1}, \boldsymbol{\theta}_{k+1}, \ldots, \boldsymbol{\theta}_B) \equiv \pi(\boldsymbol{\theta}_k|\boldsymbol{X}_n, \boldsymbol{\theta}_{-k})$, is not available. Therefore, it is difficult to generate a sample $\boldsymbol{\theta}_k$ from the conditional posterior distribution

$\pi(\boldsymbol{\theta}_k|\boldsymbol{X}_n, \boldsymbol{\theta}_{-k})$. In such a case, the Metropolis-Hastings algorithms can be used.

Similar to the Gibbs sampler, the Metropolis-Hastings algorithm sequentially generates a set of random variables $\{\boldsymbol{\theta}^{(j)}\}$, $j = 1, 2, ...L$. To generate samples from $\pi(\boldsymbol{\theta}_k|\boldsymbol{X}_n, \boldsymbol{\theta}_{-k})$, the Metropolis-Hastings algorithm requires us to specify a proposal density $p(\boldsymbol{\theta}_k^{(j+1)}, \boldsymbol{\theta}_k^{(j)})$, e.g., multivariate normal distribution. The Metropolis-Hastings algorithm then first draws a candidate parameter value $\boldsymbol{\theta}_k^{(j+1)}$ from the a proposal density $p(\boldsymbol{\theta}_k^{(j+1)}, \boldsymbol{\theta}_k^{(j)})$. The generated parameter value $\boldsymbol{\theta}_k^{(j+1)}$ will be accepted or rejected based on the acceptance probability

$$\alpha(\boldsymbol{\theta}_k^{(j)}, \boldsymbol{\theta}_k^{(j+1)}) = \min\left\{1, \frac{f(\boldsymbol{X}_n|\boldsymbol{\theta}_k^{(j+1)})\pi(\boldsymbol{\theta}_k^{(j+1)})/p(\boldsymbol{\theta}_k^{(j+1)}, \boldsymbol{\theta}_k^{(j)})}{f(\boldsymbol{X}_n|\boldsymbol{\theta}_k^{(j)})\pi(\boldsymbol{\theta}_k^{(j)})/p(\boldsymbol{\theta}_k^{(j)}, \boldsymbol{\theta}_k^{(j+1)})}\right\}.$$

If the Gibbs sampler cannot be applied, we just replace the Gibbs sampling step with the Metropolis-Hastings algorithm that samples iteratively. Therefore, the Metropolis-Hastings algorithm is widely applicable because the complete conditional posterior density need not be known in closed form.

Note also that the Gibbs sampling is a special case of Metropolis-Hasting algorithms. In the Metropolis-Hastings, let us specify a proposal density as $p(\boldsymbol{\theta}_k^{(j+1)}, \boldsymbol{\theta}_k^{(j)}) = \pi(\boldsymbol{\theta}_k^{(j+1)}|\boldsymbol{X}_n, \boldsymbol{\theta}_{-k}^{(j)})$. The acceptance probability is then

$$\begin{aligned}\alpha(\boldsymbol{\theta}_k^{(j)}, \boldsymbol{\theta}_k^{(j+1)}) &= \min\left\{1, \frac{f(\boldsymbol{X}_n|\boldsymbol{\theta}_k^{(j+1)})\pi(\boldsymbol{\theta}_k^{(j+1)})/\pi(\boldsymbol{\theta}_k^{(j+1)}|\boldsymbol{\theta}_{-k}^{(j)})}{f(\boldsymbol{X}_n|\boldsymbol{\theta}_k^{(j)})\pi(\boldsymbol{\theta}_k^{(j)})/\pi(\boldsymbol{\theta}_k^{(j)}|\boldsymbol{\theta}_{-k}^{(j)})}\right\} \\ &= \min\{1, 1\} = 1.\end{aligned}$$

Therefore, the acceptance probability is always 1, and the MCMC algorithm always moves.

In practical implementation of the Metropolis-Hasting algorithms, the choice of proposal density will greatly affect the performance of the MCMC algorithm. There are mainly two specifications of the proposal density.

Independence Metropolis-Hastings

The Independence Metropolis-Hastings algorithm draws a new candidate $\boldsymbol{\theta}_k$ from proposal density, $p(\boldsymbol{\theta}_k^{(j+1)}|\boldsymbol{\theta}_k^{(j)}) \equiv p(\boldsymbol{\theta}_k^{(j+1)})$ which does not depend on the previous parameter value $\boldsymbol{\theta}_k^{(j)}$. The acceptance probability is

$$\alpha(\boldsymbol{\theta}_k^{(j)}, \boldsymbol{\theta}_k^{(j+1)}) = \min\left\{1, \frac{f(\boldsymbol{X}_n|\boldsymbol{\theta}_k^{(j+1)})\pi(\boldsymbol{\theta}_k^{(j+1)})/p(\boldsymbol{\theta}_k^{(j+1)})}{f(\boldsymbol{X}_n|\boldsymbol{\theta}_k^{(j)})\pi(\boldsymbol{\theta}_k^{(j)})/p(\boldsymbol{\theta}_k^{(j)})}\right\}.$$

When we use the independence Metropolis-Hastings algorithm, the proposal density is commonly picked up to close the posterior distribution.

Random-walk Metropolis-Hastings

The Random-walk Metropolis-Hastings algorithm draws a new candidate $\boldsymbol{\theta}_k$ from the following random walk model,

$$\boldsymbol{\theta}_k^{(j+1)} = \boldsymbol{\theta}_k^{(j)} + \boldsymbol{\varepsilon},$$

where $\boldsymbol{\varepsilon}$ is an independent random variable with mean zero. Due to the symmetry in the proposal density, $p(\boldsymbol{\theta}_k^{(j+1)}|\boldsymbol{\theta}_k^{(j)}) = p(\boldsymbol{\theta}_k^{(j)}|\boldsymbol{\theta}_k^{(j+1)})$, the acceptance probability simplifies to

$$\alpha(\boldsymbol{\theta}_k^{(j)}, \boldsymbol{\theta}_k^{(j+1)}) = \min\left\{1, \frac{f(\boldsymbol{X}_n|\boldsymbol{\theta}_k^{(j+1)})\pi(\boldsymbol{\theta}_k^{(j+1)})}{f(\boldsymbol{X}_n|\boldsymbol{\theta}_k^{(j)})\pi(\boldsymbol{\theta}_k^{(j)})}\right\}.$$

We have to be careful to ensure enough tail of the proposal density. For example, Student-t density might be considered. Also, we have to adjust the variance of the error term $\boldsymbol{\varepsilon}$ to obtain an acceptable level of accepted draws.

4.2.3 Convergence check

The theoretical convergence of the MCMC process has been studied from both theoretical and practical aspects. In practice, a verification of MCMC convergence to the posterior distribution cannot easily be implemented, while the generated samples from MCMC after the convergence can be used for the posterior analysis. For example, we have to identify a certain simulation point, where MCMC converged to the posterior distribution, or equivalently, we have to specify the length of a burn-in period. Even after the MCMC has reached the posterior distribution, we have to decide the number of iterations to summarize the posterior distribution.

To examine the convergence of MCMC, trace plots might be useful for detecting poorly sampled Markov Chains. If the MCMC has converged to the posterior distribution, the traceplot will fluctuate the mode of the distribution. On the other hand, if we observe some trends in the MCMC output series, it is a clear sign of nonconvergence. Although the trace plot can be implemented easily, we have to be careful because the traceplot will also give a convergence sign even when the MCMC is trapped in a local region.

Geweke (1992) proposed a convergence diagnostic (CD) test statistic that measures the equality of the means of the first and last part of a Markov chain. If the samples are drawn from the stationary distribution, these two means calculated from the first and the last part of a Markov chain are equal. Therefore, a CD test statistic has an asymptotically standard normal distribution.

Gelman and Rubin (1992) proposed another approach to identify nonconvergence by simulating multiple sequences for over-dispersed starting points. They pointed out that the variance within the chains should be the same as the variance across the chains.

A related concept to the MCMC convergence is the inefficiency factor which is useful to measure the efficiency of the MCMC sampling algorithm. The inefficiency factor is defined as

$$\text{Inefficiency factor} = 1 + 2\sum_{k=1}^{\infty} \rho(k), \tag{4.1}$$

where $\rho(k)$ is the sample autocorrelation at lag k calculated from the sampled draws. A large value of inefficiency factor indicates that we need large MCMC simulation. The effective sample size, the number of MCMC output L divided by the inefficiency factor,

$$\text{Effective sample size} = \frac{L}{1 + 2\sum_{k=1}^{\infty} \rho(k)},$$

is also an useful measure to assess an efficiency of MCMC simulation. If the value of effective sample size is low, MCMC simulation chain is not fully mixed.

Many other convergence checking methods have been proposed, e.g., Raftery and Lewis (1992), Brooks and Gelman (1997), Zellner and Min (1995), and so on.

4.2.4 Example: Gibbs sampling for seemingly unrelated regression model

In many areas of economics and other sciences, the seemingly unrelated regression (SUR) model, introduced by Zellner (1962), is used as a tool to study a wide range of phenomena. As has been widely appreciated in the literature, the seemingly unrelated regression (SUR) model is useful in analyzing a broad range of problems. The linear SUR model involves a set of regression equations with cross-equation parameter restrictions and correlated error terms having differing variances.

Seemingly unrelated regression model

Algebraically, the SUR model is given by:

$$\boldsymbol{y}_{nj} = X_{nj}\boldsymbol{\beta}_j + \boldsymbol{\varepsilon}_j, \ j = 1, ..., m, \tag{4.2}$$

with

$$E[\boldsymbol{\varepsilon}_i \boldsymbol{\varepsilon}_j^T] = \begin{cases} \sigma_{ij} I, & (i \neq j) \\ \sigma_i^2 I, & (i = j) \end{cases}.$$

Here $\boldsymbol{y}_{nj}$ and $\boldsymbol{\varepsilon}_j$ are the $n \times 1$ vectors, X_{nj} is the $n \times p_j$ pre-determined matrix, and $\boldsymbol{\beta}_j$ is the p_j-dimensional vector. As shown in (4.2), the equations have different independent variables and variances. Also, the model permits error terms in different equations to be correlated.

In the matrix form, the SUR model in (4.2) is expressed as

$$\begin{pmatrix} \boldsymbol{y}_{n1} \\ \boldsymbol{y}_{n2} \\ \vdots \\ \boldsymbol{y}_{nm} \end{pmatrix} = \begin{pmatrix} X_{n1} & O & \cdots & O \\ O & X_{n_2} & \cdots & O \\ \vdots & \vdots & \ddots & \vdots \\ O & O & \cdots & X_{nm} \end{pmatrix} \begin{pmatrix} \boldsymbol{\beta}_1 \\ \boldsymbol{\beta}_2 \\ \vdots \\ \boldsymbol{\beta}_m \end{pmatrix} + \begin{pmatrix} \varepsilon_1 \\ \varepsilon_2 \\ \vdots \\ \varepsilon_m \end{pmatrix},$$

or, equivalently,

$$\boldsymbol{y}_n = X_n\boldsymbol{\beta} + \boldsymbol{\varepsilon}, \quad \boldsymbol{\varepsilon} \sim N(\mathbf{0}, \Sigma \otimes I),$$

where O is the zero matrix, $\otimes$ is the tensor product, Σ is the $m \times m$ matrix with the diagonal elements $\{\sigma_1^2, ..., \sigma_m^2\}$, and the off-diagonal ijth elements are σ_{ij}.

The maximum likelihood estimates of $\boldsymbol{\beta}$ and Σ are obtained by maximizing the likelihood function.

$$f(\boldsymbol{Y}_n|X_n, \boldsymbol{\beta}, \Sigma) = \frac{1}{(2\pi)^{nm/2}|\Sigma|^{n/2}} \exp\left[-\frac{1}{2}\text{tr}\left\{R\Sigma^{-1}\right\}\right],$$

where "tr" denotes the trace of a matrix, $|\Sigma| = \det(\Sigma)$ is the value of the determinant of Σ, the ijth elements of $m \times m$ matrix $R = (r_{ij})$ is $r_{ij} = (\boldsymbol{y}_{ni} - X_{ni}\boldsymbol{\beta}_i)^T(\boldsymbol{y}_{nj} - X_{nj}\boldsymbol{\beta}_j)$. Zellner (1962) considered the parameter estimation problem from the frequentist points of view. If Σ is known, a parameter estimate can be obtained as the generalized least squares (GLS) estimator $\hat{\boldsymbol{\beta}}$. In practice, however, Σ in $\hat{\boldsymbol{\beta}}$ is usually unknown and the feasible generalized least squares estimators have been proposed. The ordinary least squares residuals for each equation can be used to consistently estimate Σ. The maximum likelihood estimates of $\boldsymbol{\beta}$ and Σ are obtained by using the iterative SUR approach.

Prior and Posterior analysis

Zellner (1971), Box and Tiao (1973), and Percy (1992) studied the posterior distributions of parameters in the SUR model. In the absence of prior knowledge, Bayesian analysis with noninformative priors is very common in practice. One of the most widely used noninformative priors, introduced by Jeffreys (1946, 1961), is Jeffreys's invariant prior:

$$\pi_1(\boldsymbol{\beta}, \Sigma) = \pi_1(\boldsymbol{\beta})\pi_1(\Sigma) \propto |\Sigma|^{-\frac{m+1}{2}}, \tag{4.3}$$

which is proportional to the square root of the determinant of the Fisher information matrix.

The joint posterior density function for the parameters is then:

$$\pi_1(\boldsymbol{\beta}, \Sigma|\boldsymbol{Y}_n, X_n) \propto |\Sigma|^{-(n+m+1)/2} \exp\left[-\frac{1}{2}\text{tr}\left\{R\Sigma^{-1}\right\}\right].$$

However, the prior distribution just allows us to get the analytical conditional posterior densities of $\boldsymbol{\beta}$ and Σ, but also the analytical joint posterior density.

It is obvious from the form of the joint posterior density function $\pi_1(\boldsymbol{\beta}, \Sigma|\boldsymbol{Y}_n, X_n)$ that the conditional posteriors $\pi_1(\boldsymbol{\beta}|\boldsymbol{Y}_n, X_n, \Sigma)$ and $\pi(\Sigma|\boldsymbol{Y}_n, X_n, \boldsymbol{\beta})$ are

$$\pi_1 (\boldsymbol{\beta}|\boldsymbol{Y}_n, X_n, \Sigma) = N\left(\hat{\boldsymbol{\beta}}, \hat{\Omega}\right) \quad \text{and} \quad \pi_1 (\Sigma|\boldsymbol{Y}_n, X_n, \boldsymbol{\beta}) = IW(R, n), \quad (4.4)$$

with

$$\hat{\boldsymbol{\beta}} = \left\{X_n^T (\Sigma^{-1} \otimes I) X_n\right\}^{-1} X_n^T (\Sigma^{-1} \otimes I) \boldsymbol{y}_n,$$
$$\hat{\Omega} = \left(X_n^T (\Sigma^{-1} \otimes I) X_n\right)^{-1},$$

where $IW(\cdot,\cdot)$ denotes the inverse Wishart distribution.

Although the posteriors of $\boldsymbol{\beta}$ and Σ are depending upon each other, we can use the Gibbs sampler. Starting from an initial value $\boldsymbol{\beta}^{(0)}$ and $\Sigma^{(0)}$, the following steps are performed:

Gibbs sampling

Step 1. Update the coefficient vector $\boldsymbol{\beta}^{(j)}$ by drawing a new value from the conditional posterior density $\pi_1\left(\boldsymbol{\beta}|\boldsymbol{Y}_n, X_n, \Sigma^{(j-1)}\right)$ in (4.4).

Step 2. Update $\Sigma^{(j)}$ by drawing a new value from the conditional posterior density $\pi_1\left(\Sigma|\boldsymbol{Y}_n, X_n, \boldsymbol{\beta}^{(j)}\right)$ in (4.4).

The process is then repeated a large number of times $j = 1, ...,$. The initial part of generates is discarded as being unrepresentative of the posterior distribution. The remaining samples are then used for the posterior inference.

Practical implementation of MCMC

To implement the Gibbs sampling procedure with the Jeffreys's invariant prior (4.3), we simulate data sets from the $m = 2$ dimensional SUR model, corresponding to a bivariate response. The dimension of X_{nj} is set to be $p_j = 2$, $j = 1, 2$ in model (4.2). This model can thus be written as follows:

$$\begin{pmatrix} \boldsymbol{y}_1 \\ \boldsymbol{y}_2 \end{pmatrix} = \begin{pmatrix} X_{n1} & O \\ O & X_{n2} \end{pmatrix} \begin{pmatrix} \boldsymbol{\beta}_1 \\ \boldsymbol{\beta}_2 \end{pmatrix} + \begin{pmatrix} \boldsymbol{\varepsilon}_1 \\ \boldsymbol{\varepsilon}_2 \end{pmatrix}, \quad i = 1, ..., n,$$

where $\boldsymbol{y}_j$ and $\boldsymbol{u}_j$ are $n \times 1$ vectors, X_{jn} is the $n \times 2$ matrix and $\boldsymbol{\beta}_j$ is the 2-dimensional vector. Each element of Σ is set to be

$$\Sigma = \begin{pmatrix} \sigma_1^2 & \sigma_{12} \\ \sigma_{21} & \sigma_2^2 \end{pmatrix} = \begin{pmatrix} 0.2 & -0.1 \\ -0.1 & 0.4 \end{pmatrix}.$$

The design matrices X_{jn} $j = 1, 2$ were generated from a uniform density over the interval $[-4, 4]$. The coefficient vectors were set to be $\boldsymbol{\beta}_1 = (3, -2)^T$ and $\boldsymbol{\beta}_2 = (2, 1)^T$, respectively. This enabled the generation of simulated response

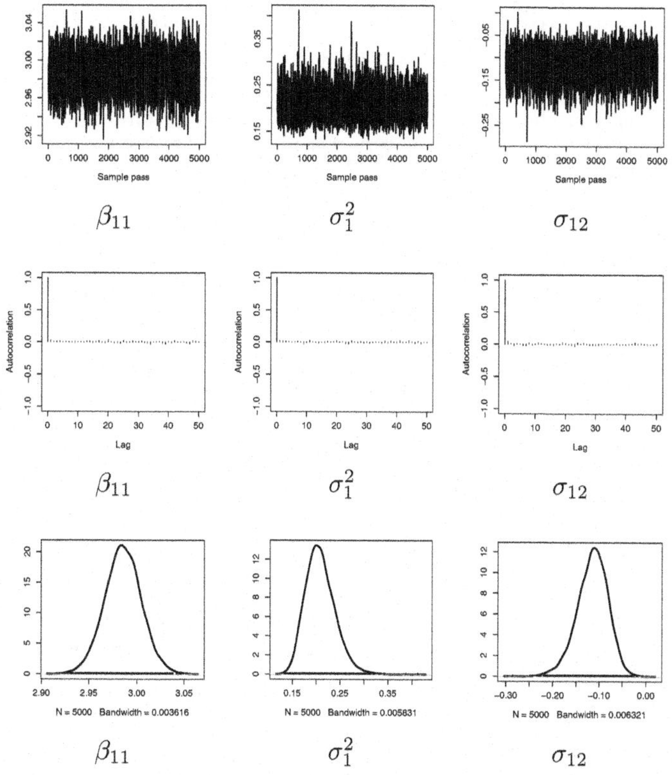

FIGURE 4.1: Sample paths, estimated posterior densities and sample autocorrelation functions for β_{11}, σ_1^2, and σ_{12}, respectively. True values are $\beta_{11} = 3$, $\sigma_1^2 = 0.2$, and $\sigma_{12} = -0.1$. Results are based on the output from the Gibbs sampling with the Jeffreys's invariant prior (4.3).

observations. In this simulation we set the number of observations to be $n = 100$.

To save computational time, the initial value of the parameter is chosen to be a generalized least squared estimate. The total number of Markov chain Monte Carlo iterations is chosen to be 6,000, of which the first 1,000 iterations are discarded. Thus, we generated 5,000 posterior samples by using the Gibbs sampling approach. It is necessary to check whether the generated posterior sample is taken from the stationary distribution. We assessed the convergence of MCMC simulation by calculating the Geweke's (1992) convergence diagnostic (CD) test statistics. All the posterior samples have passed the Geweke's (1992) convergence test at a significance level of 5% for all parameters.

Figure 4.1 shows sample paths of 5,000 draws, sample autocorrelation functions and estimated posterior densities. As shown in Figure 4.1, there was also

TABLE 4.1: Simulated data: Summary of the parameter estimates for the MCMC algorithm. The posterior means, modes, the standard deviations (SDs), 95% confidence intervals (95%CIs), Geweke's (1992) convergence diagnostic test statistic (CD) and the inefficiency factors (INEFs) are calculated.

	Mean	Mode	SDs	95%CIs		CD	INEFs
β_{11}	2.980	2.985	0.019	2.947	3.024	-0.965	0.345
β_{12}	-1.994	-1.995	0.017	-2.028	-1.961	-0.820	0.176
β_{21}	2.029	2.032	0.025	1.983	2.083	0.194	0.366
β_{22}	0.980	0.983	0.024	0.934	1.031	0.114	1.824
σ_1^2	0.195	0.208	0.030	0.157	0.275	1.115	0.703
σ_{12}	-0.115	-0.114	0.032	-0.184	-0.055	-1.653	2.203
σ_2^2	0.407	0.420	0.060	0.317	0.552	1.852	1.956

no evidence of lack of convergence based on an examination of trace plots. We can also see that the autocorrelation is very small in this simulation setting.

Table 4.1 reports the posterior means, the standard errors, 95% posterior confidence intervals. The inefficiency factor and the convergence diagnostic (CD) test statistics of MCMC algorithm were also reported. Using the posterior draws for each of the parameters, we calculated the posterior means, the standard errors and the 95% confidence intervals. The 95% confidence intervals are estimated using the 2.5th and 97.5th percentiles of the posterior samples. We have used 1,000 lags in the estimation of the inefficiency factors. It can be seen that the results appear quite reasonable. For instance, the true model is estimated with reasonable accuracy. The 95% posterior intervals include the true parameter values.

Figure 4.2 shows the estimated predictive density based on the Gibbs sampling output. By using the posterior samples $\{\boldsymbol{\beta}^{(j)}, \Omega^{(j)}; j = 1, ..., L\}$, the predictive density of the future value $\boldsymbol{z}$, given a prediction point $\boldsymbol{x}$, can be approximated as

$$\int f(\boldsymbol{z}|\boldsymbol{x}, \boldsymbol{\beta}, \Sigma)\, \pi_1(\boldsymbol{\beta}, \Sigma|\boldsymbol{Y}_n, X_n)\, d\boldsymbol{\beta} d\Sigma \approx \frac{1}{L}\sum_{j=1}^{L} f\left(\boldsymbol{z}|\boldsymbol{x}, \boldsymbol{\beta}^{(j)}, \Sigma^{(j)}\right),$$

where $\boldsymbol{x}^T = (\boldsymbol{x}_1^T, \boldsymbol{x}_2^T)$. The density is evaluated at a point $\boldsymbol{x}_1 = (0.1, -0.4)^T$ and $\boldsymbol{x}_2 = (0.2, -0.3)^T$. Because the actual predictive density is not known and so we have no bench-mark against which to compare them, we compare the estimated predictive density with the true sampling density of $\boldsymbol{y} = (y_1, y_2)^T$ given $\boldsymbol{x}_1$ and $\boldsymbol{x}_2$. For easy visual comparison, the results are also presented as contour plots, which can be produced with R software. The scales are the same for both plot and the contours join points with equal probability density. We can see that the estimated predictive density is very close to the true density.

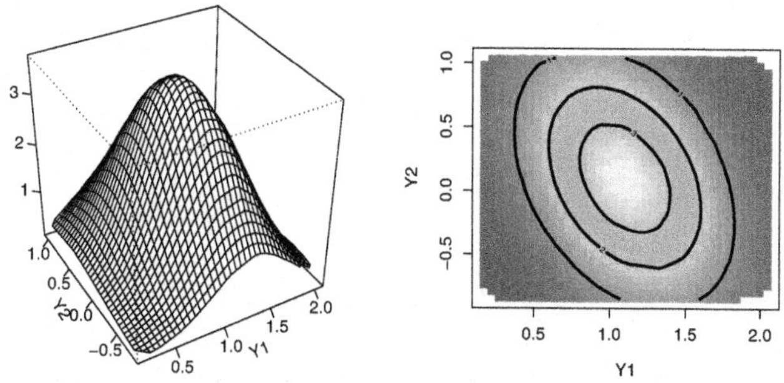

True sampling density.

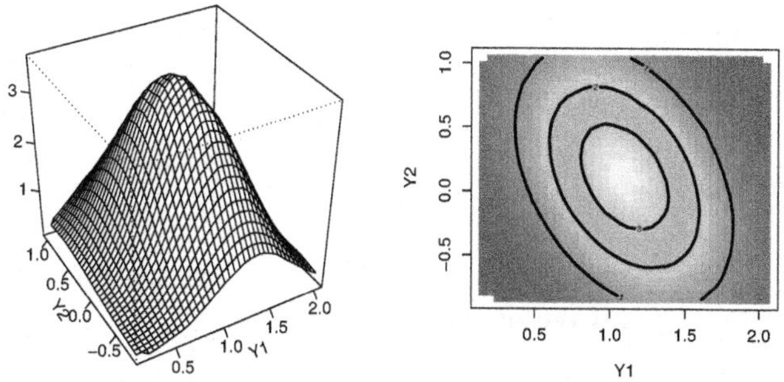

Estimated predictive density.

FIGURE 4.2: Estimated predictive density based on the Gibbs sampling output. By using the posterior samples $\{\boldsymbol{\beta}^{(j)}, \Sigma^{(j)}; j = 1, ..., L\}$, the predictive density is approximated. The density is evaluated at a point $\boldsymbol{x}_1 = (0.1, -0.4)^T$ and $\boldsymbol{x}_2 = (0.2, -0.3)^T$. We compare the estimated predictive density with the true sampling density of $\boldsymbol{y} = (y_1, y_2)^T$.

4.2.5 Example: Gibbs sampling for auto-correlated errors

In panel data analysis, we often have a set of observations with auto-correlated errors. The simplest model of autocorrelation is the p-th order auto-regressive model

$$y_\alpha = \boldsymbol{x}_\alpha^T \boldsymbol{\beta} + \varepsilon_\alpha,$$
$$\varepsilon_\alpha = \sum_{j=1}^{p} \rho_j \varepsilon_{\alpha-j} + u_\alpha,$$

for $\alpha = 1, ..., n$. Here $\boldsymbol{x}_\alpha$ is the q-dimensional predictors, u_α follows independently, identically normal with mean 0 and variance σ^2 and ρ_j, $j = 1, ..., p$ determines the dependency of the error terms ε_α. In contrast to the standard linear regression model, the error terms ε_α are correlated.

One of the common approaches for estimating this model is to transform it as follows:

$$y_\alpha^* = \boldsymbol{x}_\alpha^{*T}\boldsymbol{\beta} + u_\alpha,$$

with

$$y_\alpha^* = y_\alpha - \sum_{j=1}^{p} \rho_j y_{\alpha-j} \quad \text{and} \quad \boldsymbol{x}_\alpha^* = \boldsymbol{x}_\alpha - \sum_{j=1}^{p} \rho_j \boldsymbol{x}_{\alpha-j}.$$

Although the model inference theoretically requires the initial values $\{y_0, ..., y_{1-p}\}$ and $\{\boldsymbol{x}_0, ..., \boldsymbol{x}_{1-p}\}$, we shall follow the common practice (see for example, Bauwens et al. (1999) and Koop (2003)).

Instead of using the likelihood for $\{y_1^*, ..., y_n^*\}$, we use the likelihood function with respect to the observations $\{y_{p+1}^*, ..., y_n^*\}$. Thus, the transformed model reduces to the standard linear regression framework described in Section 2.7. Noting that the error terms u_α are independent normal, the density can be expressed as

$$f\left(\boldsymbol{y}_n^*|X_n^*, \boldsymbol{\beta}, \boldsymbol{\rho}, \sigma^2\right) = \frac{1}{(2\pi\sigma^2)^{(n-p)/2}} \exp\left[-\frac{(\boldsymbol{y}_n^* - X_n^*\boldsymbol{\beta})^T(\boldsymbol{y}_n^* - X_n^*\boldsymbol{\beta})}{2\sigma^2}\right],$$

with $\boldsymbol{\rho} = (\rho_1, ..., \rho_p)^T$, and X_n^* and $\boldsymbol{y}_n^*$ are the $(n-p) \times q$ dimensional matrix and $(n-p)$ dimensional vector, respectively, given by

$$X_n^* = \begin{pmatrix} \boldsymbol{x}_{p+1}^{*T} \\ \vdots \\ \boldsymbol{x}_n^{*T} \end{pmatrix} = \begin{pmatrix} \boldsymbol{x}_{p+1} - \sum_{j=1}^p \rho_j \boldsymbol{x}_{p+1-j} \\ \vdots \\ \boldsymbol{x}_n - \sum_{j=1}^p \rho_j \boldsymbol{x}_{n-j} \end{pmatrix} \quad \text{and} \quad \boldsymbol{y}_n^* = \begin{pmatrix} y_{p+1}^* \\ \vdots \\ y_n^* \end{pmatrix}.$$

Under the normal inverse-gamma prior $\pi(\boldsymbol{\beta}, \sigma^2) = \pi(\boldsymbol{\beta}|\sigma^2)\pi(\sigma^2)$, with

$$\pi(\boldsymbol{\beta}|\sigma^2) = N\left(\boldsymbol{\beta}_0, \sigma^2 A^{-1}\right) \quad \text{and} \quad \pi(\sigma^2) = IG\left(\frac{\nu_0}{2}, \frac{\lambda_0}{2}\right),$$

the conditional posterior distribution of $\boldsymbol{\beta}$ given $\boldsymbol{\rho}$ and σ^2 is the multivariate normal distribution. The conditional posterior distribution of σ^2 given $\boldsymbol{\rho}$ is an inverse-gamma distribution:

$$\pi\left(\boldsymbol{\beta}\middle|\sigma^2, \boldsymbol{\rho}, \boldsymbol{y}_n^*, X_n^*\right) = N\left(\hat{\boldsymbol{\beta}}_n^*, \sigma^2 \hat{A}_n^*\right), \quad \pi\left(\sigma^2\middle|\boldsymbol{\rho}, \boldsymbol{y}_n^*, X_n^*\right) = IG\left(\frac{\hat{\nu}_n^*}{2}, \frac{\hat{\lambda}_n^*}{2}\right),$$

with $\hat{\nu}_n^* = \nu_0 + n - q$ and

$$\begin{aligned}
\hat{\boldsymbol{\beta}}_n^* &= \left(X_n^{*T}X_n^* + A\right)^{-1}\left(X_n^{*T}X_n^*\hat{\boldsymbol{\beta}}^* + A\boldsymbol{\beta}_0\right), \\
\hat{\boldsymbol{\beta}}^* &= \left(X_n^{*T}X_n^* + A\right)^{-1} X_n^{*T}\boldsymbol{y}_n^*, \quad \hat{A}_n^* = \left(X_n^{*T}X_n^* + A\right)^{-1}, \\
\hat{\lambda}_n^* &= \lambda_0 + \left(\boldsymbol{y}_n^* - X_n^*\hat{\boldsymbol{\beta}}^*\right)^T\left(\boldsymbol{y}_n^* - X_n^*\hat{\boldsymbol{\beta}}^*\right) \\
&\quad + \left(\boldsymbol{\beta}_0 - \hat{\boldsymbol{\beta}}^*\right)^T\left((X_n^{*T}X_n^*)^{-1} + A^{-1}\right)^{-1}\left(\boldsymbol{\beta}_0 - \hat{\boldsymbol{\beta}}^*\right).
\end{aligned}$$

Finally, specifying the prior density for $\boldsymbol{\rho}$, we shall derive its posterior density. Here we simply restrict the region of $\boldsymbol{\rho}$ and denote Φ. Thus, the prior density of $\boldsymbol{\rho}$ is $\pi(\boldsymbol{\rho}) = I(\boldsymbol{\rho} \in \Phi) \times \text{Const.}$. Note that given value of $\boldsymbol{\beta}$, we have the actual values of the error terms ε_α. Thus, we again can apply the results of the standard linear regression models.

Let E be a $(n-p) \times q$ matrix and $\boldsymbol{\varepsilon}$ be an $(n-p)$ dimensional vector with

$$E = \begin{pmatrix} \varepsilon_p & \cdots & \varepsilon_1 \\ \vdots & \ddots & \vdots \\ \varepsilon_{n-1} & \cdots & \varepsilon_{n-p} \end{pmatrix}, \quad \boldsymbol{\varepsilon} = \begin{pmatrix} \varepsilon_{p+1} \\ \vdots \\ \varepsilon_n \end{pmatrix}.$$

The conditional posterior density of $\boldsymbol{\rho}$, given $\boldsymbol{\beta}$ and σ^2, is then

$$\pi\left(\boldsymbol{\rho} \middle| \boldsymbol{\beta}, \sigma^2, \boldsymbol{y}_n^*, X_n^*\right) = TN\left(\hat{\boldsymbol{\rho}}_n^*, \sigma^2 \hat{V}_n^*, \Phi\right),$$

with

$$\hat{\boldsymbol{\rho}}_n^* = \left(E^T E\right)^{-1} E^T \boldsymbol{\varepsilon}, \quad \hat{V}_n^* = \left(E^T E\right)^{-1},$$

where $TN(\cdot, \cdot, \Phi)$ is the truncated normal distribution with support Φ.

Based on the conditional posterior densities of $\boldsymbol{\beta}$, σ^2, $\boldsymbol{\rho}$, we can run the Gibbs sampling approach. Starting from initial values $\boldsymbol{\beta}^{(0)}$, $\sigma^{2(0)}$ and $\boldsymbol{\rho}^{(0)}$, the following steps are performed:

Gibbs sampling

Step 1. Update the coefficient vector $\boldsymbol{\beta}^{(j)}$ by drawing a new value from the conditional posterior density $\pi(\boldsymbol{\beta} | \sigma^2, \boldsymbol{\rho}, \boldsymbol{y}_n^*, X_n^*)$.

Step 2. Update $\sigma^{2(j)}$ by drawing a new value from the conditional posterior density $\pi(\sigma^2 | \boldsymbol{\rho}, \boldsymbol{y}_n^*, X_n^*)$.

Step 3. Update $\boldsymbol{\rho}^{(j)}$ by drawing a new value from the conditional posterior density $\pi(\boldsymbol{\rho} | \boldsymbol{\beta}, \sigma^2, \boldsymbol{y}_n^*, X_n^*)$.

The process is then repeated a large number of times $j = 1, ...,$. Discarding an initial number of generates as being unrepresentative of the posterior distribution, the remaining samples are then used for the posterior inference.

Remark To generate posterior samples $\boldsymbol{\rho}$ from the truncated normal distribution $TN(\hat{\boldsymbol{\rho}}_n^*, \sigma^2 \hat{V}_n^*, \Phi)$, one first draws from the untruncated normal $N(\hat{\boldsymbol{\rho}}_n^*, \sigma^2 \hat{V}_n^*)$ and then simply discards the sample which lies outside of Φ.

4.3 Data augmentation

The data augmentation is widely used in the context of MCMC framework. Generally, the data augmentation is defined as the methods for constructing iterative optimization or sampling algorithms by introducing unobserved data or latent variables (van Dyk and Meng (2001)). Under a certain model, the data augmentation allows us to develop efficient MCMC algorithms for posterior sampling. Many studies have been conducted to develop an efficient MCMC algorithm by employing the data augmentation. We refer to Gilks et al. (1996), Albert and Chib (1993). To illustrate an idea of the data augmentation for MCMC, we review the method developed by Albert and Chib (1993).

4.3.1 Probit model

Consider the probit model:

$$f(\boldsymbol{y}_n|X_n,\boldsymbol{\beta}) = \prod_{\alpha=1}^{n} \Phi\left(\boldsymbol{x}_\alpha^T\boldsymbol{\beta}\right)^{y_\alpha}\left[1-\Phi\left(\boldsymbol{x}_\alpha^T\boldsymbol{\beta}\right)\right]^{1-y_\alpha},$$

where $\Phi(\cdot)$ is the distribution function of the standard normal, y_α takes values 0 or 1, and $\boldsymbol{x}_\alpha$ is the p-dimensional predictors.

Although one can generate posterior samples from the posterior distribution $\pi(\boldsymbol{\beta}|\boldsymbol{y}_n\boldsymbol{X}_n)$ from the Metropolis-Hasting algorithm, let us consider another approach. First, we introduce the latent variable z_α as follows:

$$z_\alpha = \boldsymbol{x}_\alpha^T\boldsymbol{\beta} + \varepsilon_\alpha, \quad \varepsilon_\alpha \sim N(0,1),$$

such that

$$y_\alpha = \begin{cases} 1, & (z_\alpha \geq 0) \\ 0, & (z_\alpha < 0) \end{cases}.$$

Noting that $z_\alpha = \boldsymbol{x}_\alpha^T\boldsymbol{\beta} + \varepsilon_\alpha \geq 0$ implies $-\varepsilon_\alpha \leq \boldsymbol{x}_\alpha^T\boldsymbol{\beta}$, we have

$$\begin{aligned} P(y_\alpha = 1|\boldsymbol{x}_\alpha) &= P(z_\alpha \geq 0|\boldsymbol{x}_\alpha) \\ &= \Phi\left(\boldsymbol{x}_\alpha^T\boldsymbol{\beta}\right) \\ &= \int_{-\infty}^{\infty} \frac{1}{(2\pi)^{1/2}} \exp\left[-\frac{(z_\alpha - \boldsymbol{x}_\alpha^T\boldsymbol{\beta})^2}{2}\right] I(z_\alpha \geq 0) dz_\alpha \end{aligned}$$

and

$$1 - P(y_\alpha = 1|\boldsymbol{x}_\alpha) = \int_{-\infty}^{\infty} \frac{1}{(2\pi)^{1/2}} \exp\left[-\frac{(z_\alpha - \boldsymbol{x}_\alpha^T\boldsymbol{\beta})^2}{2}\right] I(z_\alpha < 0) dz_\alpha,$$

where $I(z_\alpha \geq 0)$ takes 1 if $z_\alpha \geq 0$ and 0 otherwise.

Specifying the prior for $\boldsymbol{\beta}$, the joint posterior distribution of latent variables $\boldsymbol{z}_n = (z_1, ..., z_n)^T$ and $\boldsymbol{\beta}$ is then

$$\begin{aligned}&\pi(\boldsymbol{\beta}, \boldsymbol{z}_n | \boldsymbol{y}_n, X_n)\\ &\propto \prod_{\alpha=1}^{n} \exp\left[-\frac{(z_\alpha - \boldsymbol{x}_\alpha^T\boldsymbol{\beta})^2}{2}\right] \times \{y_\alpha I(z_\alpha \geq 0) + (1 - y_\alpha) I(z_\alpha < 0)\} \, \pi(\boldsymbol{\beta}).\end{aligned}$$

For a simplicity of illustration, let us specify $\pi(\boldsymbol{\beta}) = \text{Const.}$. Dropping the irrelevant terms in the joint posterior distribution $\pi(\boldsymbol{\beta}, \boldsymbol{z}_n | \boldsymbol{y}_n, X_n)$ with respect to $\boldsymbol{\beta}$, we have

$$\pi(\boldsymbol{\beta} | \boldsymbol{y}_n, X_n, \boldsymbol{z}_n) \propto \exp\left[-\frac{1}{2}(\boldsymbol{z}_n - X_n\boldsymbol{\beta})^T(\boldsymbol{z}_n - X_n\boldsymbol{\beta})\right].$$

Thus, the conditional posterior distribution of $\boldsymbol{\beta}$ given $\boldsymbol{z}_n$ is

$$\pi(\boldsymbol{\beta} | \boldsymbol{y}_n, X_n, \boldsymbol{z}_n) = N\left(\left(X_n^T X_n\right)^{-1} X_n^T \boldsymbol{z}_n, \left(X_n^T X_n\right)^{-1}\right). \tag{4.5}$$

Likewise, we obtain

$$\pi(z_\alpha | y_\alpha, \boldsymbol{x}_\alpha, \boldsymbol{\beta}) \propto \exp\left[-\frac{(z_\alpha - \boldsymbol{x}_\alpha^T\boldsymbol{\beta})^2}{2}\right] \times \{y_\alpha I(z_\alpha \geq 0) + (1 - y_\alpha) I(z_\alpha < 0)\},$$

which implies that the conditional posterior distribution of z_α given $\boldsymbol{\beta}$ is

$$\pi(z_\alpha | y_\alpha, \boldsymbol{x}_\alpha, \boldsymbol{\beta}) = \begin{cases} TN\left(\boldsymbol{x}_\alpha^T\boldsymbol{\beta}, 1, +\right), & (y_\alpha = 1) \\ TN\left(\boldsymbol{x}_\alpha^T\boldsymbol{\beta}, 1, -\right), & (y_\alpha = 0) \end{cases}, \tag{4.6}$$

where $TN(\cdot, \cdot, +)$ is the truncated normal distribution with mean $\boldsymbol{x}_\alpha^T\boldsymbol{\beta}$, the variance 1 and its support is $[0, \infty)$, and $TN(\cdot, \cdot, -)$ is the truncated normal distribution support $(-\infty, 0)$.

Since we know the full conditional posterior densities of $\boldsymbol{\beta}$ and $\boldsymbol{z}_n$, we can use the Gibbs sampler. Starting from an initial value $\boldsymbol{\beta}^{(0)}$ and $\boldsymbol{z}^{(0)}$, the following steps are performed:

Gibbs sampling

Step 1. Update the coefficient vector $\boldsymbol{\beta}^{(j)}$ by drawing a new value from the conditional posterior density $\pi\left(\boldsymbol{\beta} | \boldsymbol{y}_n, X_n, \boldsymbol{z}_n^{(j-1)}\right)$ in (4.5).

Step 2. Update $z_\alpha^{(j)}$ by drawing a new value from the conditional posterior density $\pi\left(z_\alpha | y_\alpha, \boldsymbol{x}_\alpha, \boldsymbol{\beta}^{(j)}\right)$ for $\alpha = 1, ..., n$ in (4.6).

4.3.2 Generating random samples from the truncated normal density

In the Gibbs sampling algorithm for the probit model, we have to generate the random samples truncated normal distribution. We first review a general version of the inverse transform method. Let $F(x)$ be the cumulative probability distribution function of the random variable. If we have an inverse mapping of F, say F^{-1}, we can easily generate random samples using the uniform random variable $u \sim U[0,1]$. Generally, there are two steps in the inverse transform method.

Inverse transform method

Step 1. Generate the uniform random variable $u \sim U[0,1]$.
Step 2. Transform $x = F^{-1}(u)$.

The transformed variable x has a provability distribution function $F(x)$. To show this, let $G(z)$ be the cumulative probability distribution function of the transformed random variable. Then, noting that

$$x_1 \leq x_2 \rightarrow F(x_1) \leq F(x_2)$$

we have

$$\begin{aligned} G(z) &= \Pr(x \leq z) \\ &= \Pr(F(x) \leq F(z)) \\ &= \Pr(F(F^{-1}(u)) \leq F(z)) \\ &= \Pr(u \leq F(z)) \\ &= F(z). \end{aligned}$$

because u is the uniform random variable $u \sim U[0,1]$. Thus, $G(z) = F(z)$.

Generally the probability density function and the cumulative probability distribution function of the truncated normal distribution with the mean μ, the variance σ^2 and its support $S = [a, b]$ are given as

$$\begin{aligned} f(x|\mu,\sigma^2,S) &= \frac{1}{(\Phi(b^*) - \Phi(a^*))} \times \frac{1}{\sqrt{2\pi\sigma^2}} \exp\left(-\frac{(x-\mu)^2}{2\sigma^2}\right) \times I(a \leq x \leq b), \\ F(x|\mu,\sigma^2,S) &= \frac{\Phi\left(\frac{x-\mu}{\sigma}\right) - \Phi(a^*)}{\Phi(b^*) - \Phi(a^*)} \times I(a \leq x \leq b), \end{aligned}$$

where

$$a^* = \frac{a-\mu}{\sigma} \quad \text{and} \quad b^* = \frac{b-\mu}{\sigma},$$

and $\Phi(\cdot)$ is the cumulative probability distribution function of the standard normal distribution. Thus, the transformed random variable $z = (x - \mu)/\sigma$

has the probability density function $f(z)$ and the cumulative probability distribution function $F(z)$ as

$$\begin{aligned} f(z|0,1,S^*) &= \frac{1}{(\Phi(b^*)-\Phi(a^*))} \times \frac{1}{\sqrt{2\pi}} \exp\left(-\frac{z^2}{2}\right) \times I(a^* \le z \le b^*), \\ F(z|0,1,S^*) &= \frac{\Phi(z)-\Phi(a^*)}{\Phi(b^*)-\Phi(a^*)} \times I(a^* \le x \le b^*), \end{aligned}$$

which has the form of the truncated standard normal density form with support $S^* = [a^*, b^*]$.

Thus, using the inverse transform method, we first generate the random variable z from the truncated standard normal density form with support S^*. Then the transformed random variable $x = \mu + z\sigma$ follows the truncated normal distribution with the mean μ, the variance σ^2 and its support S.

4.3.3 Ordered probit model

Suppose that there are J categories and that we observe a set of n data $\boldsymbol{Y}_n = \{y_1, ..., y_n\}$ and related covariates $\boldsymbol{X}_n = \{\boldsymbol{x}_1, ..., \boldsymbol{x}_n\}$. In a similar manner of the data augmentation method for the probit model, let us introduce the latent variable z_α:

$$z_\alpha = \boldsymbol{x}_\alpha^T \boldsymbol{\beta} + \varepsilon_\alpha, \quad \varepsilon_\alpha \sim N(0,1),$$

satisfying

$$y_\alpha = j \quad (\gamma_{j-1} < z_\alpha \le \gamma_j),$$

where the cut-off points $-\infty = \gamma_0, \gamma_1, ..., \gamma_{J-1}, \gamma_J = \infty$ defines a series of ranges into which the latent variable may fall. Then the probability that the observation y_α is assigned to be the category j becomes

$$\begin{aligned} \Pr(y_\alpha = j|\boldsymbol{x}_\alpha) &= \Pr(\gamma_{j-1} < z_\alpha \le \gamma_j \big| \boldsymbol{x}_\alpha) \\ &= \Pr\left(\gamma_{j-1} - \boldsymbol{x}_\alpha^T\boldsymbol{\beta} < \varepsilon_\alpha \le \gamma_j - \boldsymbol{x}_\alpha^T\boldsymbol{\beta} \Big| \boldsymbol{x}_\alpha\right) \\ &= \Phi\left(\gamma_j - \boldsymbol{x}_\alpha^T\boldsymbol{\beta}\right) - \Phi\left(\gamma_{j-1} - \boldsymbol{x}_\alpha^T\boldsymbol{\beta}\right). \end{aligned}$$

The likelihood function for the ordered probit model is then

$$f(\boldsymbol{y}_n|X_n, \boldsymbol{\beta}, \boldsymbol{\gamma}) = \prod_{\alpha=1}^{n} \left[\sum_{j=1}^{J} I(y_\alpha = j) \times \left\{\Phi\left(\gamma_j - \boldsymbol{x}_\alpha^T\boldsymbol{\beta}\right) - \Phi\left(\gamma_{j-1} - \boldsymbol{x}_\alpha^T\boldsymbol{\beta}\right)\right\}\right],$$

where $\boldsymbol{\gamma} = (\gamma_2, ..., \gamma_{J-1})^T$.

The use of diffuse prior $\pi(\boldsymbol{\beta}, \boldsymbol{\gamma}) = \text{Const.}$ leads to the following joint posterior distribution of latent variables $\boldsymbol{z}_n = (z_1, ..., z_n)^T$ and model parameters

$\boldsymbol{\beta}$ and $\boldsymbol{\gamma}$:

$$\begin{aligned}&\pi(\boldsymbol{\beta},\boldsymbol{\gamma},\boldsymbol{z}_n|\boldsymbol{y}_n,X_n)\\&\quad\propto\prod_{\alpha=1}^{n}\exp\left[-\frac{(z_\alpha-\boldsymbol{x}_\alpha^T\boldsymbol{\beta})^2}{2}\right]\times\left\{\sum_{j=1}^{J}I(y_\alpha=j)I(\gamma_{j-1}<z_\alpha<\gamma_j)\right\}.\end{aligned}$$

Thus, the conditional posterior distribution of $\boldsymbol{\beta}$, given $\boldsymbol{z}_n$ and $\boldsymbol{\gamma}$ is

$$\pi(\boldsymbol{\beta}|\boldsymbol{y}_n,X_n,\boldsymbol{z}_n,\boldsymbol{\gamma})=N\left(\left(X_n^TX_n\right)^{-1}X_n^T\boldsymbol{z}_n,\left(X_n^TX_n\right)^{-1}\right).$$

Also,

$$\begin{aligned}&\pi(z_\alpha|y_\alpha,\boldsymbol{x}_\alpha,\boldsymbol{\beta},\boldsymbol{\gamma})\\&\quad\propto\exp\left[-\frac{(z_\alpha-\boldsymbol{x}_\alpha^T\boldsymbol{\beta})^2}{2}\right]\times\left\{\sum_{j=1}^{J}I(y_\alpha=j)I(\gamma_{j-1}<z_\alpha<\gamma_j)\right\},\end{aligned}$$

which implies that the conditional posterior distribution of z_α given $\boldsymbol{\beta}$ and $\boldsymbol{\gamma}$ is

$$\pi(z_\alpha|y_\alpha,\boldsymbol{x}_\alpha,\boldsymbol{\beta},\boldsymbol{\gamma})=TN\left(\boldsymbol{x}_\alpha^T\boldsymbol{\beta},1,S_{\gamma_{y_\alpha}}\right),$$

where $TN(\cdot,\cdot,S_{\gamma_{y_\alpha}})$ is the truncated normal distribution with mean $\boldsymbol{x}_\alpha^T\boldsymbol{\beta}$, the variance 1 and its support is $S_{\gamma_{y_\alpha}}=(\gamma_{y_\alpha-1},\gamma_{y_\alpha})$.

Finally, the conditional posterior distribution of γ_j given $\boldsymbol{z}_n$, $\boldsymbol{\beta}$ and $\boldsymbol{\gamma}_{-\gamma_j}$ is

$$\begin{aligned}&\pi(\gamma_j|\boldsymbol{y}_n,X_n,\boldsymbol{\beta},\boldsymbol{\gamma}_{-\gamma_j},\boldsymbol{z}_n)\\&\quad\propto\prod_{\alpha=1}^{n}\left\{I(y_\alpha=j)I(\gamma_{j-1}<z_\alpha<\gamma_j)+I(y_\alpha=j+1)I(\gamma_j<z_\alpha<\gamma_{j+1})\right\}.\end{aligned}$$

Thus, $\pi(\boldsymbol{\beta},\boldsymbol{\gamma},\boldsymbol{z}_n|\boldsymbol{y}_n,X_n)$ is uniform on the interval

$$\left[\max\left\{\max_\alpha\{z_\alpha|y_\alpha=j\},\gamma_{j-1}\right\},\min\left\{\min_\alpha\{z_\alpha|y_\alpha=j+1\},\gamma_{j+1}\right\}\right].$$

Starting from an initial value $\boldsymbol{\beta}^{(0)}$ and $\boldsymbol{z}^{(0)}$, we can use the following Gibbs sampling algorithm for $k=1,2,....$:

Gibbs sampling

Step 1. Update the coefficient vector $\boldsymbol{\beta}^{(k)}$ by drawing a new value from the conditional posterior density $\pi\left(\boldsymbol{\beta}|\boldsymbol{y}_n,X_n,\boldsymbol{z}_n^{(k-1)},\boldsymbol{\gamma}\right)$.

Step 2. Update the cutoff value $\gamma_j^{(k)}$ by drawing a new value from the conditional posterior density

$$\pi\left(\gamma_j^{(k)}|\boldsymbol{y}_n,X_n,\boldsymbol{z}_n^{(k-1)},\boldsymbol{\gamma}^{(k)},\gamma_2^{(k)},...,\gamma_{j-1}^{(k)},\gamma_{j+1}^{(k-1)},....,\gamma_{J-1}^{(k-1)}\right)$$

for $j = 2, ..., J-1$.

Step 3. Update $z_\alpha^{(k)}$ by drawing a new value from the conditional posterior density $\pi\left(z_\alpha | y_\alpha, \boldsymbol{x}_\alpha, \boldsymbol{\beta}^{(k)}, \boldsymbol{\gamma}^{(k)}\right)$ for $\alpha = 1, ..., n$.

We repeat the Steps 1–3 above for a sufficiently long time.

4.4 Hierarchical modeling

Advancement of information technology has made us able to collect high-dimensional data. In most applications, the use of many predictors usually leads to an over-parameterization problem, especially when the number of predictors exceeds the number of observations. Recent years have seen a development of various statistical modeling approaches; Lasso (Tibishirani (1996)), LARS (Efron, et al. (2004)), adaptive Lasso (Zou (2006)), the dantzig selector (Candes and Tao (2007)) and so forth.

Interpreting the Lasso estimate for linear regression as a Bayesian posterior mode estimate when the regression parameters have independent Laplace (i.e., double-exponential) priors, Park and Casella (2008) developed a Gibbs sampling algorithm. In this section, we cover the issue of regression under the high-dimensional number of predictors and then describe the Bayesian Lasso.

4.4.1 Lasso

Suppose that we have a set of data $\{(\boldsymbol{x}_\alpha, y_\alpha); \alpha = 1, ..., n\}$, where y_α denotes the response variable and $\boldsymbol{x}_\alpha = (x_{1\alpha}, ..., x_{p\alpha})'$ consists of p predictors. For simplicity, we assume that y_α is mean corrected. The Lasso of Tibshirani (1996) estimates linear regression

$$y_\alpha = \beta_1 x_{1\alpha} + \cdots + \beta_p x_{p\alpha} + \varepsilon_\alpha, \quad \alpha = 1, ..., n,$$

through L_1 penalized least squared methods

$$\hat{\boldsymbol{\beta}}_n = \min_{\boldsymbol{\beta}} \left(\boldsymbol{y}_n - X_n \boldsymbol{\beta}\right)^T \left(\boldsymbol{y}_n - X_n \boldsymbol{\beta}\right) + \lambda \sum_{j=1}^{p} |\beta_j|,$$

with $\lambda > 0$. Also, the independent errors $\{\varepsilon_\alpha\}$ have mean zero and variance $\sigma^2 < \infty$.

As pointed out by Tibshirani (1996) and Park and Casella (2008), we can interpret the Lasso estimates $\hat{\boldsymbol{\beta}}_n$ as posterior mode estimates when the regression parameters have independent and identical Laplace priors. In other words, the penalty term of the Lasso corresponds to a conditional Laplace prior specification of the form

$$\pi(\boldsymbol{\beta}|\sigma^2) = \prod_{j=1}^{p} \frac{\lambda}{2\sigma} \exp\left[-\lambda \frac{|\beta_j|}{\sigma}\right].$$

The next section describes the Gibbs sampling algorithm for the Lasso (Park and Casella (2008)).

4.4.2 Gibbs sampling for Bayesian Lasso

In this section, we assume that the errors $\{\varepsilon_i\}$ are independent and normally distributed with mean zero and variance $\sigma^2 < \infty$. To develop the Gibbs sampler for the Bayesian Lasso, Park and Casella (2008) exploited the following representation of the Laplace distribution as a well known scale mixture of normals (see Andrews and Mallows (1974)):

$$\frac{a}{2}\exp\{-a|z|\} = \int_0^\infty \frac{1}{2\pi s} \exp\left[-\frac{z^2}{2s}\right] \frac{a^2}{2} \exp\left[-\frac{a^2 s}{2}\right] ds, \quad a > 0.$$

This suggests the following hierarchical representation of the prior

$$\pi(\boldsymbol{\beta}|\sigma^2, \tau_1^2, ..., \tau_p^2) = N(\mathbf{0}, \sigma^2 D_\tau)$$

$$\pi(\sigma^2, \tau_1^2, ..., \tau_p^2, \lambda) = \pi(\sigma^2) \prod_{j=1}^{p} \frac{\lambda^2}{2} \exp\left(-\frac{\lambda^2 \tau_j^2}{2}\right),$$

where $D_\tau = \text{diag}(\tau_1^2, ..., \tau_p^2)$. Park and Casella (2008) considered the noninformative scale-invariant marginal prior $\pi(\sigma^2) = 1/\sigma^2$, and the class of gamma priors on λ^2 of the form

$$\pi(\lambda^2) = \frac{b^r}{\Gamma(r)} (\lambda^2)^{r-1} \exp\left(-b\lambda^2\right),$$

with $b, r > 0$.

Then the full conditional posterior distributions of $\boldsymbol{\beta}$, σ^2, $\tau_1^2, ..., \tau_p^2$, and λ are given as follows. The full conditional posterior distribution for $\boldsymbol{\beta}$ is multivariate normal with mean $(X_n^T X_n + D_\tau^{-1})^{-1} X_n^T \boldsymbol{y}_n$ and the covariance matrix $\sigma^2 (X_n^T X_n + D_\tau^{-1})^{-1}$. The full conditional distribution for σ^2 is inverse-gamma with parameters $(n+p-1)/2$ and $(\boldsymbol{y}_n - X_n\boldsymbol{\beta})^T(\boldsymbol{y}_n - X_n\boldsymbol{\beta})/2 + \boldsymbol{\beta}^T D_\tau^{-1}\boldsymbol{\beta}/2$. The full conditional posterior distribution for τ_j^2 is inverse-Gaussian with parameters $\mu = \sqrt{\lambda^2\sigma^2/\beta_j^2}$ and $\lambda' = \lambda^2$. Here the parameterization of the inverse-Gaussian density is given by

$$f(x|\mu, \lambda') = \sqrt{\frac{\lambda'}{2\pi}} x^{-3/2} \exp\left[-\frac{\lambda'(x-\mu)^2}{2\mu^2 x}\right].$$

Finally, the full posterior conditional distribution of λ^2 is gamma distribution with parameter $p+r$ and $\sum_{j=1}^{p}\tau_j^2/2+b$. Thus, starting from an initial value, we can use the following Gibbs sampling algorithm.

Gibbs sampling

Step 1. Update the coefficient vector $\boldsymbol{\beta}^{(k)}$ by drawing a new value from the conditional posterior density.

Step 2. Update the variance parameter σ^2 by drawing a new value from the conditional posterior density of σ^2.

Step 3. Update the parameter τ_j by drawing a new value from the conditional posterior posterior density, for $j=1,...,p$.

Step 4. Update the penalty parameter λ by drawing a new value from the conditional posterior density of λ.

We repeat Steps 1–3 above for a sufficiently long time.

4.5 MCMC studies for the Bayesian inference on various types of models

4.5.1 Volatility time series models

It is a well-known fact that the volatility of financial asset return changes randomly over time. This phenomenon motivates the commonly used Autoregressive Conditionally Heteroskedastic model (ARCH; Engle (1982)). The ARCH-type models have recently attracted the attention of financial economics researchers and practitioners. Let us observe a time series of observations, $\boldsymbol{y}_n=(y_1,...,y_n)^T$. Then the ARCH model of order p captures the time-varying volatility as

$$\begin{cases} y_t=h_tu_t \\ h_t^2=\mu+\beta_1y_{t-1}^2+\cdots+\beta_py_{t-p}^2 \end{cases},$$

where h_t is a volatility of y_t and $u_t\sim N(0,1)$ are Gaussian white noise sequence. There are many literatures on ARCH and its extended models. Generalized Autoregressive Conditionally Heteroskedastic model (GARCH; Bollerslev(1986)) is one of the extensions:

$$\begin{cases} y_t=h_tu_t \\ h_t^2=\mu+\beta_1y_{t-1}^2+\cdots+\beta_py_{t-p}^2+\gamma_1h_{t-1}^2+\cdots+\gamma_qh_{t-q}^2 \end{cases}.$$

For more details on the characteristics and the Bayesian inference for the ARCH-type models, we refer to Geweke (1989b), Bauwens et al. (1999), and Tsay (2002).

As an alternative to the ARCH-type models, Stochastic Volatility (SV) models, pioneered by Taylor (1982), provide useful tools to describe the evolution of asset returns, which exhibit time-varying volatility. In the context of the basic SV model, the observation equation and the system equation are specified as follows:

$$\begin{cases} y_t = \exp(h_t/2)u_t, \\ h_t = \mu + \phi(h_{t-1} - \mu) + \tau v_t, \end{cases} \quad (t = 1, ..., n).$$

Here $\boldsymbol{y}_n = (y_1, ..., y_n)^T$ is a time series of observations, h_t is an unobserved log-volatility of y_t and $u_t \sim N(0,1)$ and $v_t \sim N(0,1)$ are uncorrelated Gaussian white noise sequences. The scaling factors $\exp(h_t/2)$ and $\exp(\mu/2)$ specify the amount of volatility on time t and the model volatility, τ determines the volatility of log-volatilities and ϕ measures the autocorrelation.

A critical difference between the SV type models and ARCH type models is the difficulty level of the likelihood evaluation. Since ARCH type models specify the volatility of the current return as a nonstochastic function of past observations, it is easy to evaluate the likelihood function. On the other hand, the SV type models generally specify the volatility as a stochastic process and the likelihood function depends upon high-dimensional integrals:

$$f(\boldsymbol{y}_n|\boldsymbol{\theta}) = \prod_{t=1}^{n} f(y_t|I_{t-1}, \boldsymbol{\theta}) = \prod_{t=1}^{n} \int f(y_t|h_t, \boldsymbol{\theta}) f(h_t|I_{t-1}, \boldsymbol{\theta}) dh_t,$$

where I_{t-1} denotes the history of the information sequence up to time $t-1$, $\boldsymbol{\theta}$ is an unknown parameter vector. The two conditional density functions in the likelihood function are specified by the observation equation and the system equation, respectively. Thus it is not straightforward to construct the likelihood function of the SV type models and to implement the maximum likelihood method.

Numerous studies have been conducted to estimate the volatility process based on MCMC approach with extension of the basic SV model (Ando (2006), Berg et al. (2004), Chib et al. (2002), Jacquier et al. (2004), Meyer and Yu (2000), Tanizaki (2004), Yu (2005)), Tanizaki (2004)), Barndorff-Nielssen and Shephard (2001). Comprehensive reviews of the SV models can be found in the work of Shephard (2005). Gerlach and Tuyl (2006) studied MCMC and importance sampling techniques for volatility estimation, model misspecification testing and comparisons for general volatility models, including GARCH and SV models.

4.5.2 Simultaneous equation model

In many areas of economics and other sciences, sets of variables are often jointly generated with instantaneous feedback effects present. For instance, a fundamental feature of markets is that prices and quantities are jointly determined with current prices dependent on current quantities and current quantities dependent on current prices along with other variables appearing in these relationships. The Simultaneous Equation Model (SEM), that incorporates instantaneous feedback relationships, was put forward many years ago and has been widely employed to analyze the behavior of markets, economies and other multivariate systems.

As an intuitive example, let us consider the supply and demand of a particular economic product. The quantity supplied of a good Q_s depends upon the price P and other factors $\boldsymbol{x}_s$. In a similar manner, the quantity demand of a good Q_d depends upon the price P and other factors $\boldsymbol{x}_d$. In equilibrium, the economics theory considers that the quantity demand of a good equals to the quantity supplied of a good $Q_s = Q_d \equiv Q$. Thus, we can consider the following log-linear demand and supply model:

$$\begin{cases} \log Q = \gamma_s \log P + \boldsymbol{\beta}_s^T \boldsymbol{x}_s + \varepsilon_s, \\ \log Q = \gamma_d \log P + \boldsymbol{\beta}_d^T \boldsymbol{x}_d + \varepsilon_d, \end{cases}$$

where γ_d and γ_s are unknown coefficients that are related to a price elasticity. Since the quantity demanded Q_d and the quantity supplied Q_s are effecting each other, we cannot handle the estimation problem of this model by using the traditional linear regression model. Thus, SEM plays an important role in the economic studies.

Generally, the m equation SEM is defined as:

$$\boldsymbol{Y}_n \Gamma = X_n B + E_n,$$

where $\boldsymbol{Y}_n = (\boldsymbol{y}_1, ..., \boldsymbol{y}_m)$ is an $n \times m$ matrix of observations on m endogenous variables, the $m \times m$ nonsingular matrix Γ is a matrix coefficient for the endogenous variables, $X_n = (\boldsymbol{x}_1, ..., \boldsymbol{x}_n)^T$ is an $n \times p$ matrix of observations on the p predetermined variables, the $p \times m$ matrix $B = (\boldsymbol{b}_1, ..., \boldsymbol{b}_m)$ is the coefficient matrix for the predetermined variables, and $E_n = (\boldsymbol{\varepsilon}_1, ..., \boldsymbol{\varepsilon}_m)$ is the $n \times m$ error matrix. It is known that some restrictions on the parameters are needed for model identification.

Multiplying both sides of $\boldsymbol{Y}_n \Gamma = X_n B + E_n$, by Γ^{-1}, the unrestricted reduced form equations are

$$\boldsymbol{Y}_n = X_n \Pi + U_n, \tag{4.7}$$

where $\Pi = B\Gamma^{-1} = (\boldsymbol{\pi}_1, ..., \boldsymbol{\pi}_m)$ is a $p \times m$ reduced form coefficient matrix, and $U_n = E_n \Gamma^{-1} = (\boldsymbol{u}_1, ..., \boldsymbol{u}_m)$ is the reduced form error matrix. The n rows of U_n are generally assumed to be independently drawn from a multivariate normal distribution with zero mean vector and $m \times m$ positive definite covariance matrix

$$\Sigma = \Gamma^{-1} \text{Cov}(\boldsymbol{\varepsilon}) \Gamma,$$

where $\text{Cov}(\boldsymbol{\varepsilon})$ is the covariance matrix of the error term in the structural form of SEM.

Thus, the problem becomes how to estimate the unknown parameters in the restricted model (4.7), Π and Σ. Noting that the Bayesian inference on the unrestricted reduced form equations can be implemented by using the method of the SUR models, we can make an inference. However, usually, we are also interested in the Bayesian inference on the structural form of SEM. Using the relationship between the parameters

$$\Pi = B\Gamma^{-1} \quad \text{and} \quad \Sigma = \Gamma^{-1}\text{Cov}(\boldsymbol{\varepsilon})\Gamma,$$

we can work on the Bayesian inference after we make an inference on the unrestricted reduced form equation model. However, the number of parameters of the structural form of SEM usually larger than that of the unrestricted reduced form equation model. Therefore, some restrictions are required.

Many studies have been conducted to develop estimation, testing, prediction and other inference techniques for the SEM from both the Bayesian points of view; see, e.g., Drèze (1976), Drèze and Morales (1976), Zellner, Bauwens and van Dijk (1988), Chao and Phillips (1998), Kleibergen and Zivot (2003), Kleibergen and van Dijk (1998) and the references cited in these works as well as past and recent Bayesian and non-Bayesian econometrics textbooks. Rossi et al. (2005) applied MCMC algorithm to the Bayesian inference on the SEM.

4.5.3 Quantile regression

Quantile regression (Koenker and Bassett (1978), Koenker (2005)), a comprehensive statistical methodology for estimating models of conditional quantile functions, is now well known by its widespread applications. By complementing the focus on the conditional mean of classical linear regression, it allows us to estimate the effect of covariates not only in the center of a distribution, but also in the upper and lower tails. Now, quantile regression is used widely in empirical research and also investigated extensively from theoretical aspects. Quantile regression involves many important applications of the study of various study fields. For example, the analysis extremely low infant birthweights data (Abrevaya (2001)), the auction data (Donald and Paarsch (1993)), and identification of factors of high risk in finance (Tsay (2002)), the survival analysis (Koenker and Geling (2001)), and so on.

Suppose that we have a set of data $\{(\boldsymbol{x}_\alpha, y_\alpha); \alpha = 1, ..., n\}$, where y_α denotes the response variable and $\boldsymbol{x}_\alpha = (x_{1\alpha}, ..., x_{p\alpha})^T$ consists of p predictors. A quantile regression approach to estimate the conditional quantile regression function is suggested in Koenker and Bassett (1978). Consider the following standard linear model:

$$y_\alpha = \beta_1 x_{1\alpha} + \cdots + \beta_p x_{p\alpha} + \varepsilon_\alpha, \quad \alpha = 1, ..., n, \tag{4.8}$$

where the errors $\{\varepsilon_\alpha\}$ are independent and have mean zero and variance $\sigma^2 < \infty$. A quantile of the conditional distribution of the response variable are of interest. It should be noticed that it is not necessary to specify the distribution of the error.

Contrast to the classical theory of linear models, where a conditional expectation of the response variable is in focus, the quantile regression tries to estimate the τ-th conditional quantile of y_i given $\boldsymbol{x}_i$,

$$q_\tau(y_\alpha|\boldsymbol{x}_\alpha) = \beta_1(\tau)x_{1\alpha} + \cdots + \beta_p(\tau)x_{p\alpha}, \quad \alpha = 1, ..., n,$$

where $\boldsymbol{\beta}(\tau) = (\beta_1(\tau), ..., \beta_p(\tau))^T$ is a vector of coefficients that is dependent on the quantile τ. The τth regression quantile is obtained by minimizing the following cost function:

$$C_\tau(\boldsymbol{\beta}(\tau)) = \sum_{\alpha=1}^{n} \rho_\tau\left(y_\alpha - \boldsymbol{x}_\alpha^T \boldsymbol{\beta}(\tau)\right), \tag{4.9}$$

where the loss function is $\rho_\tau(u) = u(\tau - I(u < 0))$. When we set $\tau = 0.5$, the problem reduces to the conditional median regression, which is more robust to outliers than the conditional mean regression.

This is a linear optimization problem. Koenker and Bassett (1978) showed that a solution of (4.9) is

$$\hat{\boldsymbol{\beta}}(\tau) = X_n^{-1}(h)\boldsymbol{y}_n(h), \tag{4.10}$$

where h is a p element index set from the set $\{1, 2, ..., n\}$, $X_n(h)$ refer to rows in X_n, $\boldsymbol{y}_n(h)$ refer to elements in $\boldsymbol{y}_n$. This notation is also used in Koenker (2005).

It can be easily shown that the minimization of the loss function (4.9) is exactly equivalent to the maximization of a likelihood function formed by combining independently distributed asymmetric Laplace densities (Yu and Moyeed (2001)). The probability density of asymmetric Laplace distribution with parameter τ is given by

$$f_\tau(u) = \tau(1 - \tau)\exp\{-\rho_\tau(u)\},$$

with $0 < \tau < 1$. The estimate $\hat{\boldsymbol{\beta}}(\tau)$ can also be obtained by maximizing the following likelihood function:

$$f_\tau(\boldsymbol{y}_n|X_n, \boldsymbol{\beta}) = \tau^n(1 - \tau)^n \exp\left\{-\sum_{\alpha=1}^{n} \rho_\tau(y_\alpha - \boldsymbol{x}_\alpha^T \boldsymbol{\beta}(\tau))\right\}.$$

Yu and Moyeed (2001) developed the theoretical framework of quantile regression from a Bayesian perspective. Specifying the prior density $\pi(\boldsymbol{\beta})$, one can implement a Bayesian inference on the quantile regression models. As pointed out by Yu and Moyeed (2001), a standard conjugate prior distribution is not available for the quantile regression formulation. Therefore, MCMC methods are used for extracting the posterior distributions of unknown parameters. Yu and Stander (2007) developed a Bayesian framework for Tobit quantile regression modeling.

4.5.4 Graphical models

Graphical Gaussian model, known as covariance selection model (Dempster (1972)) allows the description of dependencies between stochastic variables. There are amounts of literature on graphical Gaussian model both in theoretical and practical aspects, together with the wide range of practical applications of the methodologies (Whittaker (1990), Cox and Wermuth (1996), Edwards (2000), Drton and Perlman (2004), Lauritzen (1996)).

Consider an undirected graph $G = (\boldsymbol{X}, E)$ with a set of p random variables $\boldsymbol{x} = (x_1, ..., x_p)^T$ and edge set $E = \{e_{ij}; i = 1, ..., p,\ j < i\}$ where $e_{ij} = 1$ is set to be one or zero according to the vertices between x_i and x_j, included in G or not. The focus of interest is how to optimize the graph structure, or equivalently, how to determine the edge set E. In the context of graphical Gaussian model, the p-variate normal distribution $N_p(\mu, \Sigma)$ is assumed for $\boldsymbol{x}$. Under the positive definiteness of Σ, it is known that the precision matrix $\Sigma_{ij}^{-1} = (\sigma^{ij})$ satisfies the following property $e_{ij} = 0 \iff \sigma^{ij} = 0$. The key idea behind Gaussian graphical models is, therefore, to use the partial correlation coefficients

$$\rho^{ij} = -\frac{\sigma^{ij}}{\sqrt{\sigma^{ii}\sigma^{jj}}},$$

as a measure of conditional independence between two variables x_i and x_j given $\boldsymbol{x}/\{x_i, x_j\}$. The partial correlations that remain significantly different from zero may be taken as indicators of a possible causal link (i.e., $\rho^{ij} >> 0 \rightarrow e_{ij} = 1$).

To reflect the uncertainty in the graph inference, great efforts for graphical Gaussian models are now being undertaken from a Bayesian point of view (Giudici and Green (1999), Roverato (2002), Dellaportas et al. (2003), Wong et al. (2003)).

4.5.5 Multinomial probit models

Discrete choice models are widely used in the social and natural sciences. The multinomial probit model is often appealing because it does not imply the independence of irrelevant alternatives property, unlike other models, such as the multinomial logit models (Nobile (1998)).

Suppose there are J categories and that $\boldsymbol{y}_\alpha = (y_{1\alpha}, .., y_{J\alpha})^T$ is a multinomial vector, with $y_{j\alpha} = 1$ if an individual α chooses alternative j, and $y_{j\alpha} = 0$ otherwise. Let $\boldsymbol{z}_\alpha = (y_{1\alpha}, .., y_{J\alpha})^T$ be the unobserved utility vector of an individual α. Each individual chooses the alternative yielding maximum utility,

$$y_{j\alpha} = 1 \quad \text{such that} \quad u_{j\alpha} = \text{argmax}_k u_{k\alpha}.$$

As well as the Bayesian analysis of probit models, it is assumed that the utility vectors $\boldsymbol{u}_\alpha = (u_{1\alpha}, ..., u_{J\alpha})^T$ are linear functions of p predictors plus an error term:

$$\boldsymbol{u}_\alpha = W_\alpha \boldsymbol{\beta} + \boldsymbol{\varepsilon}_\alpha, \quad \boldsymbol{\varepsilon}_\alpha \sim N(\mathbf{0}, \Omega), \tag{4.11}$$

where $\boldsymbol{\beta}$ is the vector of p coefficients, $W_\alpha = (\boldsymbol{w}_{1\alpha}, ..., \boldsymbol{w}_{J\alpha})^T$ is $J \times p$ matrix of covariates, and $\boldsymbol{\varepsilon}_\alpha$ is a zero mean normal error term with covariance matrix Ω.

We note that an arbitrary constant can be added to both sides of (4.11) and some restriction is needed for the model identification. This problem is commonly dealt with by subtracting the p-th equation in (4.11) from the remaining first $(p-1)$ equations. Thus we obtain

$$\boldsymbol{z}_\alpha = X_\alpha \boldsymbol{\beta} + \boldsymbol{\varepsilon}_\alpha, \quad \boldsymbol{\varepsilon}_\alpha \sim N(\mathbf{0}, \Sigma), \tag{4.12}$$

with

$$X_\alpha = \begin{pmatrix} \boldsymbol{w}_{1\alpha}^T - \boldsymbol{w}_{J\alpha}^T \\ \vdots \\ \boldsymbol{w}_{J-1,\alpha}^T - \boldsymbol{w}_{J\alpha}^T \end{pmatrix} \quad \text{and} \quad \boldsymbol{z}_n = \begin{pmatrix} u_{1\alpha} - u_{J\alpha} \\ \vdots \\ u_{J-1\alpha} - u_{J\alpha} \end{pmatrix},$$

and the covariance matrix of the new error term is

$$\Sigma = [I, -\mathbf{1}]\Omega[I, -\mathbf{1}]^T,$$

with I denoting the $(J-1)$-dimensional identity matrix and $\mathbf{1} = (1, ..., 1)^T$ a $(J-1)$ dimensional vector. The multinomial choice probability vector of an individual α is then given as the $(J-1)$-dimensional multivariate normal integrals

$$\Pr(y_{j\alpha} = 1 | X_\alpha, \boldsymbol{\beta}, \Sigma) = \int_{S_j} \frac{1}{(2\pi)^{(J-1)/2} |\Sigma|^{-1/2}} \exp\left[-\frac{1}{2} \boldsymbol{\varepsilon}_\alpha^T \Sigma^{-1} \boldsymbol{\varepsilon}_\alpha \right] d\boldsymbol{\varepsilon}_\alpha,$$

where the sets S_j are given by

$$S_j = \cap_{k \neq j} \left\{ \varepsilon_{j\alpha} - \varepsilon_{k\alpha} > (\boldsymbol{x}_{k\alpha} - \boldsymbol{x}_{j\alpha})^T \boldsymbol{\beta} \right\} \cap \left\{ \varepsilon_{j\alpha} > -\boldsymbol{x}_{j\alpha}^T \boldsymbol{\beta} \right\}.$$

Thus the likelihood function for the multinomial probit model is then

$$f(\boldsymbol{y}_n | X_n, \boldsymbol{\beta}, \Sigma) = \prod_{\alpha=1}^{n} \left[\prod_{j=1}^{J} \Pr(y_{j\alpha} = 1 | X_\alpha, \boldsymbol{\beta}, \Sigma)^{y_{j\alpha}} \right].$$

The individual's choice $y_{j\alpha} = 1$ can be re-expressed in terms of the utility differentials $\boldsymbol{z}_\alpha$ as follows:

$$y_{j\alpha} = \begin{cases} 1 & \text{if} \quad z_{j\alpha} = \max_k z_{k\alpha} > 0 \\ 0 & \text{if} \quad \max_k z_{k\alpha} < 0 \end{cases}.$$

Note that the model still lacks an identification since multiplication of both sides of the transformed model (4.12) by a positive constant leaves unaltered the distribution of $\boldsymbol{\varepsilon}$. Usually, this problem is solved by restricting the (1,1) element of Σ to be unity: $\sigma_{11} = 1$. After we specify the prior distributions $\pi(\boldsymbol{\beta}, \Sigma)$, the posterior inference can be implemented by using the MCMC algorithm.

Albert and Chib (1993), McCulloch and Rossi (1994), Nobile (1998), Imai and van Dyk (2005) have studied the Bayesian inference for the multinomial probit model. Forster et al. (2003) developed Metropolis-Hastings algorithms for exact conditional inference for multinomial logistic regression models.

4.5.6 Markov switching models

In the research of macroeconomic and financial time series, we often realize that the conventional framework with a fixed probability model may not be suitable to capture the movements in the observed time series. To treat such phenomenon, Hamilton (1989) introduced a Markov-switching model, in which the latent state variable controlling the regime shifts follows a Markov-chain. An advantage of the Markov-switching model is that it can take into account the structural shift in time series.

Suppose that we have a series of asset return process $\boldsymbol{y}_n = (y_1, ..., y_n)^T$, jointly observed with the p-dimensional covariate vector $X_n = \{\boldsymbol{x}_1, ..., \boldsymbol{x}_n\}$. To get an idea of the Markov-switching model, let us first illustrate the following simple model:

$$\begin{cases} y_t = \boldsymbol{x}_t^T \boldsymbol{\beta}_1 + \varepsilon_t, & t = 1, ..., k \\ y_t = \boldsymbol{x}_t^T \boldsymbol{\beta}_2 + \varepsilon_t, & t = k+1, ..., n \end{cases},$$

where $\boldsymbol{\beta}_j, j = 1, 2$ are the coefficient parameter vector, and k is the time point that the structural change on the coefficient parameter occurs. For simplicity, we assume that the error term follows the normal distribution

$$\begin{cases} \varepsilon_t \sim N(0, \sigma_1^2), & t = 1, ..., k \\ \varepsilon_t \sim N(0, \sigma_2^2), & t = k+1, ..., n \end{cases}.$$

Given value of k, this model can be expressed as a matrix form

$$\boldsymbol{y}_{jn} = X_{jn} \boldsymbol{\beta}_j + \boldsymbol{\varepsilon}_{jn}, \quad \boldsymbol{\varepsilon}_{jn} \sim N(0, \sigma_j^2 I),$$

with

$$X_{1n} = \begin{pmatrix} \boldsymbol{x}_1^T \\ \vdots \\ \boldsymbol{x}_k^T \end{pmatrix}, \; X_{2n} = \begin{pmatrix} \boldsymbol{x}_{k+1}^T \\ \vdots \\ \boldsymbol{x}_n^T \end{pmatrix}, \; \boldsymbol{y}_{1n} = \begin{pmatrix} y_1 \\ \vdots \\ y_k \end{pmatrix}, \; \boldsymbol{y}_{2n} = \begin{pmatrix} y_{k+1} \\ \vdots \\ y_n \end{pmatrix}.$$

To make the Bayesian inference on this model, let us specify the conjugate prior density for $(\boldsymbol{\beta}_j, \sigma_j^2)$, $j = 1, 2$, introduced in Section 2.7.

$$\pi\left(\boldsymbol{\beta}_j, \sigma_j^2\right) = \pi\left(\boldsymbol{\beta}_j | \sigma_j^2\right) \pi\left(\sigma_j^2\right),$$

with

$$\pi\left(\boldsymbol{\beta}_j|\sigma_j^2\right) = N\left(\boldsymbol{\beta}_{j0}, \sigma_j^2 A_j^{-1}\right), \quad \pi\left(\sigma_j^2\right) = IG\left(\frac{\nu_{j0}}{2}, \frac{\lambda_{j0}}{2}\right).$$

From the result of Section 2.7, given value of k, we already know that the conditional posterior distribution of $\boldsymbol{\beta}_j$ is normal, and the marginal posterior distribution of σ_j^2 is inverse-gamma distribution:

$$\pi\left(\boldsymbol{\beta}_j, \sigma_j^2\middle|\boldsymbol{y}_n, X_n, k\right) = \pi\left(\boldsymbol{\beta}_j\middle|\sigma_j^2, \boldsymbol{y}_n, X_n, k\right)\pi\left(\sigma_j^2\middle|\boldsymbol{y}_n, X_n, k\right),$$

with

$$\pi\left(\boldsymbol{\beta}_j\middle|\sigma_j^2, \boldsymbol{y}_n, X_n, k\right) = N\left(\hat{\boldsymbol{\beta}}_{jn}, \sigma_j^2\hat{A}_{jn}\right),$$
$$\pi\left(\sigma_j^2\middle|\boldsymbol{y}_n, X_n, k\right) = IG\left(\frac{\hat{\nu}_{jn}}{2}, \frac{\hat{\lambda}_{jn}}{2}\right),$$

where $\hat{\nu}_{jn} = \nu_0 + n_j$ with $n_1 = k$ and $n_2 = n - k$,

$$\hat{\boldsymbol{\beta}}_{jn} = \left(X_{jn}^T X_{jn} + A_j\right)^{-1}\left(X_{jn}\boldsymbol{y}_{jn} + A_j\boldsymbol{\beta}_{j0}\right), \quad \hat{A}_{jn} = (X_{jn}^T X_{jn} + A_j)^{-1},$$
$$\hat{\lambda}_{jn} = \lambda_{j0} + \left(\boldsymbol{y}_{jn} - X_{jn}\hat{\boldsymbol{\beta}}_{j,\mathrm{MLE}}\right)^T\left(\boldsymbol{y}_{jn} - X_{jn}\hat{\boldsymbol{\beta}}_{j,\mathrm{MLE}}\right)$$
$$+ \left(\boldsymbol{\beta}_{j0} - \hat{\boldsymbol{\beta}}_{j,\mathrm{MLE}}\right)^T\left((X_{jn}^T X_{jn})^{-1} + A_j^{-1}\right)^{-1}\left(\boldsymbol{\beta}_{j0} - \hat{\boldsymbol{\beta}}_{j,\mathrm{MLE}}\right),$$
$$\hat{\boldsymbol{\beta}}_{j,\mathrm{MLE}} = \left(X_{jn}^T X_{jn}\right)^{-1} X_{jn}^T\boldsymbol{y}_{jn},$$

Although one can easily implement the posterior Gibbs sampling, the change point, k, is usually unknown. Thus, for the change point k, we use the uniform prior on the range $[k_{\min}, k_{\max}]$:

$$\pi(k) = \frac{1}{k_{\max} - k_{\min} + 1}, \quad k = k_{\min}, \ldots, k_{\max},$$

that is independent from the priors for $(\boldsymbol{\beta}_j, \sigma_j^2)$, $j = 1, 2$. It then follows Bayes' theorem that the conditional posterior distribution of k is given as

$$\pi\left(k|\boldsymbol{y}_n, X_n, \boldsymbol{\beta}_1, \boldsymbol{\beta}_2, \sigma_1^2, \sigma_2^2\right) = \frac{\prod_{j=1}^{2}\pi\left(\boldsymbol{\beta}_j, \sigma_j^2\middle|\boldsymbol{y}_n, X_n, k\right)}{\sum_{i=k_{\min}}^{k_{\max}}\prod_{j=1}^{2}\pi\left(\boldsymbol{\beta}_j, \sigma_j^2\middle|\boldsymbol{y}_n, X_n, i\right)},$$

$k = k_{\min}, \ldots, k_{\max}$. Thus, we can easily implement the Gibbs sampling algorithm.

Let us next cover the general idea behind a class of Markov switching models. In this class of models, there is an underlying state s_t that takes one of S states $s_t \in \{1, 2, ..., S\}$, which represents the probability of being in a particular state. The state s_t is generally assumed to be governed by an S-state Markov process (Hamilton (1989)).

$$f(y_t|\boldsymbol{F}_{t-1}, s_t) = \begin{cases} f(y_t|\boldsymbol{F}_{t-1}, \boldsymbol{\theta}_1), & s_t = 1 \\ \vdots & \\ f(y_t|\boldsymbol{F}_{t-1}, \boldsymbol{\theta}_S), & s_t = S \end{cases},$$

where $\boldsymbol{\theta}_j$ is the parameter vector associated with regime j, $\boldsymbol{F}_t$ denotes the history of the information sequence up to time t, including some exogenous variables $X_t = \{\boldsymbol{x}_1, ..., \boldsymbol{x}_t\}$, observable time series $\boldsymbol{Y}_{t-1} = \{y_1, ..., y_{t-1}\}$ up to time $t-1$ and so on.

To complete the formulation of the model, we next formulate a mechanism that governs the evolution of the stochastic and unobservable regimes. In Markov switching models, the regime process s_t follows an ergodic S-state Markov process with the transition probabilities:

$$P(s_t = j|s_{t-1} = i) = p_{ij},$$

with $\sum_{j=1}^{S} p_{ij} = 1$. This stochastic structure is also expressed as a transition matrix

$$P = \begin{pmatrix} p_{11} & \cdots & p_{1S} \\ \vdots & \ddots & \vdots \\ p_{S1} & \cdots & p_{SS} \end{pmatrix}.$$

The transition probability parameter p_{ij} represents the transition probability that the next state will be state j, given that the current state is i. Therefore, the density function of $\boldsymbol{y}_t$ is a function of the underlying state s_t, which is assumed to follow an S-state Markov process.

An example of the Markov switching model is the Markov switching stochastic volatility model:

$$\begin{cases} y_t = \exp(h_t/2)\varepsilon_t, \\ h_t = \mu_{s_t} + \phi(h_{t-1} - \mu_{s_{t-1}}) + \tau v_t, \\ P(s_t = j|s_{t-1} = i) = p_{ij}. \end{cases}$$

The error terms ε_t and v_t are assumed to follow the standard normal. The scaling factor $\exp(h_t/2)$ specifies the amount of the model volatility on day t, τ determines the volatility of log-volatilities and ϕ measures the autocorrelation. As shown in this model specification, the level of the model volatility depends on the underlying state s_t. If we assume there are two regimes $S = 2$, it implies there are a low and high volatility states.

Albert and Chib (1993) developed the Gibbs-sampling for the Markov regime-switching model. Kim and Nelson (1998) employed the Gibbs-sampling for the Bayesian analysis of the state-space model with Markov switching. Fruhwirth-Schnatter (2001) used MCMC method for fully Bayesian analysis of Markov Gaussian state space models. So et al. (1998), Ando (2006) and Shibata and Watanabe (2005) conducted the Bayesian analysis of a stochastic volatility model with Markov switching using MCMC approach. Billio et al. (1999) implemented MCMC for the Bayesian inference in switching ARMA models. Chopin and Pelgrin (2004) conducted the Bayesian inference in switching regression models. They proposed a new method that delivers a joint estimation of the parameters and the number of regimes. An excellent textbook for Markov switching state space model is provided by Kim and Nelson (1999).

4.6 Noniterative computation methods for Bayesian inference

Although much recent methodological progress in Bayesian analysis has been due to use of Markov chain Monte Carlo (MCMC) methods, there are many Bayesian inference procedures that are usually based on a noniterative algorithm. Thus, it avoids the convergence diagnosis difficulties caused by the use of MCMC methods.

Suppose we wish to approximate a posterior expectation,

$$\int h(\boldsymbol{\theta})\pi\left(\boldsymbol{\theta}|\boldsymbol{X}_n\right) d\boldsymbol{\theta}.$$

Monte Carlo integration approximates the integrand by using the independent posterior samples $\boldsymbol{\theta}^{(1)}, ..., \boldsymbol{\theta}^{(L)}$. This section covers several other approaches for calculating this quantity.

4.6.1 The direct Monte Carlo

The direct Monte Carlo or Direct sampling approach (Geweke 2005, pp. 106–109) is another important and well-known procedure in posterior computing. Under the situation that the joint posterior density for the parameters is analytically converted into a product of conditional and marginal densities from which draws can be made, we can easily yield a draw from the joint posterior density. When the marginal posterior distributions of each of the parameters are obtained analytically, it is straightforward to draw posterior samples using simple simulation methods.

For simplicity of explanation, let us assume that the posterior parameter $\pi\left(\boldsymbol{\theta}|\boldsymbol{X}_n\right)$ of the parameter vector $\boldsymbol{\theta} = (\boldsymbol{\theta}_1^T, \boldsymbol{\theta}_2^T)^T$ can be decomposed as follows:

$$\pi(\boldsymbol{\theta}|\boldsymbol{X}_n) = \pi(\boldsymbol{\theta}_2|\boldsymbol{X}_n, \boldsymbol{\theta}_1) \times \pi(\boldsymbol{\theta}_1|\boldsymbol{X}_n).$$

In this case, we can easily generate the posterior samples as follows:

Step 1. Set the number of samples L to be generated.

Step 2. Draw $\boldsymbol{\theta}_1^{(j)}$ from the marginal posterior density $\pi(\boldsymbol{\theta}_1|\boldsymbol{X}_n)$, $j = 1, ..., L$, and insert the drawn values in the conditional posterior density of $\boldsymbol{\theta}_2$, $\pi\left(\boldsymbol{\theta}_2^{(j)}|\boldsymbol{X}_n, \boldsymbol{\theta}_1\right)$, $j = 1, ..., L$.

Step 3. Draw $\boldsymbol{\theta}_2^{(j)}$ from $\pi\left(\boldsymbol{\theta}_2^{(j)}|\boldsymbol{X}_n, \boldsymbol{\theta}_1\right)$, $j = 1, ..., L$.

Repeating this procedure many times provides a large sample from the joint posterior density. Therefore, Monte Carlo integration can be done by using the generated posterior samples $\boldsymbol{\theta}^{(1)}, ..., \boldsymbol{\theta}^{(L)}$. Application examples of this procedure are Zellner and Chen (2001) and Ando (2009a).

In Section 2.7, Bayesian inference on linear regression models is provided. The conditional posterior distribution of $\boldsymbol{\beta}$ is normal, and the marginal posterior distribution of σ^2 is inverse-gamma distribution:

$$\pi\left(\boldsymbol{\beta}, \sigma^2 \middle| \boldsymbol{y}_n, X_n\right) = \pi\left(\boldsymbol{\beta} \middle| \sigma^2, \boldsymbol{y}_n, X_n\right) \pi\left(\sigma^2 \middle| \boldsymbol{y}_n, X_n\right).$$

Using the direct Monte Carlo method, we can generate the posterior samples $\{\boldsymbol{\beta}^{(j)}, \sigma^{2(j)}; j = 1, ..., L\}$ using a direct Monte Carlo algorithm.

A direct Monte Carlo algorithm

Step 1. Set the number of posterior samples L to be generated.

Step 2. Draw $\sigma^{2(j)}$ from the marginal posterior density $\pi\left(\sigma^2 \middle| \boldsymbol{y}_n, X_n\right)$, $j = 1, ..., L$, and insert the drawn values in the conditional posterior density of $\boldsymbol{\beta}$, $\pi\left(\boldsymbol{\beta} \middle| \sigma^{2(j)}, \boldsymbol{y}_n, X_n\right)$, $j = 1, ..., L$.

Step 3. Draw $\boldsymbol{\beta}^{(j)}$ from $\pi\left(\boldsymbol{\beta} \middle| \sigma^{2(j)}, \boldsymbol{y}_n, X_n\right)$, $j = 1, ..., L$.

4.6.2 Importance sampling

Suppose we wish to approximate a posterior expectation,

$$\int h(\boldsymbol{\theta}) \pi(\boldsymbol{\theta}|\boldsymbol{X}_n) \, d\boldsymbol{\theta}.$$

Monte Carlo integration approximates the integrand by using the independent posterior samples $\boldsymbol{\theta}^{(1)}, ..., \boldsymbol{\theta}^{(L)}$.

Suppose we can approximate the posterior distribution $\pi(\boldsymbol{\theta}|\boldsymbol{X}_n)$ by $u(\boldsymbol{\theta})$, where $u(\boldsymbol{\theta})$ is some density from which it is easy to simulate and whose support is also that of the posterior distribution. Then Importance sampling estimates the integrand only having samples generated from $u(\boldsymbol{\theta})$ rather than the posterior distribution $\pi\left(\boldsymbol{\theta}|\boldsymbol{X}_n\right)$ as follows:

$$\begin{aligned}
\int h(\boldsymbol{\theta})\pi\left(\boldsymbol{\theta}|\boldsymbol{X}_n\right)d\boldsymbol{\theta} &= \frac{\displaystyle\int h(\boldsymbol{\theta})f\left(\boldsymbol{X}_n|\boldsymbol{\theta}\right)\pi(\boldsymbol{\theta})d\boldsymbol{\theta}}{\displaystyle\int f\left(\boldsymbol{X}_n|\boldsymbol{\theta}\right)\pi(\boldsymbol{\theta})d\boldsymbol{\theta}} \\
&= \frac{\displaystyle\int h(\boldsymbol{\theta})\frac{f\left(\boldsymbol{X}_n|\boldsymbol{\theta}\right)\pi(\boldsymbol{\theta})}{u\left(\boldsymbol{\theta}\right)}u\left(\boldsymbol{\theta}\right)d\boldsymbol{\theta}}{\displaystyle\int \frac{f\left(\boldsymbol{X}_n|\boldsymbol{\theta}\right)\pi(\boldsymbol{\theta})}{u\left(\boldsymbol{\theta}\right)}u\left(\boldsymbol{\theta}\right)d\boldsymbol{\theta}} \\
&\approx \frac{L^{-1}\displaystyle\sum_{j=1}^{L} h\left(\boldsymbol{\theta}^{(j)}\right)w\left(\boldsymbol{\theta}^{(j)}\right)}{L^{-1}\displaystyle\sum_{j=1}^{L} w\left(\boldsymbol{\theta}^{(j)}\right)},
\end{aligned}$$

where $u(\boldsymbol{\theta})$ is called the importance function, $w(\boldsymbol{\theta})$ is the weight function

$$w(\boldsymbol{\theta}) = \frac{f\left(\boldsymbol{X}_n|\boldsymbol{\theta}\right)\pi(\boldsymbol{\theta})}{u\left(\boldsymbol{\theta}\right)}$$

and $\boldsymbol{\theta}^{(1)},...,\boldsymbol{\theta}^{(L)}$ are generated from the importance function but the independent posterior distribution.

We have to be careful about the selection of the importance function. If the importance function $u(\boldsymbol{\theta})$ is a good approximation to the posterior distribution, then the weight function takes almost equal value at any point of $\boldsymbol{\theta}$. On the other hand, if the importance function $u(\boldsymbol{\theta})$ poorly approximates the posterior distribution, then many of the weights will be close to zero. As a result, we will obtain inaccurate Monte Carlo approximation. With respect to the importance sampling, we refer to Geweke (1989a).

4.6.3 Rejection sampling

Rejection sampling is a general sampling technique. In this approach, under the restriction that $f(\boldsymbol{X}_n|\boldsymbol{\theta})\pi(\boldsymbol{\theta}) < Mq(\boldsymbol{\theta})$ with $M > 0$, it generates samples from an arbitrary posterior distribution $\pi(\boldsymbol{\theta}|\boldsymbol{X}_n)$ by using a smooth density $q(\boldsymbol{\theta})$, which is called the envelope function. Using an envelope function $q(\boldsymbol{\theta})$, generated samples from an envelope function are probabilistically accepted or rejected. Rejection sampling proceeds as follows:

Step 1. Generate $\boldsymbol{\theta}$ from an envelope function $q(\boldsymbol{\theta})$.
Step 2. Generate u from an uniform density $u \in [0, 1]$.

Step 3. Accept if $f(\boldsymbol{X}_n|\boldsymbol{\theta})\pi(\boldsymbol{\theta}) > u \times Mq(\boldsymbol{\theta})$ and reject otherwise.
Step 4. Return to Step 1 and repeat the above sampling steps.

The generated samples are random variables from the posterior distribution. In practical applications, we want to set a value of M to be as small as possible for sampling efficiency (Carlin and Louis (2000)). Noting that a probability of acceptance, p, is given as

$$p = \frac{\int f\left(\boldsymbol{X}_n|\boldsymbol{\theta}\right)\pi(\boldsymbol{\theta})d\boldsymbol{\theta}}{M},$$

the probability distribution of the number of iterations k required to get one acceptable candidate is

$$P(k = j) = (1 - p)^{j-1}p, \quad j = 1, 2, ...$$

Also, the expected value of k is

$$E[k] = \frac{1}{p} = \frac{M}{\int f\left(\boldsymbol{X}_n|\boldsymbol{\theta}\right)\pi(\boldsymbol{\theta})d\boldsymbol{\theta}}.$$

Therefore, we want to minimize the value of M.

Also, it is ideal that the envelope density function $q(\boldsymbol{\theta})$ is similar to the posterior and have heavier tails so that enough candidates in the tail are generated. With respect to the rejection sampling, we refer to Ripley (1987). However, it often happens that an appropriate value is unknown, or difficult to find. In such a case, one can employ a weighted bootstrap approach. This approach will be described in the next section.

4.6.4 Weighted bootstrap

Smith and Gelfand (1992) discussed a sampling-resampling method for obtaining posterior samples. Let us have a set of samples $\boldsymbol{\theta}^{(j)}$, $j = 1, ..., L$, from some density function $q(\boldsymbol{\theta})$, which is easy to sample from. Define

$$w_j = \frac{f\left(\boldsymbol{X}_n|\boldsymbol{\theta}^{(j)}\right)\pi\left(\boldsymbol{\theta}^{(j)}\right)}{q\left(\boldsymbol{\theta}^{(j)}\right)},$$

and then calculate

$$p_j = \frac{w_j}{\sum_{j=1}^{L} w_j}, \quad j = 1, ..., L.$$

Resampling the draws $\boldsymbol{\theta}^{(k)}$, $k = 1, ..., L$ with probabilities p_k, we obtain the posterior samples.

Let us consider a case where θ is one dimensional. For the weighted bootstrap, we have

$$P(\theta \le x) = \sum_{j=1}^{L} p_j I(-\infty < \theta < x) = \frac{\sum_{j=1}^{L} w_j I(-\infty < \theta < x)}{\sum_{j=1}^{L} w_j}.$$

Thus, as $L \to \infty$

$$P(\theta \le x) \to \frac{\displaystyle\int_{-\infty}^{\infty} \frac{f(\boldsymbol{X}_n|\theta)\,\pi(\theta)}{q(\theta)} I(-\infty < \theta < x) q(\theta) d\theta}{\displaystyle\int_{-\infty}^{\infty} \frac{f(\boldsymbol{X}_n|\theta)\,\pi(\theta)}{q(\theta)} q(\theta) d\theta} = \frac{\displaystyle\int_{-\infty}^{x} f(\boldsymbol{X}_n|\theta)\,\pi(\theta)\,d\theta}{\displaystyle\int_{-\infty}^{\infty} f(\boldsymbol{X}_n|\theta)\,\pi(\theta)\,d\theta}.$$

Resampled draws $\boldsymbol{\theta}^{(k)}$, $k = 1, ..., L$ can be regarded as samples from the posterior distribution.

As well as the rejection sampling, it is ideal that the sampling density $q(\boldsymbol{\theta})$ is similar to the posterior. Also, it should have heavier tails than the posterior distribution so that enough candidates in the tail are generated.

Exercises

1. *For the Bayesian analysis of SUR model, we can also use the usual normal and the inverse Wishart priors for $\boldsymbol{\beta}$ and Σ, $\pi(\boldsymbol{\beta}, \Sigma) = \pi(\boldsymbol{\beta})\pi(\Sigma)$ with $\pi(\boldsymbol{\beta}) = N(\boldsymbol{\beta}_0, A^{-1})$ and $\pi(\Sigma) = IW(\Lambda_0, \nu_0)$. Show that this prior specification leads the following conditional posteriors $\pi(\boldsymbol{\beta}|\Sigma, \boldsymbol{Y}_n, X_n)$ and $\pi(\Sigma|\boldsymbol{\beta}, \boldsymbol{Y}_n, X_n)$:*

$$\pi(\boldsymbol{\beta}|\Sigma, \boldsymbol{Y}_n, X_n) = N\left(\hat{\boldsymbol{\beta}}_A, \hat{\Omega}_A\right),$$
$$\pi(\Sigma|\boldsymbol{\beta}, \boldsymbol{Y}_n, X_n) = IW(\Lambda_0 + R, n + \nu_0),$$

with

$$\hat{\boldsymbol{\beta}}_A = (X_n^T(\Sigma^{-1} \otimes I_n)X_n + A)^{-1}(X_n^T(\Sigma^{-1} \otimes I)X_n\hat{\boldsymbol{\beta}} + A\boldsymbol{\beta}_0),$$
$$\hat{\Omega}_A = (X_n^T(\Sigma^{-1} \otimes I_n)X_n^T + A)^{-1},$$

where $\hat{\boldsymbol{\beta}}$ is given in equation (4.4).

2. *Generate a set of* $n = 100$ *observations from the following* $m = 2$ *dimensional SUR model:*

$$\begin{pmatrix} y_1 \\ y_2 \end{pmatrix} = \begin{pmatrix} X_{n1} & O \\ O & X_{n2} \end{pmatrix} \begin{pmatrix} -1 & 2 \\ 2 & 0.5 \end{pmatrix} + \begin{pmatrix} \varepsilon_1 \\ \varepsilon_2 \end{pmatrix}, \qquad i = 1, ..., n,$$

where $\boldsymbol{y}_j$ *and* $\boldsymbol{u}_j$ *are* $n \times 1$ *vectors,* X_{jn} *is the* $n \times 2$ *matrix and* $\boldsymbol{\beta}_j$ *is the 2-dimensional vector. Each element of* Σ *is set to be*

$$\Sigma = \begin{pmatrix} \sigma_1^2 & \sigma_{12} \\ \sigma_{21} & \sigma_2^2 \end{pmatrix} = \begin{pmatrix} 0.68 & 0.33 \\ 0.33 & 0.45 \end{pmatrix}.$$

The design matrices X_{jn} $j = 1, 2$ *were generated from a uniform density over the interval* $[-3, 3]$.

Then implement Gibbs sampling approach for the Bayesian analysis of SUR model with the normal and the inverse Wishart priors for $\boldsymbol{\beta}$ *and* Σ*. The* **R** *function* **rsurGibbs** *may be useful.*

3. *Economic applications of SUR models frequently involve inequality restrictions on the coefficients. To express such restrictions, let us define a feasible region for the coefficients* $\boldsymbol{\beta}$ *by the inequality constraints, denoted by* S*, and define the indicator function*

$$I_S(\boldsymbol{\beta}) = \begin{cases} 1, & (\boldsymbol{\beta} \in S) \\ 0, & (\boldsymbol{\beta} \notin S) \end{cases}.$$

The inequality restrictions can be accommodated by setting up the following modification of the Jeffreys's invariant prior in (4.3)

$$\pi(\boldsymbol{\beta}, \Sigma) = \pi(\boldsymbol{\beta})\pi(\Sigma) \propto |\Sigma|^{-\frac{m+1}{2}} \times I_S(\boldsymbol{\beta}).$$

Show that the conditional posterior density of parameters of $\{\boldsymbol{\beta}, \Sigma\}$ *are*

$$\pi(\boldsymbol{\beta}|\boldsymbol{Y}_n, X_n, \Sigma) = TN\left(\hat{\boldsymbol{\beta}}, \hat{\Omega}, I_S(\boldsymbol{\beta})\right),$$
$$\pi(\Sigma|\boldsymbol{Y}_n, X_n, \boldsymbol{\beta}) = IW(R, n),$$

where $TN\left(\hat{\boldsymbol{\beta}}, \hat{\Omega}, I_S(\boldsymbol{\beta})\right)$ *denotes the truncated multivariate normal distribution with support* $I_S(\boldsymbol{\beta})$*. The* $\hat{\boldsymbol{\beta}}$ *and* $\hat{\Omega}$ *are given in equation (4.4).*

4. *In Section 4.3.1, a Gibbs sampling algorithm for the Bayesian analysis of the probit model was provided. Generate 100 binary observations according to a model* $Pr(Y = 1|x) = 1/[1 + \exp\{-0.05 + 0.6x\}]$*, where the design points* x_α *are uniformly distributed in* $[-1, 1]$*. Using the* **R** *function,* **MCMCoprobit***, generate a set of 1,000 posterior samples of the probit model.*

5. *In Section 4.5.5, a Gibbs sampling algorithm for the Bayesian analysis of the multinomial probit model was provided. A waveform data (Breiman et al. (1984)) consisted of three classes with 21 feature variables, and were generated from the following probability system:*

$$x_k = \begin{cases} uH_1(k) + (1-u)H_2(k) + \varepsilon_k & (g=1) \\ uH_1(k) + (1-u)H_3(k) + \varepsilon_k & (g=2) \\ uH_2(k) + (1-u)H_3(k) + \varepsilon_k & (g=3) \end{cases} \quad k = 1, \ldots, 21, \quad (4.13)$$

where u is uniform on $(0,1)$, ε_k are standard normal variables, and $H_i(k)$ are shifted triangular waveforms; $H_1(k) = \max(6 - |k - 11|, 0)$, $H_2(k) = H_1(k-4)$, and $H_3(k) = H_1(k+4)$. Generate 300 values of training data with equal prior probability for each class by using the probability system (4.13). In the same way, generate 500 values of test data to compute the prediction error. Then implement a Gibbs sampling for the multinomial probit models. An implementation of the Gibbs sampler can be done by the **R** *package* **MNP**.

6. *In Section 2.7, Bayesian inference on linear regression models is provided. Then Section 4.6.1 provided the direct Monte Carlo method for generating a set of posterior samples of $\boldsymbol{\beta}$ and σ^2. Using the results, generate a set of $L = 1,000$ posterior samples. Set $A = 10^{-5} \times I_p$ and $a = b = 10^{-10}$, which make the prior to be diffuse. Dataset can be generated from the example used in Section 2.8.1.*

Chapter 5

Bayesian approach for model selection

In the Bayesian and Non-Bayesian statistical modeling, we are naturally involved in the question of model selection. In the Bayesian linear regression modeling in Section 2.7, we studied that it is important to check how independent variables affect a response variable of interest. In other words, we have to select a set of variables that is important to predict the response variable. Also, the prior setting affected the prediction results. Although we assumed the normal error term for a simplicity of illustration of Bayesian regression modeling, we can also employ fat-tailed error terms, e.g., Student-t sampling density. The estimated Bayesian model depends on the specifications of the sampling density structure and the prior distribution of the model parameters, and thus crucial issues with Bayesian statistical modeling are the model evaluation.

Many approaches have been proposed over the years for dealing with this key issue in the Bayesian statistical modeling. In this chapter, we first provide the definition of the Bayes factor (Kass and Raftery (1995)), the Bayesian information criterion (BIC; Schwarz (1978)), the generalized Bayesian information criterion (GBIC; Konishi et al. (2004)).

5.1 General framework

Suppose we are interested in selecting a model from a set of candidate models $M_1, ..., M_r$. It is assumed that each model M_k is characterized by the probability density $f_k(\boldsymbol{x}|\boldsymbol{\theta}_k)$, where $\boldsymbol{\theta}_k$ $(\in \Theta_k \subset R^{p_k})$ is a p_k-dimensional vector of unknown parameters. Let $\pi_k(\boldsymbol{\theta}_k)$ be the prior distribution for parameter vector $\boldsymbol{\theta}_k$ under model M_k. The posterior probability of the model M_k for a particular data set $\boldsymbol{X}_n = \{\boldsymbol{x}_1, ..., \boldsymbol{x}_n\}$ is then given by

$$P(M_k|\boldsymbol{X}_n) = \frac{P(M_k)\displaystyle\int f_k(\boldsymbol{X}_n|\boldsymbol{\theta}_k)\pi_k(\boldsymbol{\theta}_k)d\boldsymbol{\theta}_k}{\displaystyle\sum_{\alpha=1}^{r} P(M_\alpha)\int f_\alpha(\boldsymbol{X}_n|\boldsymbol{\theta}_\alpha)\pi_\alpha(\boldsymbol{\theta}_\alpha)d\boldsymbol{\theta}_\alpha}, \tag{5.1}$$

where $f_k(\boldsymbol{X}_n|\boldsymbol{\theta}_k)$ and $P(M_k)$ are the likelihood function and the prior probability for model M_k. The prior probabilities $P(M_k)$ and $\pi_k(\boldsymbol{\theta}_k|M_k)$ for the model M_k specify an initial view of model uncertainty. Having observed information $\boldsymbol{X}_n$, we then update the view of model uncertainty based on the posterior model probability $P(M_k|\boldsymbol{X}_n)$.

In principal, the Bayes approach for selecting a model is to choose the model with the largest posterior probability among a set of candidate models. Therefore, the posterior model probability $P(M_1|\boldsymbol{X}_n)$,...,$P(M_r|\boldsymbol{X}_n)$ is the fundamental object of interest for model selection. This is equivalent to choosing the model that maximizes

$$P(M_k)\int f_k(\boldsymbol{X}_n|\boldsymbol{\theta}_k)\pi_k(\boldsymbol{\theta}_k)d\boldsymbol{\theta}_k. \tag{5.2}$$

The quantity

$$P(\boldsymbol{X}_n|M_k) = \int f_k(\boldsymbol{X}_n|\boldsymbol{\theta}_k)\pi_k(\boldsymbol{\theta}_k)d\boldsymbol{\theta}_k \tag{5.3}$$

obtained by integrating over the parameter space Θ_k is the marginal probability of the data $\boldsymbol{X}_n$ under the model M_k. This quantity measures how well the specified prior distributions fit to the observed data.

With respect to the prior model probabilities $P(M_1),...,P(M_r)$, a simple and popular choice is the uniform prior

$$P(M_k) = \frac{1}{r}, \quad k = 1, ..., r.$$

Thus this prior is noninformative in the sense of favoring all models equally. Under this prior, the quantity (5.2) is proportional to the marginal likelihood, and the posterior model probabilities reduce to

$$P(M_k|\boldsymbol{X}_n) = \frac{\displaystyle\int f_k(\boldsymbol{X}_n|\boldsymbol{\theta}_k)\pi_k(\boldsymbol{\theta}_k)d\boldsymbol{\theta}_k}{\displaystyle\sum_{\alpha=1}^{r}\int f_\alpha(\boldsymbol{X}_n|\boldsymbol{\theta}_\alpha)\pi_\alpha(\boldsymbol{\theta}_\alpha)d\boldsymbol{\theta}_\alpha}.$$

Although uniform prior might be convenient, we often have non-uniform prior model probabilities based on the model characteristics such as model complexicity, or, the number of parameters. For example, for the linear regression model $y = \sum_{j=1}^{p}\beta_j x_j + \varepsilon$, we might want to place higher probability on simpler models. In such a case, we will not assign prior probability uniformly to each of the models. To specify the prior model probabilities for each of the models, Denison et al. (1998) used a Poisson distribution

$$P(M_k) \propto \lambda^{p_k}\exp(-\lambda),$$

where p_k is the number of predictors in the model M_k, and the parameter λ

adjust the expectation of the number of predictors included in the models. As an alternative approach, we can also use

$$P(M_k) \propto \prod_{j=1}^{p} \pi_j^{\gamma_j}(1-\pi_j)^{1-\gamma_j},$$

where $\pi_j\ j = 1, ..., p$ are the prior probability that the j-th predictor is included in the model, and $\gamma_j = 1$ if the j-th predictor is included in the model and $\gamma_j = 0$ otherwise (see example, Smith and Kohn (1996)).

5.2 Definition of the Bayes factor

The Bayes factor, a quantity for comparing models and for testing hypotheses in the Bayesian framework, has played a major role in assessing the goodness of fit of competing models. It allows us to consider a pairwise comparison of models, say M_k and M_j based on the posterior probabilities $P(M_k|\boldsymbol{X}_n)$ in (5.1). The Bayes factor is defined as the odds of the marginal likelihood of the data $\boldsymbol{X}_n$:

$$\text{Bayes factor}(M_k, M_j) \equiv \frac{P(\boldsymbol{X}_n|M_k)}{P(\boldsymbol{X}_n|M_j)}, \tag{5.4}$$

which measures the evidence for model M_k versus model M_j based on the data information. The Bayes factor chooses the model with the largest value of marginal likelihood among a set of candidate models.

Noting that

$$\frac{P(M_k|\boldsymbol{X}_n)}{P(M_j|\boldsymbol{X}_n)} = \frac{P(\boldsymbol{X}_n|M_k)}{P(\boldsymbol{X}_n|M_j)} \times \frac{P(M_k)}{P(M_j)}, \tag{5.5}$$

$$\left[\text{Posterior odds}(M_k, M_j) = \text{Bayes factor}(M_k, M_j) \times \text{Prior odds}(M_k, M_j)\right]$$

the Bayes factor is also given as the ratio of posterior odds and prior odds

$$\text{Bayes factor}(M_k, M_j) = \frac{\text{Posterior odds}(M_k, M_j)}{\text{Prior odds}(M_k, M_j)}.$$

When the prior model probabilities $P(M_k)$ and $P(M_j)$ are equal, the Bayes factor reduces to the Posterior odds(M_k, M_j).

Jeffreys (1961) recommended interpreting the Bayes factors as a scale of evidence. Table 5.1 gives Jeffreys scale. Although these partitions seem to be somewhat arbitrary, it provides some descriptive statements. Kass and Raftery (1995) also give guidelines for interpreting the evidence from the Bayes factor.

TABLE 5.1: Jeffreys' scale of evidence for Bayes factor(M_k, M_j).

Bayes factor	Interpretation
$B_{kj} < 1$	Negative support for M_k
$1 < B_{kj} < 3$	Barely worth mentioning evidence for M_k
$3 < B_{kj} < 10$	Substantial evidence for M_k
$10 < B_{kj} < 30$	Strong evidence for M_k
$30 < B_{kj} < 100$	Very strong evidence for M_k
$100 < B_{kj}$	Decisive evidence for M_k

The Bayes factor can reduce to the classical likelihood ratio. Let $\hat{\boldsymbol{\theta}}_{\mathrm{MLE},k}$ and $\hat{\boldsymbol{\theta}}_{\mathrm{MLE},j}$ be the maximum likelihood estimates for the models M_k and M_j, respectively. Also, suppose that the prior densities for both models $\pi_k(\boldsymbol{\theta}_k)$ and $\pi_k(\boldsymbol{\theta}_k)$ consist of a point mass at the maximum likelihood values $\hat{\boldsymbol{\theta}}_{\mathrm{MLE},k}$ and $\hat{\boldsymbol{\theta}}_{\mathrm{MLE},j}$. Then from (5.3) and (5.5), the Bayes factor reduces to the classical likelihood ratio

$$\text{Bayes factor}(M_k, M_j) = \frac{f_k(\boldsymbol{X}_n|\boldsymbol{\theta}_{\mathrm{MLE},k})}{f_j(\boldsymbol{X}_n|\boldsymbol{\theta}_{\mathrm{MLE},j})}.$$

Bayes factor has been discussed in recent years. Note that if the model specific prior $\pi_k(\boldsymbol{\theta}_k)$ is improper, then the marginal likelihood is not well defined. Of course, numerous approaches have been proposed to solve this problem, ranging from the use of various "pseudo Bayes factor" approaches, such as the posterior Bayes factor (Aitkin (1991)), the intrinsic Bayes factor (Berger and Pericchi (1996)), the fractional Bayes factor (O'Hagan (1995)), the pseudo Bayes factors based on cross validation (Gelfand et al. (1992)) and so on. In Section 5.7, these criteria are discussed. The Bayesian information criterion (Schwarz (1978)) covered in this chapter is one of solutions. Kass and Raftery (1995) provides a comprehensive review of asymptotic methods for approximating Bayes' factors. We also refer to the review papers by Wasserman (2000) and Kadane and Lazar (2004) and the references therein.

5.2.1 Example: Hypothesis testing 1

Consider a simple linear regression model

$$y_\alpha = \beta_0 + \beta_1 x_{1\alpha} + \varepsilon_\alpha, \qquad \alpha = 1, ..., n,$$

where errors ε_α are independently, normally distributed with mean zero and variance σ^2. Suppose we are interested in testing:

$$H_0 : \ \beta_1 = 0 \quad \text{versus} \quad H_1 : \ \beta_1 \neq 0.$$

These hypotheses can be considered as two different models:

$$M_0: \; y_\alpha = \beta_0 + \varepsilon_\alpha,$$
$$M_1: \; y_\alpha = \beta_0 + \beta_1 x_{1\alpha} + \varepsilon_\alpha.$$

We first choose a prior probability $P(M_1)$ for M_1, which assigns probability $P(M_0) = 1 - P(M_1)$ to the null hypothesis. We then use a conjugate normal inverse-gamma prior $\pi(\boldsymbol{\beta}, \sigma^2)$, introduced in Section 2.6, for the parameters in each model. Let us have a set of n observations $\{y_\alpha, \boldsymbol{x}_\alpha\}$, $\alpha = 1, ..., n$. From (2.5), the marginal likelihood $P(\boldsymbol{y}_n | X_n, M_j)$, $j = 1, 2$, under the null hypothesis and the alternative hypothesis can be obtained analytically. The posterior probability of the alternative hypothesis is then

$$P(M_1 | \boldsymbol{y}_n, X_n) = \frac{P(M_1) P\left(\boldsymbol{y}_n \middle| X_n, M_1\right)}{P(M_0) P\left(\boldsymbol{y}_n \middle| X_n, M_0\right) + P(M_1) P\left(\boldsymbol{y}_n \middle| X_n, M_1\right)}.$$

Dividing the posterior odds by the prior odds, we obtain the Bayes factor

$$\begin{aligned}
\text{Bayes factor}(M_1, M_0) &= \frac{\text{Posterior odds}(M_k, M_j)}{\text{Prior odds}(M_k, M_j)} \\
&= \frac{\dfrac{P(M_1) P\left(\boldsymbol{y}_n \middle| X_n, M_1\right)}{P(M_0) P\left(\boldsymbol{y}_n \middle| X_n, M_0\right) + P(M_1) P\left(\boldsymbol{y}_n \middle| X_n, M_1\right)}}{\dfrac{P(M_0) P\left(\boldsymbol{y}_n \middle| X_n, M_0\right)}{P(M_0) P\left(\boldsymbol{y}_n \middle| X_n, M_0\right) + P(M_1) P\left(\boldsymbol{y}_n \middle| X_n, M_1\right)}} \times \frac{P(M_0)}{P(M_1)} \\
&= \frac{P\left(\boldsymbol{y}_n \middle| X_n, M_1\right)}{P\left(\boldsymbol{y}_n \middle| X_n, M_0\right)},
\end{aligned}$$

which is simply the ratio of the marginal likelihoods. The next example considers the hypothesis testing for checking the parameter region.

5.2.2 Example: Hypothesis testing 2

Consider again the simple example used in the above section. We test the hypotheses

$$H_0: \; \beta_1 \in \Theta_0 \quad \text{versus} \quad H_1: \; \beta_1 \in \Theta_1,$$

where Θ_0 and Θ_1 form a partition of the parameter space of β_1. In a similar manner, the corresponding models are:

$$M_0: \; y_\alpha = \beta_0 + \beta_1 x_{1\alpha} + \varepsilon_\alpha, \quad \beta_1 \in \Theta_0,$$
$$M_1: \; y_\alpha = \beta_0 + \beta_1 x_{1\alpha} + \varepsilon_\alpha, \quad \beta_1 \in \Theta_1.$$

For simplicity, assume that we know the parameter values β_0 and σ^2 accurately. Assigning a proper prior density $\pi(\beta_1)$, we can compare the two hypotheses a priori based on the prior odds:

$$\frac{P(M_1)}{P(M_0)} = \frac{P(\beta_1 \in \Theta_1)}{P(\beta_1 \in \Theta_0)} = \frac{\int_{\Theta_1} \pi(\beta_1)d\beta_1}{\int_{\Theta_0} \pi(\beta_1)d\beta_1}.$$

Similarly, the posterior odds based on a set of n observations $\{y_\alpha, \boldsymbol{x}_\alpha\}$, $\alpha = 1, ..., n$ is

$$\frac{P(M_1|\boldsymbol{y}_n, X_n)}{P(M_0|\boldsymbol{y}_n, X_n)} = \frac{P(M_1)\int_{\Theta_1} f(\boldsymbol{y}_n|X_n, \beta_1)\pi(\beta_1)d\beta_1}{P(M_0)\int_{\Theta_0} f(\boldsymbol{y}_n|X_n, \beta_1)\pi(\beta_1)d\beta_1}.$$

The Bayes factor, the ratio of the posterior odds to the prior odds of the hypotheses, has again given as the ratio of the marginal likelihood:

$$\text{Bayes factor}(M_k, M_j) = \frac{\int_{\Theta_1} f(\boldsymbol{y}_n|X_n, \beta_1)\pi(\beta_1)d\beta_1}{\int_{\Theta_0} f(\boldsymbol{y}_n|X_n, \beta_1)\pi(\beta_1)d\beta_1}.$$

In a practical situation, we have to specify the prior density $\pi(\beta_1)$. If we have a prior knowledge about β_1, one might use the truncated normal prior for β_1 with the mean μ_β, the variance σ_β^2, and its support Θ_j, $j = 1, 2$. Then the Bayes factor can be evaluated.

5.2.3 Example: Poisson models with conjugate priors

Suppose that we have a set of n independent observations $\boldsymbol{X}_n = \{x_1, ..., x_n\}$ from the Poisson distribution with parameter λ, where λ is the expected value for x. The joint probability density function for $\boldsymbol{X}_n$ is the product of the individual probability density functions:

$$f(\boldsymbol{X}_n|\lambda) = \prod_{\alpha=1}^{n} \frac{\exp(-\lambda)\lambda^{x_\alpha}}{x_\alpha!} = \frac{\exp(-n\lambda)\lambda^{n\bar{x}_n}}{\prod_{\alpha=1}^{n} x_\alpha!},$$

where $\bar{x}_n = \sum_{\alpha=1}^{n} x_\alpha / n$. The conjugate prior of the Poisson distribution, the Gamma distribution with parameter α and β

$$\pi(\lambda|\alpha, \beta) = \frac{\beta^\alpha}{\Gamma(\alpha)}\lambda^{\alpha-1}\exp(-\beta\lambda),$$

leads to a posterior distribution

$$\begin{aligned}\pi\left(\lambda\middle|\boldsymbol{X}_n\right) &\propto f(\boldsymbol{X}_n|\lambda)\times\pi(\lambda|\alpha,\beta)\\ &\propto \left\{\exp(-n\lambda)\lambda^{n\bar{x}_n}\right\}\times\left\{\lambda^{\alpha-1}\exp\left(-\beta\lambda\right)\right\}\\ &\propto \lambda^{n\bar{x}_n+\alpha-1}\exp\left(-\lambda(n+\beta)\right),\end{aligned}$$

which is a Gamma distribution with parameters $n\bar{x}_n+\alpha$ and $n+\beta$. Noting that the marginal likelihood is the denominator of the posterior distribution, we have

$$\begin{aligned}P(\boldsymbol{X}_n) &= \frac{f(\boldsymbol{X}_n|\lambda)\pi(\lambda)}{\pi(\lambda|\boldsymbol{X}_n)}\\ &= \frac{\left(\dfrac{\exp(-n\lambda)\lambda^{n\bar{x}_n}}{\prod_{\alpha=1}^{n}x_\alpha!}\right)\times\left(\dfrac{\beta^\alpha}{\Gamma(\alpha)}\lambda^{\alpha-1}\exp\left(-\beta\lambda\right)\right)}{\dfrac{(n+\beta)^{(n\bar{x}_n+\alpha)}}{\Gamma\left(n\bar{x}_n+\alpha\right)}\lambda^{n\bar{x}_n+\alpha-1}\exp\left(-(n+\beta)\lambda\right)}\\ &= \frac{1}{\prod_{\alpha=1}^{n}x_\alpha!}\frac{\Gamma\left(n\bar{x}_n+\alpha\right)}{\Gamma\left(\alpha\right)}\frac{\beta^\alpha}{(n+\beta)^{(n\bar{x}_n+\alpha)}}.\end{aligned}$$

Suppose we randomly collect the number of car accidents in a year for 8 Japanese drivers. Assuming that each of the drivers $\boldsymbol{X}_8=\{x_1,...,x_8\}$ independently, identically follow Poisson distribution with parameter λ. Noting that the mean and variance of a gamma distribution with parameter α and β are mean $=\alpha/\beta$, and variance $=\alpha/\beta^2$, we consider the following two priors for λ:

M_1: λ is Gamma distribution with parameter $\alpha=2$ and $\beta=2$. This prior reflects one's belief that the mean of λ is 1.

M_2: λ is Gamma distribution with parameter $\alpha=10$ and $\beta=10$. This prior also has the same mean as the model M_1. However, reflecting stronger information about λ, the variance of this prior is much more confident than that of M_1.

After the collection of data for the 8 drivers, we found that 6 had no accidents, 4 had exactly one accident, and 1 had three accidents. In the Figure 5.1, we display the likelihood function, the prior density, and the posterior density. We can see that the mode of prior under M_1 is consistent with the likelihood. The prior density under M_2 is flat relative to the likelihood. However, the constructed posterior densities are very similar.

We can compare these two models with the use of Bayes factors. We first compute the marginal likelihood for each of the models. These quantities are given as $f(\boldsymbol{X}_n|M_1)=0.000015$ and $f(\boldsymbol{X}_n|M_2)=0.000012$. Thus the Bayes factor is

$$\text{Bayes factor}(M_1,M_2)=\frac{P(\boldsymbol{X}_n|M_1)}{P(\boldsymbol{X}_n|M_2)}=\frac{0.000015}{0.000012}=1.25.$$

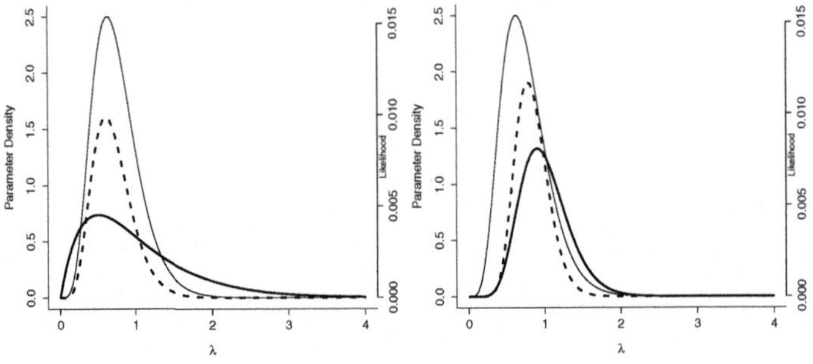

(a): M_1 (b): M_2

FIGURE 5.1: Comparison of the the likelihood function (—), the prior density (—), and the posterior density (- - -), under (a): M_1 and (b): M_b. M_1: Gamma prior with parameter $\alpha = 2$ and $\beta = 2$ is used for λ. M_2: Gamma prior with parameter $\alpha = 1$ and $\beta = 1$ is used for λ.

This means that the model M_1 is about 125% times as likely as the model M_2.

5.3 Exact calculation of the marginal likelihood

We have seen that the model evaluation of the Bayesian models based on the traditional Bayesian approach requires the marginal likelihood evaluation. This section provides several examples that allow us to evaluate the exact marginal likelihood.

5.3.1 Example: Binomial model with conjugate prior

Let us have a set of n independent samples $\boldsymbol{X}_n = \{x_1, x_2, ..., x_n\}$ from the Bernoulli distribution with parameter p. We know that a conjugate prior for p in a binomial distribution is Beta distribution with parameter α and β. The marginal likelihood can be calculated as follows:

$$
\begin{aligned}
f(\boldsymbol{X}_n) &= \int_0^1 \left[\begin{pmatrix} n \\ y_n \end{pmatrix} p^{y_n}(1-p)^{n-y_n} \times \frac{\Gamma(\alpha+\beta)}{\Gamma(\alpha)\Gamma(\beta)} p^{\alpha-1}(1-p)^{\beta-1} \right] dp \\
&= \begin{pmatrix} n \\ y_n \end{pmatrix} \frac{\Gamma(\alpha+\beta)}{\Gamma(\alpha)\Gamma(\beta)} \frac{\Gamma(y_n+\alpha)\Gamma(n+\beta-y_n)}{\Gamma(n+\alpha+\beta)} \\
&\quad \times \int_0^1 \left[\frac{\Gamma(n+\alpha+\beta)}{\Gamma(y_n+\alpha)\Gamma(n+\beta-y_n)} p^{y_n+\alpha-1}(1-p)^{n-y_n+\beta-1} \right] dp \\
&= \begin{pmatrix} n \\ y_n \end{pmatrix} \frac{\Gamma(\alpha+\beta)}{\Gamma(\alpha)\Gamma(\beta)} \frac{\Gamma(y_n+\alpha)\Gamma(n+\beta-y_n)}{\Gamma(n+\alpha+\beta)},
\end{aligned}
$$

where $y_n = \sum_{\alpha=1}^{n} x_\alpha$. Thus the posterior distribution of p is the Beta density with parameter $y_n + \alpha$ and $n - y_n + \beta$. Note also that this is the normalizing constant term of the posterior density of p, Beta distribution with parameter $y_n + \alpha$ and $n - y_n + \beta$.

Suppose we bet a game $n = 10$ times. Assuming that each of the games $\boldsymbol{X}_{10} = \{x_1, ..., x_{10}\}$ independently, identically follow Bernoulli distribution with parameter p. If you are very confident in the game, your expectation to the probability of winning p would be larger than 0.5. On the other hand, if your confidence is weak, the value of p would be much smaller than 0.5. Noting that the mean and variance of a beta distribution with parameter α and β are mean $= \alpha/(\alpha + \beta)$, let us consider the following two priors for p:

M_1: p follows beta distribution with parameter $\alpha = 0.1$ and $\beta = 4$.
M_2: p follows beta distribution with parameter $\alpha = 2$ and $\beta = 4$.
M_3: p follows beta distribution with parameter $\alpha = 4$ and $\beta = 4$.
M_4: p follows beta distribution with parameter $\alpha = 8$ and $\beta = 4$.

After the game, we won the game just 2 times. In the Figure 5.2, we display the likelihood function, the prior density, and the posterior density. We can see that the mode of prior under M_2 is consistent with the likelihood. As shown in this figure, the constructed posterior densities are very different. The marginal likelihood values are

$$\begin{cases} f(\boldsymbol{X}_{10}|M_1) = 0.0277 \\ f(\boldsymbol{X}_{10}|M_2) = 0.1648 \\ f(\boldsymbol{X}_{10}|M_3) = 0.0848 \\ f(\boldsymbol{X}_{10}|M_4) = 0.0168 \end{cases}.$$

Thus, we can see that the model M_2 is the most favored prior from the Bayes factor. Also, we can find the pair of (α, β) that maximizes the marginal likelihood. This approach is so-called empirical Bayes approach.

5.3.2 Example: Normal regression model with conjugate prior and Zellner's g-prior

Suppose we have n independent observations $\{(y_\alpha, \boldsymbol{x}_\alpha); \alpha = 1, 2, ..., n\}$, where y_α are random response variables and $\boldsymbol{x}_\alpha$ are vectors of p-dimensional explanatory variables. The problem to be considered is how to estimate the relationship between the response variable and the explanatory variables from the observed data. We use the Gaussian linear regression model

$$f\left(\boldsymbol{y}_n|X_n, \boldsymbol{\beta}, \sigma^2\right) = \frac{1}{(2\pi\sigma^2)^{n/2}} \exp\left[-\frac{(\boldsymbol{y}_n - X_n\boldsymbol{\beta})^T(\boldsymbol{y}_n - X_n\boldsymbol{\beta})}{2\sigma^2}\right],$$

with a conjugate normal inverse-gamma prior $\pi(\boldsymbol{\beta}, \sigma^2) = \pi(\boldsymbol{\beta}|\sigma^2)\pi(\sigma^2)$. Here

$$\pi(\boldsymbol{\beta}|\sigma^2) = N\left(\boldsymbol{0}, \sigma^2 A\right) \quad \text{and} \quad \pi(\sigma^2) = IG\left(\frac{\nu_0}{2}, \frac{\lambda_0}{2}\right).$$

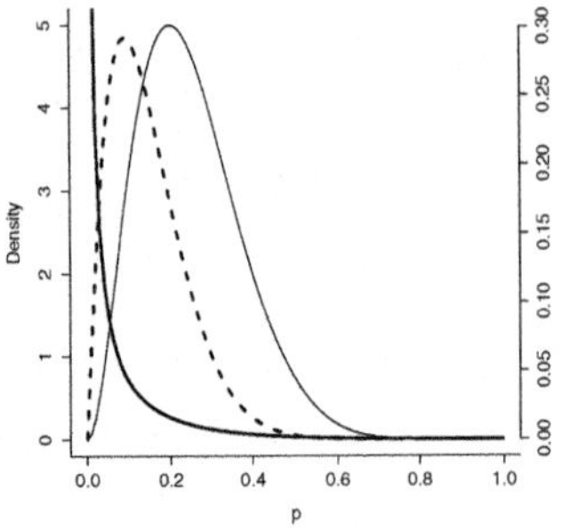

(a): M_1

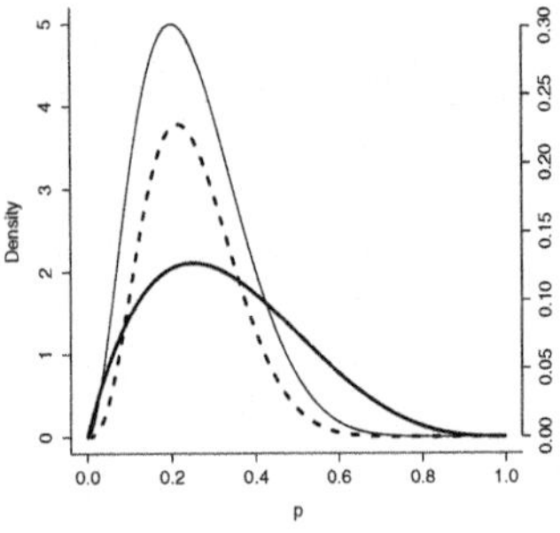

(b): M_2

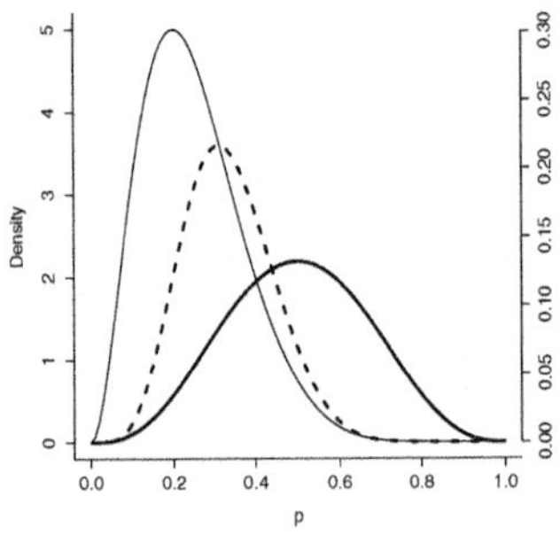

(c): M_3

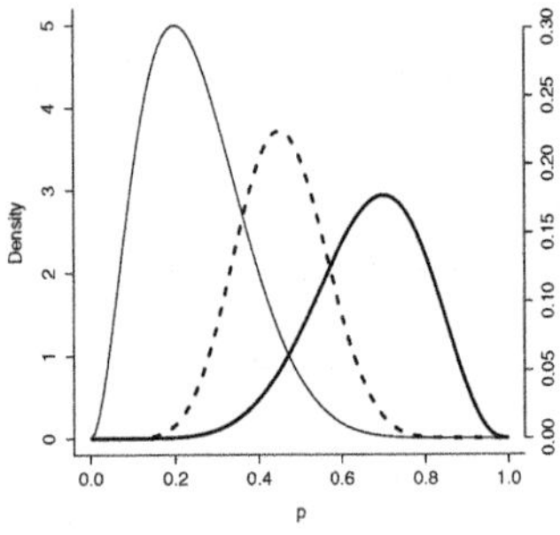

(d): M_4

FIGURE 5.2: Comparison of the likelihood function (—), the prior density (—), and the posterior density (- - -), under (a): M_1 (b): M_b (c): M_3 and (d): M_4. M_1: p follows beta distribution with parameter $\alpha = 1$ and $\beta = 4$. M_2: p follows beta distribution with parameter $\alpha = 2$ and $\beta = 4$. M_3: p follows beta distribution with parameter $\alpha = 4$ and $\beta = 4$. M_4: p follows beta distribution with parameter $\alpha = 8$ and $\beta = 4$.

Let the model M_k and M_j be one of the model specifications. Depending on the specification of the explanatory variables $\boldsymbol{x}$ in X_n, the number of predictors ranges from 0 to p. Also, the prediction for the future observation $\boldsymbol{z}_n$ depends on the values of A, ν and λ in the prior distributions. In Section 2.7, we have the marginal likelihood of the model (2.5). The marginal likelihood is given as

$$P\left(\boldsymbol{y}_n \middle| X_n, M\right) = \frac{\left|\hat{A}_n\right|^{1/2} |A|^{1/2} \left(\frac{\lambda_0}{2}\right)^{\frac{\nu_0}{2}} \Gamma\left(\frac{\hat{\nu}_n}{2}\right)}{\pi^{\frac{n}{2}} \Gamma\left(\frac{\nu_0}{2}\right)} \left(\frac{\hat{\lambda}_n}{2}\right)^{-\frac{\hat{\nu}_n}{2}},$$

where $\hat{A}_n$, $\hat{\nu}_n$ and $\hat{\lambda}_n$ are defined in Section 2.7. Thus, we can optimize the optimal combinations to select the best model among a set of candidate models.

As an alternative prior specification, Zellner (1986) suggested a particular form of the conjugate normal-gamma prior, a g-prior:

$$\pi(\boldsymbol{\beta}|\sigma^2) = N\left(\boldsymbol{\beta}_0, g\sigma^2(X_n^T X_n)^{-1}\right) \quad \text{and} \quad \pi(\sigma^2) \propto 1/\sigma^2,$$

where the prior covariance matrix of $\boldsymbol{\beta}$ is a scalar multiple g of the Fisher information matrix, which depends on the observed data through the design matrix X_n. As we can see, the amount of subjectivity in Zellner's g-prior is limited to the choice of g. Although the prior looks like an improper prior, a major advantage of Zellner's g-prior is it also provides to the closed form expression of all marginal likelihood

$$P\left(\boldsymbol{y}_n \middle| X_n, M\right) = \frac{\Gamma(n/2)}{\pi^{n/2}}(1+g)^{-p}S^{-n}$$

with

$$S^2 = \boldsymbol{y}_n^T \boldsymbol{y}_n - \frac{g}{1+g}\boldsymbol{y}_n^T X_n (X_n^T X_n)^{-1} X_n^T \boldsymbol{y}_n - \frac{1}{1+g}\boldsymbol{\beta}_0^T (X_n^T X_n)^{-1}\boldsymbol{\beta}_0.$$

The choice of g controls the estimation results. One can select the value of g by using the marginal likelihood. Kass and Wasserman (1995) recommended to choose the priors with the amount of information about the parameter equal to the amount of information contained in one observation by taking $g = n$. Foster and George (1994) recommended the use of $g = p^2$ from a minimax perspective. Fernandez et al. (2001) suggested the value $g = \max(n, p^2)$ by studying various choices of g with dependence on the sample size n and the model dimension p. Liang et al. (2008) studied mixtures of g-priors as an alternative to default g-priors.

5.3.3 Example: Multi-response normal regression model

Suppose we have a set of n independent observations $\{(\boldsymbol{y}_1, \boldsymbol{x}_1), ..., (\boldsymbol{y}_n, \boldsymbol{x}_n)\}$, where $\boldsymbol{y}_\alpha$ are m-dimensional response variables and $\boldsymbol{x}_\alpha$ are p-dimensional explanatory variables. Let us consider the multivariate regression model:

$$\boldsymbol{y}_\alpha = \boldsymbol{\alpha} + \Gamma^T \boldsymbol{x}_\alpha + \boldsymbol{\varepsilon}_\alpha, \quad \alpha = 1, ..., n, \tag{5.6}$$

where $\boldsymbol{x}_\alpha = (x_{1\alpha}, ..., x_{p\alpha})^T$ is a p-dimensional vector of explanatory variables, and $\boldsymbol{\varepsilon}_\alpha = (\varepsilon_{1\alpha}, ..., \varepsilon_{m\alpha})^T$ is an m-dimensional noise vector (normal, with mean $\mathbf{0}$ and variance Σ). The vector $\boldsymbol{\alpha} = (\alpha_1, ..., \alpha_m)^T$ and matrix $\Gamma = (\boldsymbol{\beta}_1, ..., \boldsymbol{\beta}_m)$ consist of unknown parameters; $\boldsymbol{\beta}_j = (\beta_{j1}, ..., \beta_{jp})'$ is the p-dimensional vector of coefficients.

The model in (5.6) can be expressed in matrix form:

$$\boldsymbol{Y}_n = X_n B + E,$$

where $\boldsymbol{Y}_n = (\boldsymbol{y}_1, ..., \boldsymbol{y}_n)^T$, $X_n = (\mathbf{1}_n, W_n)$, $W_n = (\boldsymbol{x}_1, ..., \boldsymbol{x}_n)^T$, $B = (\boldsymbol{\alpha}, \Gamma^T)^T$, and $E = (\boldsymbol{\varepsilon}_1, ..., \boldsymbol{\varepsilon}_n)^T$ with $\boldsymbol{\varepsilon}_\alpha \sim N(\mathbf{0}, \Sigma)$. The likelihood function is then

$$
\begin{aligned}
& f(\boldsymbol{Y}_n|X_n, B, \Sigma) \\
&= \prod_{\alpha=1}^{n} f(\boldsymbol{y}_\alpha|\boldsymbol{x}_\alpha, B, \Sigma) \\
&= \prod_{\alpha=1}^{n} \det(2\pi\Sigma)^{-\frac{1}{2}} \exp\left\{-\frac{1}{2}(\boldsymbol{y}_\alpha - \boldsymbol{\alpha} - \Gamma^T\boldsymbol{x}_\alpha)^T\Sigma^{-1}(\boldsymbol{y}_\alpha - \boldsymbol{\alpha} - \Gamma^T\boldsymbol{x}_\alpha)\right\} \\
&= (2\pi)^{-\frac{nm}{2}}|\Sigma|^{-\frac{n}{2}} \exp\left[-\frac{1}{2}\mathrm{tr}\left\{\Sigma^{-1}\left(\boldsymbol{Y}_n - X_nB\right)^T\left(\boldsymbol{Y}_n - X_nB\right)\right\}\right],
\end{aligned}
$$

where $\boldsymbol{\theta} = (\mathrm{vec}(B), \mathrm{vech}(\Sigma))^T$ contains all the unknown model parameters. (The vectorization operator *vec* converts the matrix B into a column vector by stacking its columns on top of one another; the half-vectorization operator *vech* does the same with only the lower triangular part of Σ.)

Assuming $\pi(\boldsymbol{\theta}) = \pi(B|\Sigma)\pi(\Sigma)$, we use a matricvariate normal and an inverted Wishart for the prior distributions:

$$
\begin{aligned}
\pi(B|B_0, \Sigma, A) &\propto |\Sigma|^{-\frac{m}{2}}|A|^{-\frac{p+1}{2}} \exp\left[-\frac{1}{2}\mathrm{tr}\left\{\Sigma^{-1}\left(B - B_0\right)^T A^{-1}\left(B - B_0\right)\right\}\right], \\
\pi(\Sigma|\Lambda_0, \nu_0) &= \frac{|\Lambda_0|^{\frac{\nu_0}{2}}}{2^{\frac{m\nu_0}{2}}\Gamma_m(\frac{\nu_0}{2})}|\Sigma|^{-\frac{\nu_0+m+1}{2}} \exp\left[-\frac{1}{2}\mathrm{tr}\left(\Lambda_0\Sigma^{-1}\right)\right],
\end{aligned}
$$

with $m \geq \nu_0$, $|\Sigma| > 0$. Here Λ_0, A and B_0 are $m \times m$, $(p+1) \times (p+1)$ and $(p+1) \times m$ matrices, respectively. The matrix B_0 specifies the prior mean of B, and the matrix A adjusts the strength of prior information. Based on some perspective, one might adjust A to specify that a prior is particularly informative by putting small values into A. On the other hand, one could also weaken or remove entirely the influence of prior information.

The posterior probability of Σ, $\pi(\Sigma|\boldsymbol{Y}_n, X_n)$, is modeled using the inverted Wishart distribution $W(\Sigma|S+\Lambda_0, n+\nu_0)$. That of B given Σ, $\pi(B|\Sigma, \boldsymbol{Y}_n, X_n)$, is modeled using the matricvariate normal distribution $N(B|\bar{B}, \Sigma, X_n^TX_n + A^{-1})$. See for example, Rossi et al. (2005) and Ando (2009b). The posterior means of Σ and B given Σ are $(S + \Lambda_0)/(\nu_0 + n - m - 1)$ and $\bar{B}$, respectively. Here S and $\bar{B}$ are defined below:

$$
\begin{aligned}
S &= (\boldsymbol{Y}_n - X_n\bar{B})^T(\boldsymbol{Y}_n - X_n\bar{B}) + (\bar{B} - B_0)^TA^{-1}(\bar{B} - B_0), \\
\bar{B} &= (X_n^TX_n + A^{-1})^{-1}\left(X_n^TX_n\hat{B} + A^{-1}B_0\right) := (\bar{\boldsymbol{\alpha}}, \bar{\Gamma}^T)^T,
\end{aligned}
$$

with $\hat{B} = (X_n^TX_n)^{-1}X_n^T\boldsymbol{Y}_n$.

It is well known that the predictive distribution for $\boldsymbol{z}$ given value of $\boldsymbol{x}$, is the multivariate t-distribution $Mt(\bar{B}^T\boldsymbol{x}^*, \Sigma^*, \nu^*)$, with $\boldsymbol{x}_* = (1, \boldsymbol{x}^T)^T$:

$$\begin{aligned} & f(\boldsymbol{z}|\boldsymbol{x}, \boldsymbol{Y}_n, X_n) \\ &= \int f(\boldsymbol{z}|\boldsymbol{x}, B, \Sigma)\pi(B|\Sigma, D)\pi(\Sigma|D)dBd\Sigma \\ &= \frac{\Gamma\left(\frac{\nu_*+m}{2}\right)}{\Gamma\left(\frac{\nu_*}{2}\right)\pi^{\frac{m}{2}}}|\Sigma^*|^{-\frac{1}{2}}\left\{1 + (\boldsymbol{z} - \bar{\boldsymbol{\alpha}} - \bar{\Gamma}^T\boldsymbol{x})^T\Sigma^{*-1}(\boldsymbol{z} - \bar{\boldsymbol{\alpha}} - \bar{\Gamma}^T\boldsymbol{x})\right\}^{-\frac{\nu_*+m}{2}}, \end{aligned}$$

where $\nu_* = n + \nu_0 - m + 1$ and

$$\Sigma^* = \frac{1 + \boldsymbol{x}_*^T(X_n^TX_n + A^{-1})^{-1}\boldsymbol{x}_*}{n + \nu_0 - m + 1}(S + \Lambda_0).$$

By rearranging the posterior distribution, we can evaluate the marginal likelihood as follows:

$$P(\boldsymbol{Y}_n|X_n) = \frac{f(\boldsymbol{Y}_n|X_n, B^*, \Sigma^*)\pi(B^*, \Sigma^*)}{\pi(B^*, \Sigma^*|\boldsymbol{Y}_n, X_n)},$$

for any values of B^* and Σ^*. Since every term on the right-hand side of the equation is available, we can easily evaluate the marginal likelihood:

$$P(\boldsymbol{Y}_n|X_n) = \pi^{-\frac{nm}{2}} \times \frac{|\Lambda_0|^{\frac{\nu_0}{2}} \times \Gamma_m(\frac{\nu_0+n}{2}) \times |X_n^TX_n + A^{-1}|^{\frac{p+1}{2}}}{|\Lambda_0 + S|^{\frac{\nu_0+n}{2}} \times \Gamma_m(\frac{\nu_0}{2}) \times |A|^{\frac{p+1}{2}}}. \quad (5.7)$$

5.4 Laplace's method and asymptotic approach for computing the marginal likelihood

Evaluating the posterior probability of the model M_k involves the calculation of the marginal likelihood. Unfortunately, in almost all situations, an exact analytical expression does not always exist, while the marginal likelihood calculation is an essential point in the Bayes approach for model selection. Therefore, calculating posterior probabilities for a collection of competing models has been investigated extensively.

There are mainly two approaches for calculating the marginal likelihood; (1) an asymptotic approximation approach, (2) a simulation based approach. The asymptotic approximation method is mainly based on Laplace for multidimensional integrals. To ease the computational burden, the Laplace approximation is very useful to obtain an analytic approximation for the marginal likelihood.

It may be preferable to use a computable approximation for the marginal likelihood when exact expressions are unavailable. A useful asymptotic approximation to the marginal likelihood is obtained by Laplace's method (Tierney and Kadane (1986)). Laplace's method is commonly used to obtain an analytical approximation to integrals. The use of Laplace's method for integrals has been extensively investigated as a useful tool for approximating Bayesian predictive distributions, Bayes factors and Bayesian model selection criteria (Davison, 1986; Clarke and Barron, 1994; Kass and Wasserman, 1995; Kass and Raftery, 1995; O'Hagan, 1995; Pauler, 1998).

Let us assume that the posterior density $\pi(\boldsymbol{\theta}|\boldsymbol{X}_n)$ is sufficiently well-behaved, (e.g., highly peaked at the posterior mode $\hat{\boldsymbol{\theta}}_n$), and define $s_n(\boldsymbol{\theta}) = \log\{f(\boldsymbol{X}_n|\boldsymbol{\theta})\pi(\boldsymbol{\theta})\}$. Noting that the first derivale of $s_n(\boldsymbol{\theta})$ evaluated at the posterior mode $\hat{\boldsymbol{\theta}}_n$ equals to zero, a Taylor series expansion of the $s_n(\boldsymbol{\theta})$ about the posterior mode $\hat{\boldsymbol{\theta}}_n$

$$s_n(\boldsymbol{\theta}) = s_n(\hat{\boldsymbol{\theta}}_n) - \frac{n}{2}(\boldsymbol{\theta} - \hat{\boldsymbol{\theta}}_n)^T S_n(\hat{\boldsymbol{\theta}}_n)(\boldsymbol{\theta} - \hat{\boldsymbol{\theta}}_n)$$

and then taking an exponentiation yields an approximation to

$$\exp\{s_n(\boldsymbol{\theta})\} \approx \exp\left\{s_n(\hat{\boldsymbol{\theta}}_n)\right\} \times \exp\left\{-\frac{n}{2}(\boldsymbol{\theta} - \hat{\boldsymbol{\theta}}_n)^T S_n(\hat{\boldsymbol{\theta}}_n)(\boldsymbol{\theta} - \hat{\boldsymbol{\theta}}_n)\right\}$$

that has a form of a multivariate normal density with mean posterior mode $\hat{\boldsymbol{\theta}}_n$ and covariance matrix $n^{-1}S_n(\hat{\boldsymbol{\theta}}_n)^{-1}$ with

$$S_n(\hat{\boldsymbol{\theta}}_n) = -\frac{1}{n}\left.\frac{\partial^2 \log\{f(\boldsymbol{X}_n|\boldsymbol{\theta})\pi(\boldsymbol{\theta})\}}{\partial\boldsymbol{\theta}\partial\boldsymbol{\theta}^T}\right|_{\boldsymbol{\theta}=\hat{\boldsymbol{\theta}}_n},$$

which is minus the Hessian of $n^{-1}\log\{f(\boldsymbol{X}_n|\boldsymbol{\theta})\pi(\boldsymbol{\theta})\}$ evaluated at the posterior mode $\hat{\boldsymbol{\theta}}_n$. Integrating this approximation yields

$$\begin{aligned} P(\boldsymbol{X}_n|M) &= \int \exp\{s_n(\boldsymbol{\theta})\}\, d\boldsymbol{\theta} \\ &\approx \exp\left\{s_n(\hat{\boldsymbol{\theta}}_n)\right\} \times \int \exp\left\{-\frac{n}{2}(\boldsymbol{\theta} - \hat{\boldsymbol{\theta}}_n)^T S_n(\hat{\boldsymbol{\theta}}_n)(\boldsymbol{\theta} - \hat{\boldsymbol{\theta}}_n)\right\} d\boldsymbol{\theta} \\ &= f(\boldsymbol{X}_n|\hat{\boldsymbol{\theta}}_n)\pi(\hat{\boldsymbol{\theta}}_n) \times \frac{(2\pi)^{\frac{p}{2}}}{n^{\frac{p}{2}}\left|S_n(\hat{\boldsymbol{\theta}}_n)\right|^{\frac{1}{2}}}, \end{aligned}$$

where p is the dimension of $\boldsymbol{\theta}$. When the sample size is moderate, this approximation provides accurate results. For general regularity conditions for the Laplace's method for integrals, we refer to Barndorff-Nielsen and Cox (1989).

The Bayes factor in (5.5) is then approximated as

$$\begin{aligned} &\text{Bayes factor}(M_k, M_j) \\ &\approx \frac{f_k(\boldsymbol{X}_n|\hat{\boldsymbol{\theta}}_{kn})}{f_j(\boldsymbol{X}_n|\hat{\boldsymbol{\theta}}_{jn})} \times \frac{\pi_k(\hat{\boldsymbol{\theta}}_{kn})}{\pi_j(\hat{\boldsymbol{\theta}}_{jn})} \times \frac{\left|S_{jn}(\hat{\boldsymbol{\theta}}_{jn})\right|^{\frac{1}{2}}}{\left|S_{kn}(\hat{\boldsymbol{\theta}}_{kn})\right|^{\frac{1}{2}}} \times \left(\frac{2\pi}{n}\right)^{\frac{p_k - p_j}{2}}, \end{aligned}$$

where $\hat{\boldsymbol{\theta}}_{kn}$ is the posterior mode of the model M_k, $f_k(\boldsymbol{X}_n|\hat{\boldsymbol{\theta}}_{kn})$, and $\pi_k(\hat{\boldsymbol{\theta}}_{kn})$ are the likelihood and prior evaluated at the posterior mode $\hat{\boldsymbol{\theta}}_{kn}$. This quantity depends on the ratio of the likelihood function evaluated at the posterior modes, the ratio of the prior evaluated at the posterior modes, on the difference between the dimensions of the models M_k and M_j, and on the Hessian terms.

Recently, Ando (2007) and Konishi et al. (2004) showed that the order of the prior distribution has a large influence on the calculation of the marginal likelihood based on Laplace's method. In the following sections, we consider the two cases (a): $\log \pi(\boldsymbol{\theta}) = O_p(1)$ and (b): $\log \pi(\boldsymbol{\theta}) = O_p(n)$. In the first case (a), the prior information can be ignored for a sufficiently large n, Schwarz's (1978) Bayesian information criterion is derived. The second case leads us to the Generalized Bayesian information criterion (Konishi et al. (2004)).

5.5 Definition of the Bayesian information criterion

In this section, we consider the case $\log \pi(\boldsymbol{\theta}) = O_p(1)$. Assuming that we have a set of n independent observations $\boldsymbol{X}_n = \{\boldsymbol{x}_1, ..., \boldsymbol{x}_n\}$, the marginal likelihood is approximated as

$$P(\boldsymbol{X}_n|M) \approx f(\boldsymbol{X}_n|\hat{\boldsymbol{\theta}}_{\text{MLE}})\pi(\hat{\boldsymbol{\theta}}_{\text{MLE}}) \times \frac{(2\pi)^{p/2}}{n^{p/2}|J_n(\hat{\boldsymbol{\theta}}_{\text{MLE}})|^{1/2}}, \tag{5.8}$$

where $\hat{\boldsymbol{\theta}}_{\text{MLE}}$ is the maximum likelihood estimate, p is the dimension of $\boldsymbol{\theta}$, and

$$J_n(\hat{\boldsymbol{\theta}}) = -\frac{1}{n} \left.\frac{\partial^2 \log f(\boldsymbol{X}_n|\boldsymbol{\theta})}{\partial \boldsymbol{\theta} \partial \boldsymbol{\theta}^T}\right|_{\boldsymbol{\theta}=\hat{\boldsymbol{\theta}}_{\text{MLE}}} = -\frac{1}{n}\sum_{\alpha=1}^{n} \left.\frac{\partial^2 \log f(\boldsymbol{x}_\alpha|\boldsymbol{\theta})}{\partial \boldsymbol{\theta} \partial \boldsymbol{\theta}^T}\right|_{\boldsymbol{\theta}=\hat{\boldsymbol{\theta}}_{\text{MLE}}}.$$

By ignoring the terms in (5.8) that are constant in large samples, and then taking the logarithm of the resulting formula yields the Schwarz's (1978) Bayesian information criterion,

$$\text{BIC} = -2 \log f(\boldsymbol{X}_n|\hat{\boldsymbol{\theta}}_{MLE}) + p \log n. \tag{5.9}$$

The BIC is a criterion for evaluating models estimated by the maximum likelihood method. Derivation can be found in Section 7.3.

Note that the Schwarz criterion gives a rough approximation to the logarithm of the Bayes factor (5.5)

$$\begin{aligned} \log\left[\text{Bayes factor}(M_k, M_j)\right] &= \log\left[P(\boldsymbol{X}_n|M_k)\right] - \log\left[P(\boldsymbol{X}_n|M_j)\right] \\ &\approx \left(\text{BIC}_j - \text{BIC}_k\right)/2, \end{aligned}$$

because the approximation error is $O_p(1)$. Although the approximation by

BIC does not achieve the correct value of the Bayes factor, asymptotically, as the sample size increases, we obtain

$$\frac{(\mathrm{BIC}_j - \mathrm{BIC}_k)/2 - \log\{\text{Bayes factor}(M_k, M_j)\}}{\log\{\text{Bayes factor}(M_k, M_j)\}} \to 0.$$

Thus, the approximation error relative to the true Bayes factor tends to 0.

5.5.1 Example: Evaluation of the approximation error

In example 5.3.1, we have considered the conjugate prior analysis of the binomial distribution. With a set of n independent samples $\boldsymbol{X}_n = \{x_1, x_2, ..., x_n\}$ from the Bernoulli distribution with parameter p, the marginal likelihood was given as

$$P(\boldsymbol{X}_n) = \left(\begin{array}{c} n \\ y_n \end{array} \right) \frac{\Gamma(\alpha+\beta)}{\Gamma(\alpha)\Gamma(\beta)} \frac{\Gamma(y_n+\alpha)\Gamma(n+\beta-y_n)}{\Gamma(n+\alpha+\beta)},$$

where $y_n = \sum_{\alpha=1}^n x_\alpha$.

This section investigates the approximation error of BIC as an estimator of the marginal likelihood $P(\boldsymbol{X}_n)$. Here, the BIC score is given as

$$-\frac{1}{2}\mathrm{BIC} = \log \left(\begin{array}{c} n \\ y_n \end{array} \right) + y_n \log(\hat{p}_{\mathrm{MLE}}) + (n - y_n) \log(\hat{p}_{\mathrm{MLE}}) - \frac{1}{2}\log(n),$$

where $\hat{p}_{\mathrm{MLE}} = \sum_{\alpha=1}^n x_\alpha / n$ is the maximum likelihood estimator. Note also that the number of parameters is just 1.

Using a set of n independent samples $\boldsymbol{X}_n = \{x_1, x_2, ..., x_n\}$, we can evaluate the approximation error, e.g., by using the absolute difference between the exact marginal likelihood and BIC as

$$\text{Apprximation error} = \left| P(\boldsymbol{X}_n) - \exp\left[-\frac{1}{2}\mathrm{BIC} \right] \right|.$$

Given value of n, we repeatedly generated a set of n independent samples $\boldsymbol{X}_n = \{x_1, x_2, ..., x_n\}$ and calculated the approximation error over 10,000 Monte carlo trials. The values of hyperparameter are set to be $\alpha = 2$ and $\beta = 4$, respectively. Table 5.2 shows the summary of approximation errors. The mean and the standard deviations of approximation errors are given for each n. As shown in Table 5.2, the error becomes smaller as the sample size increases. It implies that BIC is a useful tool as an estimator of the marginal likelihood.

5.5.2 Example: Link function selection for binomial regression

BIC score (5.14) is applied to the link function selection problem for binomial regression analysis of O-ring failure data. O-ring failure data is famous

TABLE 5.2: Summary of the absolute difference between the exact marginal likelihood and BIC score. The mean and the standard deviations of approximation errors over 10,000 Monte Carlo trials are given for each n.

n	Error (Standard deviations)
10	0.0725 (0.0555)
25	0.0306 (0.0140)
50	0.0173 (0.0048)
75	0.0122 (0.0027)
100	0.0095 (0.0018)

because of the Challenger disaster, January 28, 1986. A focus is the probability of primary O-ring damage conditional on 31° fahrenheit, the temperature that morning. Six primary O-rings are used in the space shuttles, and thus it is assmed that the response variable y follows binomial distribution with size parameter 6 and the probability of primary O-ring damage $p(t)$

$$f\left(y_\alpha|p(t_\alpha)\right) = \begin{pmatrix} 6 \\ y_\alpha \end{pmatrix} p(t_\alpha)^{y_\alpha}\left(1 - p(t_\alpha)\right)^{6-y_\alpha},$$

where $y_\alpha \in \{0, 1, 2, ..., 6\}$ is the number of failures and $p(t_\alpha)$ is the conditional probability at temperature t_α. In total, a set of $n = 23$ observations are used to estimate the probability of a distress to the field-joint primary O-rings. The lower temperature increases the probability of a distress.

Several options are available for expressing the conditional probability. Here we use the logistic function and the probit function, respectively, given as

$$\begin{aligned} \text{Logistic} &: \quad p(t;\boldsymbol{\beta}) = \frac{\exp\left(\beta_0 + \beta_1 t_\alpha\right)}{1 + \exp\left(\beta_0 + \beta_1 t\right)}, \\ \text{Probit} &: \quad p(t;\boldsymbol{\beta}) = \Phi\left(\beta_0 + \beta_1 t\right), \end{aligned}$$

where $\boldsymbol{\beta} = (\beta_0, \beta_1)^T$, and $\Phi(\cdot)$ is the cumulative distribution function of the standard normal distribution.

The maximum likelihood estimate can be found by maximizing the likelihood function. The **R** function **glm** might be useful for implementing the maximum likelihood approach. The BIC score is

$$\text{BIC} = -2\sum_{\alpha=1}^{23} \log f\left(y_\alpha|p(t_\alpha)\right) + \log(n) \times 2$$

where the conditional probability $p(t_\alpha)$ is given above. As a result, the BIC scores for each of the models are

$$\begin{aligned} \text{Logistic} &: \quad \text{BIC} = 39.744, \\ \text{Probit} &: \quad \text{BIC} = 39.959, \end{aligned}$$

respectively. We therefore select the model with the logistic function. The estimated coefficients and standard deviations are $\hat{\boldsymbol{\beta}} = (5.166, -0.116)^T$ and $(3.030, 0.046)^T$, respectively. Therefore, we can conclude that the temperature t affects the probability of the failure of primary O-rings.

Figure 5.3 shows the estimated conditional probabilities based on these three models. If we can assume that the estimated structures are close to the true conditional probability, we can see that the probability of the primary O-rings is very large at the temperature, 31° fahrenheit.

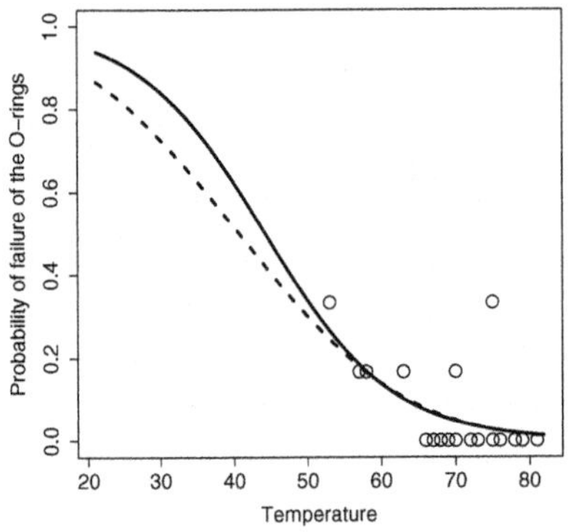

FIGURE 5.3: Comparison of the estimated conditional probabilities based on two models. Logistic: (—) and Probit (- - -). Circles are observed data.

5.5.3 Example: Selecting the number of factors in factor analysis model

It is common in econometric and financial applications to employ large panel data sets. This is particularly so in recent years because the advances in information technology make it possible to collect and process huge panel data sets. On the other hand, traditional statistical methods such as the vector autoregressive model of multivariate time series analysis often fare poorly in data analysis when the dimensions are high, and dimension reduction becomes a necessity. Factor analysis models are perhaps the most commonly used statistical tool to simplify the analysis of huge panel data sets.

This section applies the factor analysis model to select the number of factors in Japanese stock markets. Here, Japanese market return data using monthly reports from January 1980 through December 2006. Value-weighted monthly returns for the Japanese stock market indices were taken from a database at the Japan Securities Research Institute (JSRI).[1] The returns on

[1]http://www.jsri.or.jp/. This foundation serves the public interest, under the superintendence of the Financial Services Agency.

these indices, as well as details on their construction, can be obtained from the JSRI database directly.

In particular, we base decisions on the following 28 industries: Agriculture, Mining Construction, Food, Textile, Pulp, Chemicals, Coal and Oil, Steel Works etc., Electrical Equipment, Rubber, Glass, Steel manufacture, Non-steel manufacture, Metal manufacture, Machinery, Electric equipment, Transportation, Precision instruments. Other products, Commerce, Banking and Insurance, Real Estate, Land transportation, Shipping transportation, Freight transportation, Warehouse transportation, Information and communication, Utilities, and Services.

Suppose a set of n observations $\boldsymbol{X}_n = \{\boldsymbol{x}_1, ..., \boldsymbol{x}_n\}$ are generated from the following r-factor model:

$$\boldsymbol{x}_\alpha = \Lambda \boldsymbol{f}_\alpha + \boldsymbol{\varepsilon}_\alpha, \quad \alpha = 1, ..., n,$$

where $\boldsymbol{x}_\alpha = (x_{1\alpha}, ..., x_{m\alpha})^T$ is a m-dimensional random vector, $\boldsymbol{f}_\alpha = (f_{1\alpha}, ..., f_{r\alpha})^T$ is the r-dimensional random vector of factors, $\boldsymbol{\varepsilon}_\alpha = (\varepsilon_{1\alpha}, ..., \varepsilon_{m\alpha})^T$ is the m-dimensional random noise vector, and $\Lambda = (\boldsymbol{\lambda}_1, ..., \boldsymbol{\lambda}_m)^T$ is the $m \times r$ matrix of factor loadings.

The common assumptions of the maximum likelihood factor analysis are as follows: (a) the number of factors r is lower than m, (b) the noise follows a multivariate normal distribution with mean $\mathbf{0}$ and covariance matrix Ψ, which is diagonal, (c) the factor follows a multivariate normal distribution with mean $\mathbf{0}$ and unit covariance matrix I_r, and (d) $\boldsymbol{f}_i$ and $\boldsymbol{\varepsilon}_i$ are independent.

The maximum likelihood estimates unknown parameters $\boldsymbol{F}_n = (\boldsymbol{f}_1, ..., \boldsymbol{f}_n)^T$, Λ and Ψ by maximizing the likelihood function

$$\begin{aligned} & f(\boldsymbol{X}_n|\Lambda, \boldsymbol{F}_n, \Psi) \\ &= \frac{1}{(2\pi)^{nm/2}|\Psi|^{n/2}} \exp\left[-\frac{1}{2}\text{tr}\left\{(\boldsymbol{X}_n - \boldsymbol{F}_n\Lambda)^T(\boldsymbol{X}_n - \boldsymbol{F}_n\Lambda)\Psi^{-1}\right\}\right] \end{aligned}$$

with $\boldsymbol{X}_n = (\boldsymbol{x}_1, ..., \boldsymbol{x}_n)^T$. It is well known that the further constraints are required to define a unique model. For example, the model is invariant under transformations of the factor $\boldsymbol{f}^* = Q\boldsymbol{f}$, where Q is any orthogonal matrix. With regard to this problem, the author refers to Geweke and Singleton (1980), Seber (1984) and Lopes and West (2004).

The BIC score for the r-factor model is

$$\text{BIC}(r) = -2\log f(\boldsymbol{X}_n|\hat{\Lambda}, \hat{\boldsymbol{F}}_n, \hat{\Psi}) + \log(n) \times \dim(\boldsymbol{\theta}, r), \tag{5.10}$$

where $\hat{\Lambda}$, $\hat{\boldsymbol{F}}_n$ and $\hat{\Psi}$ are the maximized likelihood estimates and $\dim(\boldsymbol{\theta}, r) = m(r+1) - r(r-1)/2$ is the number of parameters included in the r-factor model.

Table 5.3 shows the relationship between the number of factors r and the BIC score (5.10). We can see that the $r = 7$ factor model can be selected as the minimizer of the BIC score. Table 5.3 also provides the maximum log-likelihood values, the number of parameters, and the penalty term $\log(n) \times$

$\dim(\boldsymbol{\theta}, r)$ for each of the r-factor models. It indicates that the maximum log-likelihood value becomes large as the number of r becomes larger, while the penalty term also becomes large. So, we can see the trade off between the fitness to the data and model complexity. The BIC score can be used to adjust this trade-off.

TABLE 5.3: The relationship between the number of factors r and the BIC score (5.10). We can see that the $r = 7$ factor model can be selected as the minimizer of the BIC score. The maximum log-likelihood values (MLLs), the number of parameters $\dim(\boldsymbol{\theta}, r)$, and the penalty term $\log(n) \times \dim(\boldsymbol{\theta}, r)$ (Penalty) for each of the r-factor models are also shown.

r	BIC	$\dim(\boldsymbol{\theta}, r)$	MLLs	Penalty
1	18038.20	56	−8857.24	323.72
2	16956.43	83	−8238.31	479.80
3	16778.00	109	−8073.95	630.10
4	16612.59	134	−7918.98	774.61
5	16531.52	158	−7809.08	913.35
6	16521.72	181	−7737.70	1046.31
7	16514.56	203	−7670.53	1173.49
8	16534.93	224	−7620.02	1294.88
9	16588.58	244	−7589.04	1410.50
10	16648.50	263	−7564.08	1520.33

Fama and French (1993) suggested that the asset return model on a stock index can be constructed using three different weighted averages of the portfolio values: one based on size (SMB), another based on the book-to-market ratio (HML), and the third based on excess return (ER) on the market. The Russell/Nomura Japan Index was used to evaluate these three factors in the Japanese stock market. So that the characteristics of factors are similar in the two markets, we computed the HML and SMB as follows:

$$\text{HML} = \{\text{LCV} + \text{SCV}\}/2 - \{\text{LCG} + \text{SCG}\}/2,$$
$$\text{SMB} = \{\text{LCV} + \text{LCG}\}/2 - \{\text{SCV} + \text{SCG}\}/2,$$

where the indices LCV, SCV, LCG, SCG refer to the Russell/Nomura Large Cap Value, Small Cap Value, Large Cap Growth, and Small Cap Growth indices. The excess return on the market is computed as the Russell/Nomura total Market Index minus the one-month call rate. (Details on these indices can be found in the website of Nomura Securities Global Quantitative Research.)

We computed the correlation between the estimated $r = 7$ factors and these three factors. The factors are estimated using Bartlett's approach. Table 5.4 provides the fitting results. We can see that the first factor $\boldsymbol{f}_1$ and the ER has a relatively large correlation. Therefore, we can suspect that the first factor represents the ER. Similarly, the fourth factor $\boldsymbol{f}_4$ and SMB, and fifth factor $\boldsymbol{f}_5$ and HML have a large correlation. Thus, we can consider that Fama and French (1993)'s three factors (ER, SMB, and HML) are important factors to explain the variation of the Japanese stock market.

TABLE 5.4: The correlation between the estimated $r = 7$ factors and Fama and French (1993)'s three factors (ER, SMB, and HML). The factors are estimated using Bartlett's approach.

r	ER	SMB	HML
$\boldsymbol{f}_1$	0.503	−0.290	−0.186
$\boldsymbol{f}_2$	0.287	0.227	0.386
$\boldsymbol{f}_3$	0.416	0.363	−0.347
$\boldsymbol{f}_4$	0.284	−0.539	0.101
$\boldsymbol{f}_5$	0.014	0.093	0.554
$\boldsymbol{f}_6$	−0.016	−0.085	−0.072
$\boldsymbol{f}_7$	0.199	0.055	−0.119

5.5.4 Example: Survival analysis

In many fields of study, the information of primary interest is how long it takes for a certain event to occur. Some examples are the time it takes for a customer to withdraw from their first purchase, the time it takes for a firm to default, the time required for a given therapy to cure the patient. These random phenomena are usually characterized as a distribution in the *time-to-event* variable, also known as survival data. The BIC is applied to statistical models and predictors of time-to-event relationships.

In survival data analysis, it is assumed that the time-to-event T is a realization of some random process. It follows that T is a random variable with probability distribution $f(t)$, where t has units of time. The probability distribution of T can also be expressed as a cumulative function denoted $F(t) = P(T \leq t)$, which represents the probability that the time-to-event will be less than or equal to any given value t.

More commonly, the distribution is treated using a survival function:

$$S(t) = \Pr(T > t) = 1 - F(t).$$

It is obvious that $S(t)$ is monotonically decreasing. We can also describe the distribution using a hazard function:

$$h(t) = \lim_{\Delta t \to 0} \frac{\Pr(t \leq T < t + \Delta t | T \geq t)}{\Delta t} = \frac{f(t)}{S(t)},$$

which measures the instantaneous risk that the event will occur at time t.

One can include the effects of p covariates $\boldsymbol{x} = (x_1, \ldots, x_p)^T$ in the survival probability through the hazard function. A famous example is the proportional hazards model (Cox, 1972), where the hazard function is expressed as

$$h(t|\boldsymbol{x};\boldsymbol{\beta}) = h_0(t)\exp\{\beta_1 x_1 + \cdots + \beta_p x_p\} = h_0(t)\exp\left\{\boldsymbol{\beta}^T\boldsymbol{x}\right\}. \quad (5.11)$$

Here $\boldsymbol{\beta} = (\beta_1, \ldots, \beta_p)^T$ is a vector of regression parameters, and $h_0(t)$ is an unknown baseline hazard function depending only on the time t. The baseline function is not specified, but can take the form of any parametric probability distribution such as a Weibull distribution or log-logistic distribution. Many different types of hazard functions can therefore be considered depending on the assumptions made about the baseline hazard function.

If the baseline hazard function is a Weibull distribution $h_0(t;\alpha) = \alpha t^{\alpha-1}$ with shape parameter α, the hazard function, its corresponding survival function, and the probability density function are given by

$$\begin{cases} h\left(t|\boldsymbol{x},\boldsymbol{\theta}\right) = \alpha t^{\alpha-1}\exp\left(\boldsymbol{\beta}^T\boldsymbol{x}\right), \\ S\left(t|\boldsymbol{x},\boldsymbol{\theta}\right) = \exp\left\{-t^{\alpha}\exp\left(\boldsymbol{\beta}^T\boldsymbol{x}\right)\right\}, \\ f\left(t|\boldsymbol{x},\boldsymbol{\theta}\right) = \alpha t^{\alpha-1}\exp\left(\boldsymbol{\beta}^T\boldsymbol{x}\right)\exp\left\{-t^{\alpha}\exp\left(\boldsymbol{\beta}^T\boldsymbol{x}\right)\right\}, \end{cases} \quad (5.12)$$

respectively. Here $\boldsymbol{\theta} = (\alpha, \boldsymbol{\beta}^T)^T$ is the unknown parameter vector. Note that the probability density function is also expressed as

$$f\left(t|\boldsymbol{x},\boldsymbol{\theta}\right) = h\left(t|\boldsymbol{x},\boldsymbol{\theta}\right) \times S\left(t|\boldsymbol{x},\boldsymbol{\theta}\right).$$

Suppose that we have a set of n independent observations $\{(t_\alpha, u_\alpha, \boldsymbol{x}_\alpha); \alpha = 1, \ldots, n\}$, where $\boldsymbol{x}_\alpha = (x_{1\alpha}, \ldots, x_{p\alpha})^T$ is a p-dimensional predictor, t_α is the survival time, and u_α is a censor function equal to one or zero. Some observations continue to exist until the end of the observation period without the event occurring; these data are censored by setting u_α to zero.

Under the Weibull model (5.12), the log-likelihood of a sample of n independent observations is given by

$$\begin{aligned} &\log f\left(\boldsymbol{T}_n|X_n, U_n, \boldsymbol{\theta}\right) \\ &= \sum_{\alpha=1}^{n}\left[u_\alpha \log f\left(t_\alpha|\boldsymbol{x}_\alpha, \boldsymbol{\theta}\right) + (1-u_\alpha)\log S\left(t_\alpha|\boldsymbol{x}_\alpha, \boldsymbol{\theta}\right)\right] \\ &= \sum_{\alpha=1}^{n}\left[u_\alpha\left\{\log\alpha + (\alpha-1)\log t_\alpha + \boldsymbol{\beta}^T\boldsymbol{x}_\alpha\right\} + t_\alpha^\alpha \exp\left(\boldsymbol{\beta}^T\boldsymbol{x}_\alpha\right)\right], \end{aligned} \quad (5.13)$$

where $\boldsymbol{T}_n = \{t_\alpha; \alpha = 1, \ldots, n\}$, $X_n = \{\boldsymbol{x}_\alpha; \alpha = 1, \ldots, n\}$, and $\boldsymbol{U}_n = \{u_\alpha; \alpha = 1, \ldots, n\}$. The maximum likelihood estimate $\hat{\boldsymbol{\theta}}$ can be obtained numerically.

The critical issue is how to optimize a set of predictors. To determine the best model among the candidates, we use the BIC score:

$$\text{BIC}(\boldsymbol{x}) = -2\log f(\boldsymbol{T}_n|X_n, U_n, \hat{\boldsymbol{\theta}}) + \log(n) \times \dim\{\boldsymbol{\theta}\},$$

where $\boldsymbol{x}$ is a possible combination of predictors.

In this example, the Bayesian analysis method is applied to ovarian cancer survival data. The data are taken from a randomized trial comparing two treatments for ovarian cancer. Some of the patients are censored because they survived for the entire observation period. In addition to the survival data itself, we have four predictors: age (in years), residual disease present (0=no, 1=yes), treatment group (0=group1, 1=group2), and ECOG performance status (0 is better than 1). Thus, there are 15 possible prior models: one for each combination of predictors.

We simply apply a stepwise variable selection method to identify an optimal subset of predictors. The best model is that based on the three predictors: x_1: age (in years), x_2: residual disease present (0=no, 1=yes), x_3: treatment group (0=group1, 1=group2), respectively. The values of estimated coefficients are $\hat{\beta}_1 = -0.069$, $\hat{\beta}_2 = -0.537$, and $\hat{\beta}_3 = 0.594$, respectively. Predictor x_1 thus has a negative impact to the time-to-event. Also, the residual disease present has a negative impact.

In the same way, we fitted the following three parametric models by applying a stepwise variable selection method.

$$\begin{aligned}
\text{Exponential}: \quad & h(t|\beta_1) = \exp\left(\beta_1 x_1\right), \\
\text{Extreme value}: \quad & h(t|\beta_1, \alpha) = \alpha \exp(\alpha t) \exp\left(\beta_1 x_1\right), \\
\text{Log logistic}: \quad & h(t|\beta_1, \alpha) = \frac{\alpha t^{\alpha-1} \exp\left(\alpha \beta_1 x_1\right)}{\left[1 + t^{\alpha} \exp\left(\alpha \beta_1 x_1\right)\right]}.
\end{aligned}$$

To compare the models, we calculated their BIC scores. The results are given in Table 5.5. We can see that the log-logistic model achieved the best fitting result.

Finally, we would like to point out that Volinsky and Raftery (2000) proposed a revision of the penalty term in BIC. They defined it in terms of the number of uncensored events instead of the number of observations. For example, in the context of Cox proportional hazards regression model, they proposed defining BIC in terms of the maximized partial likelihood with the use of the number of deaths rather than the number of individuals in the BIC penalty term. See also Kass and Wasserman (1995).

TABLE 5.5: BIC scores for various models.

Model	BIC
Weibull	76.773
Exponential	79.550
Extreme value	91.909
Log logistic	76.167

5.5.5 Consistency of the Bayesian information criteria

We have seen that the Bayesian information criteria allows us to select the best model among candidates. The candidate models may be nested, nonnested, overlapping, linear or nonlinear, and correctly specified or misspecified. Sin and White (1996) provided general conditions on the true model under which the use of several information criteria leads to the selection of model with lowest Kullback-Leibler (1951) divergence from the true model with probability one or with probability approaching one, so called "(weak) consistency of selection." We discuss the consistency of BIC and refer to Sin and White (1996) for precise assumptions.

Consider a situation that the candidate models are misspecified. In this case, we define the best model $f_0(\boldsymbol{z}|\boldsymbol{\theta}_0)$ that has the lowest Kullback-Leibler (1951) divergence from the true model, or equivalently, the maximum expected log-likelihood

$$\int \log f_0(\boldsymbol{z}|\boldsymbol{\theta}_0)dG(\boldsymbol{z}) = \max_k \left\{ \sup_{\boldsymbol{\theta}_k} \int \log f_k(\boldsymbol{z}|\boldsymbol{\theta}_k)dG(\boldsymbol{z}) \right\}.$$

To select the best model, one can consider an information criterion

$$\mathrm{IC} = -2\log f(\boldsymbol{X}_n|\hat{\boldsymbol{\theta}}) + c_{n,p}, \tag{5.14}$$

where $\hat{\boldsymbol{\theta}}$ is the maximum likelihood estimator and $c_{n,p}$ imposes a penalty to encourage the selection of a parsimonious model. It can be a sequence of nonstochastic numbers such as AIC: $c_{n,p} = 2p$ and BIC: $c_{n,p} = \log(n)p$. It is also a known fact that, under a certain condition, the maximum likelihood estimator $\hat{\boldsymbol{\theta}}_k$ converges in probability or almost surely to $\boldsymbol{\theta}_{k0}$, which achieves the maximum expected log-likelihood that minimizes the Kullback-Leibler divergence (see for example, White (1982)).

As shown in Sin and White (1996), for the consistency in picking the model $f_0(\boldsymbol{z}|\boldsymbol{\theta}_0)$, it is required that

$$\liminf_n \left[\frac{1}{n}\int \log f_0(\boldsymbol{X}_n|\boldsymbol{\theta}_0)dG(\boldsymbol{X}_n) - \frac{1}{n}\int \log f_k(\boldsymbol{X}_n|\boldsymbol{\theta}_{k0})dG(\boldsymbol{X}_n) \right] > 0$$

for $f_k(\boldsymbol{X}_n|\boldsymbol{\theta}_{k0})$ that does not achieve the maximum expected log-likelihood. Under the situation that $f_0(\boldsymbol{z}|\boldsymbol{\theta}_0)$ is a unique density and $c_{n,p} = o_p(n)$, the consistency of selection holds. As a consequence, we immediately see that both AIC and BIC satisfy this condition. Thus, both AIC and BIC are consistent in this situation.

However, suppose now that there are two or more candidate models that achieve the maximum expected log-likelihood. It is naturally considered that the simplest model (the model having fewest number of parameters) among these models is most preferred. This is often referred to as "parsimony." For the consistency in picking the simplest model that achieve the maximum expected log-likelihood, we require, for the models $f_j(\boldsymbol{X}_n|\boldsymbol{\theta}_{j0})$ and $f_k(\boldsymbol{X}_n|\boldsymbol{\theta}_{k0})$

that achieve the maximum expected log-likelihood, the log-likelihood ratio sequence of two models is bounded

$$\log\left[\frac{f_j(\boldsymbol{X}_n|\boldsymbol{\theta}_{j0})}{f_k(\boldsymbol{X}_n|\boldsymbol{\theta}_{k0})}\right] = O_p(1).$$

Under the situation $c_{n,p} \to \infty$ $(n \to \infty)$, the consistency of selection holds. As a consequence, the penalty term of BIC satisfies this condition, while that of AIC does not. Thus, it implies that BIC will select the simplest model, while AIC fails.

5.6 Definition of the generalized Bayesian information criterion

This section considers the case $\log \pi(\boldsymbol{\theta}) = O_p(n)$. In this case, under the i.i.d assumption on $\boldsymbol{X}_n$, we obtain the Laplace approximation to the marginal likelihood in the form

$$P(\boldsymbol{X}_n|M) \approx f(\boldsymbol{X}_n|\hat{\boldsymbol{\theta}}_n)\pi(\hat{\boldsymbol{\theta}}_n) \times \frac{(2\pi)^{p/2}}{n^{p/2}|S_n(\hat{\boldsymbol{\theta}}_n)|^{1/2}}, \tag{5.15}$$

where $\hat{\boldsymbol{\theta}}_n$ is the posterior mode, and

$$S_n(\hat{\boldsymbol{\theta}}_n) = -\frac{1}{n}\frac{\partial^2 \log\{f(\boldsymbol{X}_n|\boldsymbol{\theta})\pi(\boldsymbol{\theta})\}}{\partial\boldsymbol{\theta}\partial\boldsymbol{\theta}^T}\bigg|_{\boldsymbol{\theta}=\hat{\boldsymbol{\theta}}_n} = J_n(\hat{\boldsymbol{\theta}}_n) - \frac{1}{n}\frac{\partial^2 \log \pi(\boldsymbol{\theta})}{\partial\boldsymbol{\theta}\partial\boldsymbol{\theta}^T}\bigg|_{\boldsymbol{\theta}=\hat{\boldsymbol{\theta}}_n}.$$

Multiplying (-2) on the logarithm of the resulting formula (5.15), we obtain the generalized Bayesian information criterion (Konishi et al. (2004))

$$\text{GBIC} = -2\log f(\boldsymbol{X}_n|\hat{\boldsymbol{\theta}}_n) - 2\log \pi(\hat{\boldsymbol{\theta}}_n) + p\log n + \log|S_n(\hat{\boldsymbol{\theta}}_n)| - p\log 2\pi \tag{5.16}$$

Choosing the model with the largest posterior probability among a set of candidate models is equivalent to choosing the model that minimizes the criterion.

Remark

If the second derivative of the log posterior density

$$\ell(\boldsymbol{\theta}) \equiv \log\{f(\boldsymbol{X}_n|\boldsymbol{\theta})\pi(\boldsymbol{\theta})\}$$

in $S_n(\hat{\boldsymbol{\theta}}_n)$ is difficult to determine analytically, we can use their numerical derivatives.

At any specified point $\boldsymbol{\theta} = (\theta_1, ..., \theta_p)^T$, each component of the first derivative $\partial\ell(\boldsymbol{\theta})/\partial\boldsymbol{\theta}$ is approximated by

$$\frac{\partial\ell(\boldsymbol{\theta})}{\partial\theta_j} \approx \frac{\ell(\boldsymbol{\theta}+\boldsymbol{\delta}_j) - \ell(\boldsymbol{\theta}-\boldsymbol{\delta}_j)}{(2\delta)}, \quad j = 1, \cdots, p.$$

Here δ is a small value, and $\boldsymbol{\delta}_j$ is the p-dimensional vector whose j^{th} component is δ and whose other elements are zero.

Similarly, each component of the second derivative $\partial \ell(\boldsymbol{\theta})/\partial\boldsymbol{\theta}\partial\boldsymbol{\theta}^T$ can be calculated by

$$\frac{\partial^2 \ell(\boldsymbol{\theta})}{\partial\theta_j\partial\theta_k} \approx \frac{\ell(\boldsymbol{\theta}+\boldsymbol{\delta}_j+\boldsymbol{\delta}_k)-\ell(\boldsymbol{\theta}+\boldsymbol{\delta}_j-\boldsymbol{\delta}_k)-\ell(\boldsymbol{\theta}-\boldsymbol{\delta}_j+\boldsymbol{\delta}_k)+\ell(\boldsymbol{\theta}-\boldsymbol{\delta}_j-\boldsymbol{\delta}_k)}{4\delta^2},$$

for $j,k=1,\cdots,p$. Following the suggestion of Gelman et al. (1995, page 273), $\delta = 0.0001$ is low enough to approximate the derivative and high enough to avoid truncation error.

5.6.1 Example: Nonlinear regression models using basis expansion predictors

Suppose we have n independent observations $\{(y_\alpha, \boldsymbol{x}_\alpha);\alpha=1,2,...,n\}$, where y_α are random response variables and $\boldsymbol{x}_\alpha$ are vectors of p-dimensional explanatory variables. In order to draw information from the data, we use the Gaussian nonlinear regression model

$$y_\alpha = u(\boldsymbol{x}_\alpha)+\varepsilon_\alpha, \qquad \alpha=1,...,n, \tag{5.17}$$

where $u(\cdot)$ is an unknown smooth function and errors ε_α are independently, normally distributed with mean zero and variance σ^2. The problem to be considered is to estimate the function $u(\cdot)$ from the observed data, for which we use the basis expansion predictors:

$$u(\boldsymbol{x}_\alpha)=\sum_{k=1}^m w_k b_k(\boldsymbol{x}_\alpha)+w_0, \tag{5.18}$$

where $b_k(\boldsymbol{x})$ are basis functions.

There are various types of basis functions. Under a situation that the dimension of the predictor is $p=1$, we can consider

1. **Linear model**

$$u(x)=w_0+w_1x$$

2. **Polynomial model**

$$u(x)=w_0+w_1x+w_2x^2+\cdots+w_mx^m$$

3. ***B*-spline model**

$$u(x)=\sum_{k=1}^m w_k\phi_k(x)$$

4. **Cubic spline model**

$$u(x) = w_0 + w_1 x + w_2 x^2 + w_3 x^3 + \sum_{k=1}^{p} w_k |x - \kappa_k|_+^3$$

Here $\phi_j(x)$ is B-spline basis function. Each basis function $\phi_j(x)$ can be calculated using de Boor's recursion formula (de Boor (1978)). Given $m+r+1$ equally spaced knots $t_1 < ... < t_{m+r+1}$, each B-spline basis with the degree r, $\phi_k(t;r)$, can be calculated as

$$\begin{aligned}\phi_k(t,0) &= \begin{cases} 1, & t_k \le t < t_{k+1}, \\ 0, & otherwise, \end{cases} \\ \phi_j(t;r) &= \frac{t - t_k}{t_{k+r} - t_k}\phi_k(t;r-1) + \frac{t_{k+r+1} - t}{t_{k+r+1} - t_{k+1}}\phi_{k+1}(t;r-1).\end{aligned}$$

Since a zero-degree B-spline basis is just a constant on one interval between two knots, it is simple to compute the B-spline basis of any degree.

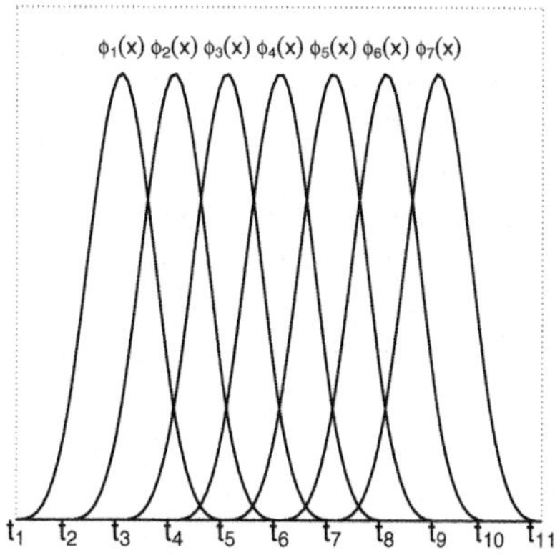

FIGURE 5.4: B-spline basis functions with degree 3.

The function $|a|_+$ in the cubic spline model is defined as $|a|_+ = a$ if $a > 0$ and o otherwise. Also, κ_k is a knot that determines the point that allows the function $|x - \kappa_k|_+$ takes a positive value.

When the dimension of predictors is $p > 1$, we can consider

1. **Linear model**

$$u(\boldsymbol{x}) = w_0 + \sum_{k=1}^{p} w_k x_k$$

2. **Additive model**

$$u(\boldsymbol{x}) = \sum_{k=1}^{p} h_k(x_k)$$

3. **Kernel expansion**

$$u(\boldsymbol{x}) = \sum_{k=1}^{m} w_k b_k(\boldsymbol{x})$$

Hastie and Tibshirani (1990) and Hastie et al. (2009) provided a nice summary of the additive models and Kernel expansion approach. We also refer to Craven and Wahba (1979) and Imoto and Konishi (2003) for nonlinear regression with spline functions. Here, for basis functions $b_j(\boldsymbol{x})$, we use a Gaussian radial basis with a hyperparameter (Ando et al. (2008)):

$$b_j(\boldsymbol{x}) = \exp\left(-\frac{||\boldsymbol{x} - \boldsymbol{\mu}_j||^2}{2\nu\sigma_j^2}\right), \qquad j = 1, 2, \ldots, m, \tag{5.19}$$

where $\boldsymbol{\mu}_j$ is a p-dimensional vector determining the location of the basis function, σ_j^2 is the scale parameter, and ν is the spread parameter that adjusts the smoothness of the regression surface. Note that the augment provided below can apply to other uses of basis functions.

Ando et al. (2008) estimated the centers $\boldsymbol{\mu}_k$ and width parameters σ_k are determined by using only the input data set $\{\boldsymbol{x}_\alpha; \alpha = 1, ..., n\}$ for explanatory variables. Among possible strategies for determining the centers and widths of the basis functions we use a k-means clustering algorithm. This algorithm divides the input data set $\{\boldsymbol{x}_\alpha; \alpha = 1, ..., n\}$ into m clusters $A_1, ..., A_m$ that correspond to the number of the basis functions. The centers and width parameters are then determined by

$$\boldsymbol{c}_k = \frac{1}{n_k} \sum_{\alpha \in A_k} \boldsymbol{x}_\alpha, \qquad s_k^2 = \frac{1}{n_k} \sum_{\alpha \in A_k} ||\boldsymbol{x}_\alpha - \boldsymbol{c}_k||^2, \tag{5.20}$$

where n_k is the number of the observations which belong to the k-th cluster A_k. Replacing $\boldsymbol{\mu}_k$ and σ_k in the Gaussian basis by $\boldsymbol{c}_k$ and s_k, respectively, we have a set of m basis functions

$$b_k(\boldsymbol{x}) = \exp\left(-\frac{||\boldsymbol{x} - \boldsymbol{c}_k||^2}{2\nu s_k^2}\right), \qquad k = 1, ..., m. \tag{5.21}$$

It follows from Equations (5.17), (5.19) and (5.21) that the Gaussian nonlinear regression model based on the radial basis functions may be expressed as

$$y_\alpha = \boldsymbol{w}^T \boldsymbol{b}(\boldsymbol{x}_\alpha) + \varepsilon_\alpha, \qquad \varepsilon_\alpha \sim N(0, \sigma^2), \tag{5.22}$$

where $\boldsymbol{w} = (w_0, ..., w_m)^T$ and $\boldsymbol{b}(\boldsymbol{x}_\alpha) = (1, b_1(\boldsymbol{x}_\alpha), ..., b_m(\boldsymbol{x}_\alpha))^T$. Then the data $\{y_1, ..., y_n\}$ are summarized by a model from a class of probability densities

$$f(y_\alpha|\boldsymbol{x}_\alpha;\boldsymbol{\theta}) = \frac{1}{\sqrt{2\pi}\sigma} \exp\left[-\frac{1}{2\sigma^2}\left\{y_\alpha - \boldsymbol{w}^T\boldsymbol{b}(\boldsymbol{x}_\alpha)\right\}^2\right], \quad \alpha = 1, ..., n, \quad (5.23)$$

where $\boldsymbol{\theta} = (\boldsymbol{w}^T, \sigma^2)^T$.

Using a singular multivariate normal prior density (Konishi et al. (2004), Ando (2007)):

$$\pi(\boldsymbol{\theta}) = \left(\frac{n\lambda}{2\pi}\right)^{(m-d)/2} |R|_+^{1/2} \exp\left\{-n\lambda\frac{\boldsymbol{\theta}^T R\boldsymbol{\theta}}{2}\right\}, \quad (5.24)$$

we can make the posterior inference. Here λ is a smoothing parameter, m is the number of basis functions, $R = \text{diag}\{D, 0\}$ is a block diagonal matrix and $|R|_+$ is the product of $(m-d)$ nonzero eigenvalues of R. Konishi et al. (2004) used the following matrix $D_2^T D_2$ for D, where D_k is a $(m-k) \times m$ matrix

$$D_k = \begin{pmatrix} (-1)^0{}_kC_0 & \cdots & (-1)^k{}_kC_k & 0 & \cdots & 0 \\ 0 & (-1)^0{}_kC_0 & \cdots & (-1)^k{}_kC_k & \cdots & 0 \\ \vdots & \ddots & \ddots & \ddots & \ddots & \vdots \\ 0 & \cdots & 0 & (-1)^0{}_kC_0 & \cdots & (-1)^k{}_kC_k \end{pmatrix},$$

with ${}_pC_k = p!/\{k!(p-k)!\}$. In this case, $\boldsymbol{w}^T D\boldsymbol{w} = \sum_{j=k}^m (\Delta^k w_j)^2$ with Δ is a difference operator such as $\Delta w_j = w_j - w_{j-1}$. The use of difference penalties has been investigated by Whittaker (1923), Green and Yandell (1985) and O'Sullivan et al. (1986). As proposed by Lang and Brezger (2004), one can also put hierachcal prior on the smoothing parameter λ, $\pi(\lambda)$.

The posterior mode $\hat{\boldsymbol{\theta}}_n$ can be found by maximizing the penalized log-likelihood function

$$\hat{\boldsymbol{\theta}}_n = \text{argmax}_\theta \left[\sum_{\alpha=1}^n \log f(y_\alpha|\boldsymbol{x}_\alpha;\boldsymbol{\theta}) - \frac{n\lambda}{2}\boldsymbol{\theta}^T R\boldsymbol{\theta}\right]. \quad (5.25)$$

Then the posterior modes of $\boldsymbol{w}$ and σ^2 are explicitly given by

$$\hat{\boldsymbol{w}}_n = \left(B^T B + n\beta D_2^T D_2\right)^{-1} B^T \boldsymbol{y}_n \text{ and } \hat{\sigma}_n^2 = \frac{1}{n}\sum_{\alpha=1}^n \left\{y_\alpha - \hat{\boldsymbol{w}}^T\boldsymbol{b}(\boldsymbol{x}_\alpha)\right\}^2 (5.26)$$

where $\beta = \lambda\hat{\sigma}_n^2$, and

$$B = \begin{pmatrix} \boldsymbol{b}(\boldsymbol{x}_1)^T \\ \vdots \\ \boldsymbol{b}(\boldsymbol{x}_n)^T \end{pmatrix} = \begin{pmatrix} b_1(\boldsymbol{x}_1) & \cdots & b_m(\boldsymbol{x}_1) \\ \vdots & \ddots & \vdots \\ b_1(\boldsymbol{x}_n) & \cdots & b_m(\boldsymbol{x}_n) \end{pmatrix}$$

is the $n \times (m+1)$ design matrix. The predictive mean z at a point $\boldsymbol{x}$ is then given as

$$\hat{z}_n = \hat{\boldsymbol{w}}_n^T \boldsymbol{b}(\boldsymbol{x}).$$

We first illustrate this Bayesian model by fitting surface to the simulated data. The data $\{y_\alpha, (x_{1\alpha}, x_{2\alpha}); \alpha = 1, ..., 400\}$ are generated from the true model $y_\alpha = \sin(2\pi x_{1\alpha}) + \cos(2\pi x_{2\alpha}) + \varepsilon_\alpha$ with Gaussian noise $N(0,1)$, where the design points are uniformly distributed in $[0,2] \times [0,2]$. Figures 5.6.1 (a) and (b) show the true surface and the interpolated surface to the simulated data, respectively. Figures 5.6.1 (c) and (d) give the smoothed surfaces for the hyperparameters $\nu = 1$ and $\nu = 5$. The smoothed surface in Figure 5.6.1 (c) is obviously undersmoothed, while the one in (d) gives a good representation of the underlying function over the region $[0,2] \times [0,2]$. We set the number of basis functions and the value of smoothing parameter to be $m = 30$ and $\lambda = 0.0001$. We observe that by appropriate choice of ν, this Bayesian nonlinear regression modeling strategy can capture the true structure generating the data.

Therefore, the estimated statistical model depends on the hyperparameter ν, the smoothing parameter λ and also the number of basis functions m. The problem is how to choose these adjusted parameters by a suitable criterion. Konishi et al. (2004) proposed a generalized Bayesian information criterion for evaluating the estimated statistical model

$$\begin{aligned}\mathrm{GBIC} &= (n+m-1)\log\hat{\sigma}_n^2 + n\lambda\hat{\boldsymbol{w}}_n^T D\hat{\boldsymbol{w}}_n + n + (n-3)\log(2\pi) \\ &\quad +3\log n + \log\left|S_n(\hat{\boldsymbol{\theta}}_n)\right| - \log\left|D_2^T D_2\right|_+ - (m-1)\log\left(\lambda\hat{\sigma}_n^2\right).\end{aligned} \tag{5.27}$$

Here a $(m+2) \times (m+2)$ matrix $S_n(\hat{\boldsymbol{\theta}}_n)$ is given as

$$\begin{aligned} S_n(\hat{\boldsymbol{\theta}}_n) &= -\frac{1}{n}\begin{pmatrix} \dfrac{\partial^2 \log\{f(\boldsymbol{y}_n|X_n,\hat{\boldsymbol{\theta}}_n)\pi(\hat{\boldsymbol{\theta}}_n)\}}{\partial\boldsymbol{w}\partial\boldsymbol{w}^T} & \dfrac{\partial^2 \log\{f(\boldsymbol{y}_n|X_n,\hat{\boldsymbol{\theta}}_n)\pi(\hat{\boldsymbol{\theta}}_n)\}}{\partial\boldsymbol{w}\partial\sigma^2} \\ \dfrac{\partial^2 \log\{f(\boldsymbol{y}_n|X_n,\hat{\boldsymbol{\theta}}_n)\pi(\hat{\boldsymbol{\theta}}_n)\}}{\partial\boldsymbol{w}^T\partial\sigma^2} & \dfrac{\partial^2 \log\{f(\boldsymbol{y}_n|X_n,\hat{\boldsymbol{\theta}}_n)\pi(\hat{\boldsymbol{\theta}}_n)\}}{\partial\sigma^2\partial\sigma^2} \end{pmatrix} \\ &= \frac{1}{n\hat{\sigma}_n^2}\begin{pmatrix} B^TB + n\lambda\hat{\sigma}_n^2 D_2^T D_2 & B^T\boldsymbol{e}/\hat{\sigma}_n^2 \\ \boldsymbol{e}^T B/\hat{\sigma}_n^2 & n/2\hat{\sigma}_n^2 \end{pmatrix}, \end{aligned}$$

with $f(\boldsymbol{y}_n|X_n,\boldsymbol{\theta}) = \prod_{\alpha=1}^n f(y_\alpha|\boldsymbol{x}_\alpha;\boldsymbol{\theta})$ and

$$\boldsymbol{e} = \{y_1 - \hat{\boldsymbol{w}}_n\boldsymbol{b}(\boldsymbol{x}_1), ..., y_n - \hat{\boldsymbol{w}}_n\boldsymbol{b}(\boldsymbol{x}_n)\}$$

is an n-dimensional vector.

Konishi et al. (2004) applied the above method to Robot arm data. Andrieu et al. (2001) proposed a hierarchical full Bayesian model for radial

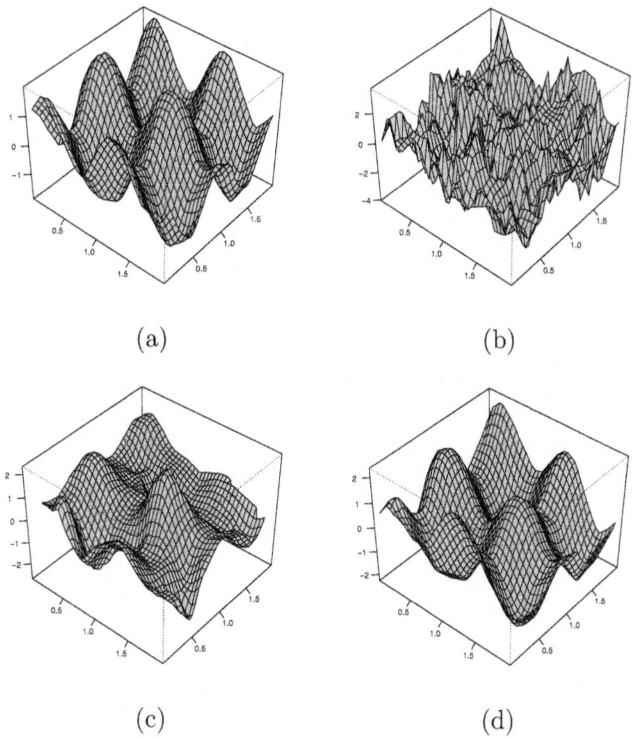

FIGURE 5.5: (From Ando, T. et al., *J. Stat. Plan. Infer.*, 138, 2008. With permission.) Comparison of the true surface and the smoothed surfaces for radial basis functions with $\nu = 1$ and $\nu = 6.38$. (a) and (b) show the true and interpolated surfaces, respectively. (c) and (d) show the smoothed surfaces for radial basis functions with $\nu = 1$ and $\nu = 5$, respectively.

basis function networks with Gaussian noise, in which the model dimension, model parameters, regularization parameters and also noise parameters are treated as unknown random variables. They developed a reversible-jump Markov chain Monte Carlo simulation algorithm for radial basis networks for computing the joint posterior distribution of the parameters. The developed Bayesian method can be regarded as an approximate Bayesian methodology, and Konishi et al. (2004) compared the method with the full Bayesian approach, by analyzing the robot arm data which is often used as a benchmark dataset in the neural network literature (Andrieu et al. (2001), Holmes and Mallick (1998), MacKay (1992), Neal (1996), Rios Insua and Müller (1998)). MacKay (1992) originally introduced the use of the Bayesian approach in the neural network literature. The dataset, created by D.J.C. MacKay and available at http://wol.ra.phy.cam.ac.uk/mackay/bigback/dat/, is a set of four-dimensional data $\{(x_{1\alpha}, x_{2\alpha}, y_{1\alpha}, y_{2\alpha}); \alpha = 1, ..., n\}$ generated from the following model:

$$y_{1\alpha} = 2\cos(x_{1\alpha}) + 1.3\cos(x_{1\alpha} + x_{2\alpha}) + \varepsilon_{1\alpha},$$
$$y_{2\alpha} = 2\sin(x_{1\alpha}) + 1.3\sin(x_{1\alpha} + x_{2\alpha}) + \varepsilon_{2\alpha},$$

where $\varepsilon_{1\alpha}$ and $\varepsilon_{2\alpha}$ are normal noise variables with means 0 and variances $(0.05)^2$. Figure 5.6 shows true surfaces.

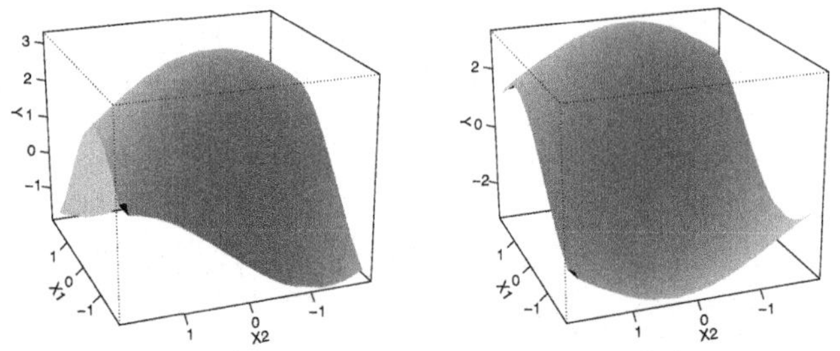

(a) (b)

FIGURE 5.6: Robot arm data: True surfaces (a) $u(\boldsymbol{x}) = 2\cos(x_1) + 1.3\cos(x_1 + x_2)$, and (b) $u(\boldsymbol{x}) = 2\sin(x_1) + 1.3\sin(x_1 + x_2)$.

The first 200 observations are used to estimate the model and the last 200 observations are used to evaluate the prediction accuracy. The values of λ, ν and m are chosen as the minimizers of GBIC. As a result, Konishi et al. (2004) obtained $\hat{m} = 20$, $\hat{\nu} = 31.81$ and $\hat{\lambda} = 2.46 \times 10^{-7}$C and the corresponding average squared error was 0.00509.

Table 5.6 summarizes the results obtained by various techniques. The implemented strategy and the full Bayesian approach yield almost the same results, and both give fitted functions that capture the true structure. An advantage of the implemented Bayesian procedure is that it is easily implemented in both its Gaussian and non-Gaussian versions.

Remarks: The posterior modes $\hat{\boldsymbol{w}}_n$ and $\hat{\sigma}_n^2$ depend on each other. Note that given a value of $\lambda_0 = \lambda\sigma^2$, we first estimate the coefficient parameter $\boldsymbol{w}$ by $\hat{\boldsymbol{w}}_n = (B^T B + n\lambda_0 D)^{-1} B^T \boldsymbol{y}$. We then obtain the estimate $\hat{\sigma}_n^2 = n^{-1} \sum_{\alpha=1}^n \{y_\alpha - \hat{\boldsymbol{w}}_n^T \boldsymbol{b}(\boldsymbol{x}_\alpha)\}^2$. Thus, the smoothing parameter λ is given as $\lambda = \lambda_0 / \hat{\sigma}_n^2$.

5.6.2 Example: Multinomial logistic model with basis expansion predictors

One well-known statistical method of multiclass classification is based on linear logistic discriminant models (Seber (1984), Hosmer and Lemeshow

TABLE 5.6: *Source:* Konishi, S., Ando, T. and Imoto, S. *Biometrika*, 91, 2004. With permission. Comparison of the average squared errors for robot arm data. The results are due to Andrieu et al. (2001), Holmes and Mallick (1998), Konishi et al. (2004), MacKay (1992), Neal (1996), Rios Insua and Müller (1998). ARD, automatic relevance determination; GA, Gaussian approximation; MLP, multilayer perceptron; RJMCMC, reversible-jump Markov Chain Monte Carlo; RBF, radial basis function; MDL, minimum description length; GBIC, generalized Bayesian information criteria.

Methods	ASE
MacKay's (1992) GA with highest evidence	0.00573
MacKay's (1992) GA with lowest test error	0.00557
Neal's (1996) hybrid Monte Carlo	0.00554
Neal's (1996) hybrid Monte Carlo with ARD	0.00549
Rios Insua and Müller's (1998) MLP with RJMCMC	0.00620
Holmes and Mallick's (1998) RBF with RJMCMC	0.00535
Andrieu et al.'s RJMCMC with Bayesian model	0.00502
Andrieu et al.'s RJMCMC with MDL	0.00512
Andrieu et al.'s RJMCMC with AIC	0.00520
Konishi et al.'s GBIC	0.00509

(1989)), which assume that the log-odds ratios of the posterior probabilities can be expressed as linear combinations of the p-dimensional feature variables $\boldsymbol{x} = (x_1, \ldots, x_p)^T$:

$$\log\left\{\frac{\Pr(g = k|\boldsymbol{x})}{\Pr(g = G|\boldsymbol{x})}\right\} = w_{k0} + \sum_{j=1}^{p} w_{kj}x_j, \qquad k = 1, \ldots, G-1. \tag{5.28}$$

Here G is the number of groups, the categorical variable $g \in \{1, \ldots, G\}$ is an indicator of the class label, and $\Pr(g = k|\boldsymbol{x})$ is the posterior probability of $g = k$ given the feature variables $\boldsymbol{x}$. When the unknown parameters $\{w_{kj}; j = 0, \ldots, p, k = 1, \ldots, G-1\}$ are estimated by the maximum likelihood method, a future observation is generally classified into one of several groups that gives the maximum posterior probability.

Although linear logistic discriminant models have become a standard tool for multiclass classification, this method has some disadvantages. First, linear decision boundaries are often too crude for complex data, and therefore nonlinear decision boundaries would be more attractive (Hastie et al. (1994)). Second, a large number of predictors relative to the sample size leads to unstable maximum likelihood parameter estimates. In addition, the existence of multicollinearity may result in infinite maximum likelihood parameter estimates and, consequently, incorrect classification results.

In such a case we can consider nonlinear models by replacing the linear predictor with a linear combination of basis functions:

$$\log\left\{\frac{\Pr(g=k|\boldsymbol{x})}{\Pr(g=G|\boldsymbol{x})}\right\} = w_{k0} + \sum_{j=1}^{m} w_{kj} b_j(\boldsymbol{x}), \tag{5.29}$$

where $\{b_j(\boldsymbol{x}); j = 1,\ldots,m\}$ are a set of basis functions and $\{w_{kj}; j = 0,\ldots,m, k = 1,\ldots,G-1\}$ are a set of unknown parameters to be estimated. For basis functions $b_j(\boldsymbol{x})$ in (5.29), Ando and Konishi (2009) used Ando et al. (2008)'s Gaussian radial basis with a hyperparameter in (5.19).

It may easily be seen that log-posterior-odds models of the form (5.29) can be rewritten in terms of the following posterior probabilities:

$$\begin{aligned}\Pr(g=k|\boldsymbol{x}) &= \frac{\exp\left\{\boldsymbol{w}_k^T \boldsymbol{b}(\boldsymbol{x})\right\}}{1+\sum_{j=1}^{G-1}\exp\left\{\boldsymbol{w}_j^T \boldsymbol{b}(\boldsymbol{x})\right\}}, \qquad k=1,\ldots,G-1,\\ \Pr(g=G|\boldsymbol{x}) &= \frac{1}{1+\sum_{k=1}^{G-1}\exp\left\{\boldsymbol{w}_j^T \boldsymbol{b}(\boldsymbol{x})\right\}},\end{aligned} \tag{5.30}$$

where $\boldsymbol{w}_k = (w_{k0},\ldots,w_{km})^T$ is an $(m+1)$-dimensional parameter vector and $\boldsymbol{b}(\boldsymbol{x}) = (1, b_1(\boldsymbol{x}),\ldots,b_m(\boldsymbol{x}))^T$ is an $(m+1)$-dimensional vector of basis functions. These posterior probabilities $\Pr(g=k|\boldsymbol{x})$ depend on a set of parameters $\boldsymbol{w} = (\boldsymbol{w}_1^T,\ldots,\boldsymbol{w}_{G-1}^T)^T$, and so we denote these posterior probabilities as $\Pr(g=k|\boldsymbol{x}) := \pi_k(\boldsymbol{x};\boldsymbol{w})$.

We now define the G-dimensional vector $\boldsymbol{y} = (y_1,\ldots,y_G)^T$ that indicates group membership. The kth element of $\boldsymbol{y}$ is set to be one or zero according to whether $\boldsymbol{x}$ belongs or does not belong to the kth group as follows:

$$\boldsymbol{y} = (y_1,\ldots,y_G)^T = (0,\ldots,\overset{(k-1)}{0},\overset{(k)}{1},\overset{(k+1)}{0},\ldots,0)^T \quad \text{if} \quad g=k.$$

This implies that $\boldsymbol{y}$ is the kth unit column vector if $g=k$. Assuming that the random variable $\boldsymbol{y}$ is distributed according to a multinomial distribution with probabilities $\pi_k(\boldsymbol{x};\boldsymbol{w})$ $(k=1,\ldots,G)$, our model (5.29) can be expressed in the following probability density form:

$$f(\boldsymbol{y}|\boldsymbol{x};\boldsymbol{w}) = \prod_{k=1}^{G} \pi_k(\boldsymbol{x};\boldsymbol{w})^{y_k}, \tag{5.31}$$

where $\pi_k(\boldsymbol{x};\boldsymbol{w})$ are the posterior probabilities given in (5.30).

Suppose that we have a set of n independent observations $\{(\boldsymbol{x}_\alpha, g_\alpha); \alpha = 1,\ldots,n\}$, where the $\boldsymbol{x}_\alpha$ are the vectors of p feature variables and g_α are the class labels. Ando and Konishi (2009) proposed the estimation procedure that consists of two stages. In the first stage, a set of Gaussian radial basis functions $\{b_j(\boldsymbol{x}); j=1,\ldots,m\}$ are constructed or, equivalently, the centers $\boldsymbol{\mu}_j$ and the scale parameters σ_j^2 in the Gaussian radial basis (5.19) are determined. In the second stage, the unknown parameter vector $\boldsymbol{w}$ is estimated.

Ando and Konishi (2009) determined the centers $\boldsymbol{\mu}_j$ and the scale parameters σ_j^2 in the Gaussian radial basis by using the k-means clustering algorithm. This algorithm divides a set of observations $\{\boldsymbol{x}_\alpha; \alpha = 1, \ldots, n\}$ into m clusters $A_1, \ldots, A_m$ that correspond to the number of basis functions. The centers and the scale parameters are then determined by $\boldsymbol{\mu}_j = \sum_{\alpha \in A_j} \boldsymbol{x}_\alpha / n_j$ and $\sigma_j^2 = \sum_{\alpha \in A_j} ||\boldsymbol{x}_\alpha - \boldsymbol{c}_j||^2 / n_j$, respectively, where n_j is the number of observations which belong to the jth cluster A_j. Using an appropriate value of the hyperparameter ν, we then obtain a set of m Gaussian radial basis functions.

The likelihood function is then

$$f(\boldsymbol{Y}_n | X_n, \boldsymbol{w}) = \sum_{\alpha=1}^{n} \log f(\boldsymbol{y}_\alpha | \boldsymbol{x}_\alpha; \boldsymbol{w}) = \sum_{\alpha=1}^{n} \sum_{k=1}^{G} y_{k\alpha} \log \pi_k(\boldsymbol{x}_\alpha; \boldsymbol{w}),$$

where $\boldsymbol{Y}_n = \{\boldsymbol{y}_1, ..., \boldsymbol{y}_n\}$, $\boldsymbol{y}_\alpha = (y_{1\alpha}, \ldots, y_{G\alpha})^T$ indicates the class label of the αth observation, and $X_n = \{\boldsymbol{x}_1, ..., \boldsymbol{x}_n\}$. Specifying the prior distributions $\pi(\boldsymbol{w})$ of the parameters of the model to be a $(G-1)(m+1)$-variate normal distribution

$$\pi(\boldsymbol{w}) = \left(\frac{2\pi}{n\lambda}\right)^{\frac{p}{2}} \exp\left\{-n\lambda \frac{\boldsymbol{w}^T \boldsymbol{w}}{2}\right\},$$

with mean zero and covariance matrix $I/(n\lambda)$, we can make a Bayesian inference on this model.

Ando and Konishi (2009) illustrated some characteristics of the model by means of a simulation study. They showed that (a) the smoothness of the decision boundary is mainly controlled by ν, and (b) the smoothing parameter λ has the effect of reducing the variances of the parameter estimates $\hat{\boldsymbol{w}}_n$ or, equivalently, it controls the stability of the decision boundary.

A set of simulated data $\{(x_{1\alpha}, x_{2\alpha}, g_\alpha), \alpha = 1, \ldots, 100\}$ were generated from equal mixtures of normal distributions with centers $(0.3, -0.7)$ and $(0.3, 0.3)$ in class 1 and $(0.7, 0.2)$ and $(0.7, 0.3)$ in class 2, with a common covariance matrix $\Sigma = 0.03 I_2$, where I_2 is a two-dimensional identity matrix. Figure 5.7 shows the true decision boundary obtained from the Bayes rule. As shown in Figure 5.7, the Bayes decision boundary $\{\boldsymbol{x}; P(g = 1|\boldsymbol{x}) = P(g = 2|\boldsymbol{x}) = 0.5\}$ represents a nonlinear structure. It is clear that the linear logistic discriminant model (5.28) cannot capture the true structure well.

Ando and Konishi (2009) investigated the effect of the smoothing parameter. The Bayesian model was estimated with various values of the smoothing parameter λ. In this experiment, the values of the smoothing parameter were specified as $\log_{10}(\lambda) = -1, -3, -5$, and -7, respectively. The number of basis functions and the value of hyperparameter were set to be $m = 20$ and $\nu = 10$. Figure 5.8 shows the estimated decision boundaries $\{\boldsymbol{x}; \pi_1(\boldsymbol{x}; \hat{\boldsymbol{w}}) = \pi_2(\boldsymbol{x}; \hat{\boldsymbol{w}}) = 0.5\}$ obtained from 50 Monte Carlo simulations. It can be seen from Fig. 5.8 that the stability of our model is closely related to

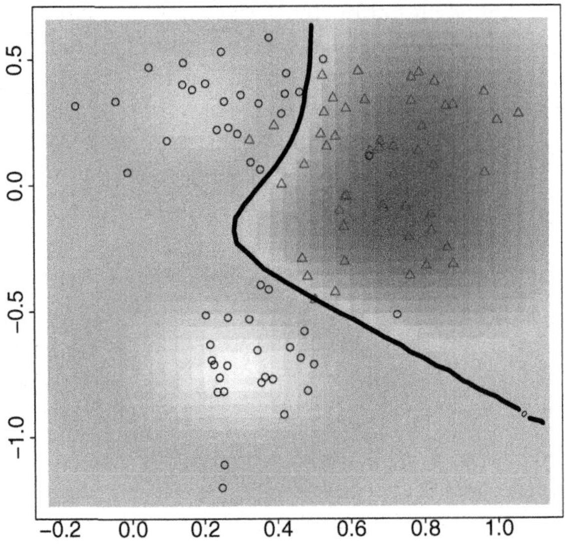

FIGURE 5.7: (From Ando, T. and Konishi, S., *Ann. I. Stat. Math.*, 61, 2009. With permission.) The Bayes boundary (—). Samples are marked by open circles ($g_\alpha = 1$) and open triangles ($g_\alpha = 2$). As the posterior probability $P(g = 2|\boldsymbol{x})$ becomes larger, the color becomes green.

the value of the smoothing parameter; as the value of the smoothing parameter becomes smaller, the variance of the estimated decision boundary becomes large. The variance of the decision boundary can be reduced by using a relatively large smoothing parameter. However, too large a smoothing parameter leads to a linear decision boundary, which cannot capture the nonlinear structure well.

Boxplots of the training errors and prediction errors obtained from 50 Monte Carlo simulations are also shown in Figure 5.9. As the smoothing parameter becomes smaller, the training error becomes small. Note that we cannot use the training error as a measure of the prediction ability of the estimated model, since we can make the training error small by using a more complicated model. In fact, the smallest value of the smoothing parameter $\log_{10}(\lambda) = -7$ gives the smallest median value of the training error, whereas it does not minimize the median value of the prediction error. On the other hand, an appropriate choice of $\log_{10}(\lambda) = -3$ gives the smallest median value of the prediction error.

Next, the effect of the hyperparameter ν in the radial basis function is illustrated. Using the penalized maximum likelihood method, we fitted the proposed model (5.31) with $\log_{10}(\nu) = 0, 1,$ and 2, respectively. In this simulation, we fixed the number of basis functions and the value of the smoothing

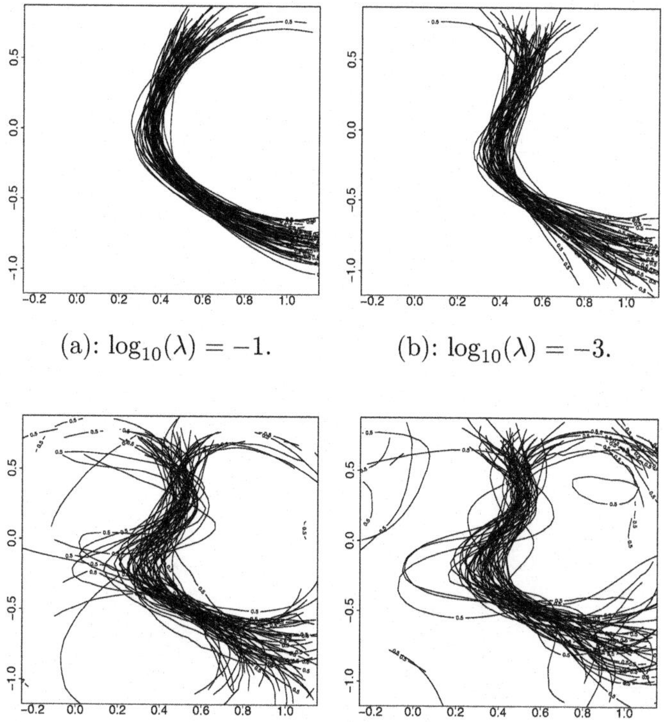

(a): $\log_{10}(\lambda) = -1$.

(b): $\log_{10}(\lambda) = -3$.

(c): $\log_{10}(\lambda) = -5$.

(d): $\log_{10}(\lambda) = -7$.

FIGURE 5.8: (From Ando, T. and Konishi, S., *Ann. I. Stat. Math.*, 61, 2009. With permission.) Comparison of the variances of the estimated decision boundaries obtained through 50 Monte Carlo simulations.

parameter at $m = 20$ and $\log_{10}(\lambda) = -3$. Figure 5.10 compares the Bayes decision boundary and the estimated decision boundaries. The estimated decision boundaries in Figures 5.10(a) and (c) are obviously undersmoothed and oversmoothed, respectively. We can see from Figure 5.10 (b) that an appropriate choice of ν gives a good approximation to the system underlying the data.

These simulation studies indicate that the crucial issue in the model building process is the choice of λ and ν. Additionally, the number of basis functions m should be optimized. A tailor-made version of the generalized Bayesian information criterion (Konishi et al. (2004)) in (5.16) that evaluates the estimated model is proposed by Ando and Konishi (2009). The score is

$$\begin{aligned} \text{GBIC} = & -2\sum_{\alpha=1}^{n}\sum_{k=1}^{G} y_{k\alpha} \log \pi_k(\boldsymbol{x}_\alpha; \hat{\boldsymbol{w}}_n) + n\lambda \hat{\boldsymbol{w}}_n^T \hat{\boldsymbol{w}}_n \\ & + \log |S_n(\hat{\boldsymbol{w}}_n)| - (G-1)(m+1)\log \lambda, \end{aligned} \tag{5.32}$$

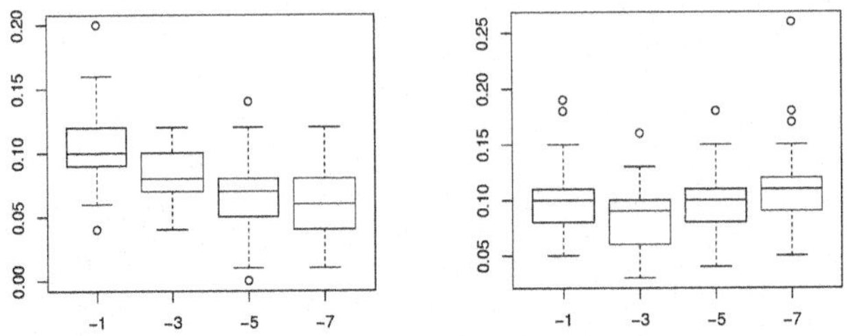

(a) (b)

FIGURE 5.9: (From Ando, T. and Konishi, S., *Ann. I. Stat. Math.*, 61, 2009. With permission.) Boxplots of (a) the training errors and (b) the prediction errors obtained from various values of the smoothing parameter $\log_{10}(\lambda)$.

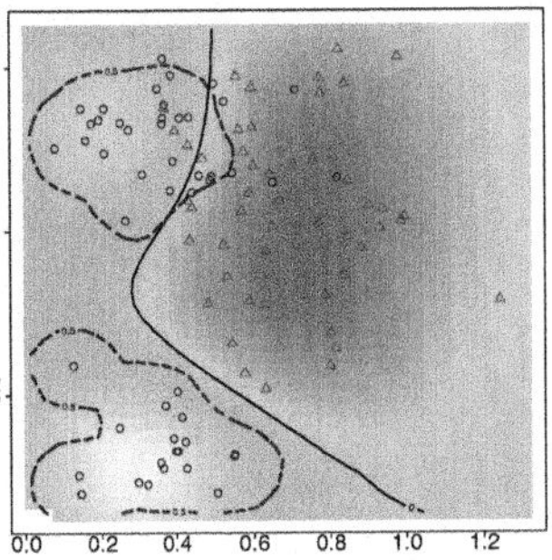

(a): $\log_{10}(\nu) = 0$.

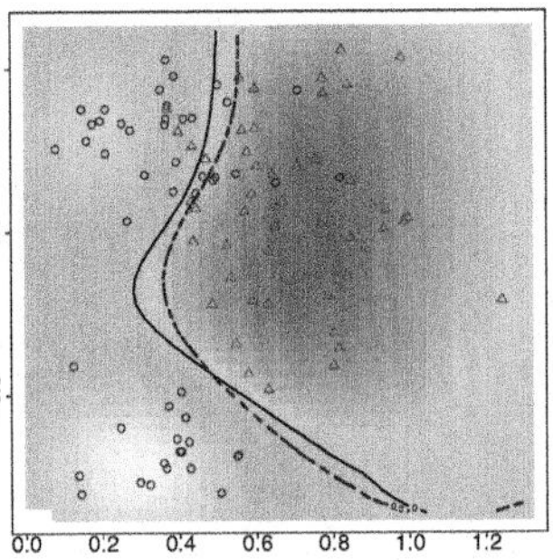

(b): $\log_{10}(\nu) = 1$.

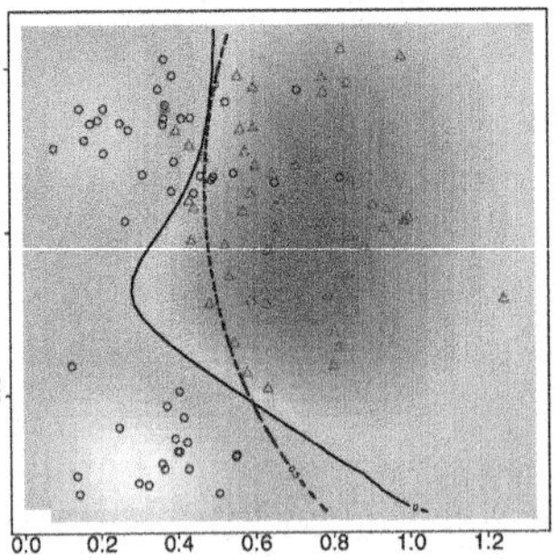

(c): $\log_{10}(\nu) = 2$.

FIGURE 5.10: (From Ando, T. and Konishi, S., *Ann. I. Stat. Math.*, 61, 2009. With permission.) The effect of hyperparameter ν. The dashed lines (- - -) and solid lines (—) represent the estimated decision boundaries and Bayes decision boundary, respectively.

where $\hat{\boldsymbol{w}}_n$ is the posterior mode. Calculating the second derivative of $\log\{f(\boldsymbol{Y}_n|\boldsymbol{X}_n,\boldsymbol{w})\pi(\boldsymbol{w})\}$ in the $(G-1)(m+1)$-dimensional matrix $S_n(\hat{\boldsymbol{w}}_n)$ is given as

$$S_n(\hat{\boldsymbol{w}}_n) = -\frac{1}{n}\begin{pmatrix} S_{11} & \cdots & S_{1,G-1} \\ \vdots & \ddots & \vdots \\ S_{G-1,1} & \cdots & S_{G-1,G-1} \end{pmatrix}$$

with

$$S_{ml} = \frac{\partial^2 \log\{f(\boldsymbol{Y}_n|X_n,\boldsymbol{w})\pi(\boldsymbol{w})\}}{\partial \boldsymbol{w}_m \partial \boldsymbol{w}_l^T} = \begin{cases} B^T\Gamma_m(\Gamma_m - I)B - n\lambda I, & (l=m) \\ B^T\Gamma_m\Gamma_l B - n\lambda I, & (l \neq m) \end{cases},$$

where

$$\Gamma_m = \mathrm{diag}\left\{\pi_m(\boldsymbol{x}_1,\boldsymbol{w}),\ldots,\pi_m(\boldsymbol{x}_n,\boldsymbol{w})\right\}.$$

Ando and Konishi (2009) applied the generalized BIC to the multiclass classification problem of optical recognition of handwritten digits (Alpaydin and Kaynak 1998). Figure 5.11 shows a set of examples. In the analysis, 32×32 bitmaps were divided into non-overlapping blocks of 4×4, and the number of pixels was counted in each block. As shown in Fig. 5.12, this handling generates an 8×8 feature matrix, where each element is an integer.

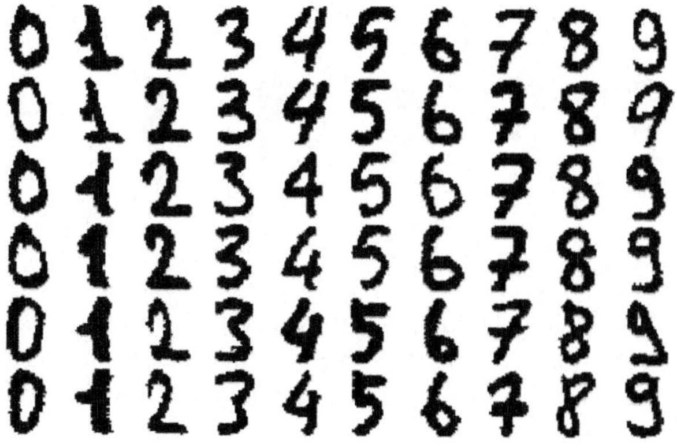

FIGURE 5.11: (From Ando, T. and Konishi, S., *Ann. I. Stat. Math.*, 61, 2009. With permission.) Examples of optical recognition of handwritten digits data.

The model was constructed using 3823 estimation data and then evaluated the prediction performance by using 1797 test data. Minimization of the generalized BIC (5.32) chose the adjusted parameters. The candidate values of m were in the range from 30 to 100. The candidates for the smoothing parameter were chosen on a geometrical grid with 100 knots between $\log_{10}(\lambda) = -5$

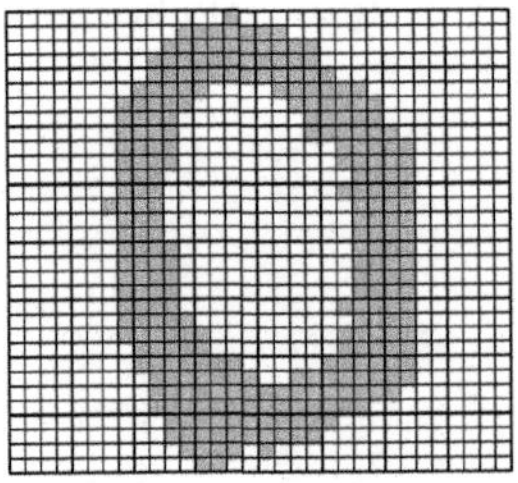

(*a*): Preprocessing data

0	0	6	13	10	0	0	0
0	2	14	5	10	12	0	0
0	4	11	0	1	12	7	0
0	5	8	0	0	9	8	0
0	4	12	0	0	8	8	0
0	3	15	2	0	11	8	0
0	0	13	15	10	15	5	0
0	0	5	13	9	1	0	0

(*b*): A transformed data.

FIGURE 5.12: (From Ando, T. and Konishi, S., *Ann. I. Stat. Math.*, 61, 2009. With permission.) An example of dimension reduction procedure.

and $\log_{10}(\lambda) = -7$. The candidates for the hyperparameter ν were chosen on a geometrical grid with 100 knots between $\log_{10}(\nu) = 0$ and $\log_{10}(\nu) = 0.8$. The use of BIC selected a model with $(m, \lambda, \nu) = (35, -5.10, 3.16)$, and the corresponding prediction error was 5.73%.

Figure 5.13 shows some examples that are classified incorrectly. Under the characters, the true label and the classified label are shown. Ando and Konishi (2009) pointed out that one reason for misclassification could be that it would be difficult even for humans to recognize these characters.

7 (9) 7 (8) 8 (5) 8 (9) 9 (5) 3 (9)

FIGURE 5.13: (From Ando, T. and Konishi, S., *Ann. I. Stat. Math.*, 61, 2009. With permission.) A set of examples, classified incorrectly. Under each character, the true label information is provided. The estimated labels are also shown in parentheses.

Remark: Finding the posterior mode

We can easily see that the posterior mode of $\boldsymbol{w}$ can be found by maximizing the penalized log-likelihood function

$$\begin{aligned} \ell(\boldsymbol{w}) &= \log\{f(\boldsymbol{Y}_n|X_n, \boldsymbol{w})\pi(\boldsymbol{w})\} \\ &\propto \sum_{\alpha=1}^{n} \log f(\boldsymbol{y}_\alpha|\boldsymbol{x}_\alpha; \boldsymbol{w}) - \frac{n\lambda}{2}\boldsymbol{w}^T\boldsymbol{w}, \end{aligned}$$

where λ is the smoothing parameter.

In practice, the posterior mode $\hat{\boldsymbol{w}}_n$ is given by the solution of $\partial \ell(\boldsymbol{w})/\partial \boldsymbol{w} = \mathbf{0}$, which is obtained by employing a Newton–Raphson algorithm. Using the first and second derivatives of $\ell(\boldsymbol{w})$, given by

$$\frac{\partial \ell(\boldsymbol{w})}{\partial \boldsymbol{w}_k} = \sum_{\alpha=1}^{n} \{y_{k\alpha} - \pi_k(\boldsymbol{x}_\alpha; \boldsymbol{w})\} \boldsymbol{b}(\boldsymbol{x}_\alpha) - n\lambda \boldsymbol{w}_k, \qquad k = 1, \ldots, G-1,$$

$$\frac{\partial \ell(\boldsymbol{w})}{\partial \boldsymbol{w}_m \partial \boldsymbol{w}_l^T} = \begin{cases} \displaystyle\sum_{\alpha=1}^{n} \pi_m(\boldsymbol{x}_\alpha; \boldsymbol{w})(\pi_m(\boldsymbol{x}_\alpha; \boldsymbol{w}) - 1)\boldsymbol{b}(\boldsymbol{x}_\alpha)\boldsymbol{b}(\boldsymbol{x}_\alpha)^T - n\lambda I_{m+1}, \\ \displaystyle\sum_{\alpha=1}^{n} \pi_m(\boldsymbol{x}_\alpha; \boldsymbol{w})\pi_l(\boldsymbol{x}_\alpha; \boldsymbol{w})\boldsymbol{b}(\boldsymbol{x}_\alpha)\boldsymbol{b}(\boldsymbol{x}_\alpha)^T, \quad (l \neq m), \end{cases}$$

respectively, we optimize the parameter vector $\boldsymbol{w}$ by use of the following iterative system:

$$\boldsymbol{w}^{\text{new}} = \boldsymbol{w}^{\text{old}} - \left\{ \frac{\partial^2 \ell(\boldsymbol{w}^{\text{old}})}{\partial \boldsymbol{w} \partial \boldsymbol{w}^T} \right\}^{-1} \frac{\partial \ell(\boldsymbol{w}^{\text{old}})}{\partial \boldsymbol{w}},$$

where I_{m+1} is an $(m+1) \times (m+1)$ identity matrix. The parameter vector $\boldsymbol{w}$ is updated until a suitable convergence criterion is satisfied.

GBIC has been applied to various Bayesian model selection problems. Fujii and Konishi (2006) employed GBIC for nonlinear regression modeling via regularized wavelets and smoothing parameter selection. For functional multivariate regression modeling, GBIC was used by Matsui et al. (2008).

5.7 Bayes factor with improper prior

We have seen that there are various methods for evaluating the marginal likelihood, or equivalently, the Bayes factors. The most serious difficulty in the use of the Bayes factor is its sensitivity to prior distributions. Unfortunately, it is generally known that the use of improper priors for the parameters in alternative models results in Bayes factors that are not well defined. Many attempts have been made to define a convenient Bayes factor in the case of non-informative priors, for example, Aitkin (1991), Berger and Pericchi (1996, 1998), Gelfand and Dey (1994), Kass and Wasserman (1995), O'Hagan (1995, 1997), Pauler (1998), Perez and Berger (2002), and Santis and Spezzaferri (2001).

Recall that improper priors are only defined up to a constant;

$$\pi(\boldsymbol{\theta}) \propto h(\boldsymbol{\theta}),$$

where $\int h(\boldsymbol{\theta})d\boldsymbol{\theta} = \infty$. Thus, with an arbitrary positive constant C, we can also use $q(\boldsymbol{\theta}) = C\pi(\boldsymbol{\theta})$ as a prior. The posterior density is

$$\pi(\boldsymbol{\theta}|\boldsymbol{X}_n) = \frac{f(\boldsymbol{X}_n|\boldsymbol{\theta})q(\boldsymbol{\theta})}{\int f(\boldsymbol{X}_n|\boldsymbol{\theta})q(\boldsymbol{\theta})d\boldsymbol{\theta}} = \frac{f(\boldsymbol{X}_n|\boldsymbol{\theta})\pi(\boldsymbol{\theta})}{\int f(\boldsymbol{X}_n|\boldsymbol{\theta})\pi(\boldsymbol{\theta})d\boldsymbol{\theta}}.$$

Even if we use an improper prior $q(\boldsymbol{\theta}) = C\pi(\boldsymbol{\theta})$, the posterior is considered well defined so long as the integral $\int f(\boldsymbol{X}_n|\boldsymbol{\theta})\pi(\boldsymbol{\theta})d\boldsymbol{\theta}$ converges.

However, the Bayes factor using these priors $q_k(\boldsymbol{\theta}_k)$ is

$$\text{Bayes factor}(M_k, M_j) = \frac{\int f_k(\boldsymbol{X}_n|\boldsymbol{\theta}_k)\pi_k(\boldsymbol{\theta}_k)d\boldsymbol{\theta}_k}{\int f_j(\boldsymbol{X}_n|\boldsymbol{\theta}_j)\pi_j(\boldsymbol{\theta}_j)d\boldsymbol{\theta}_j} \times \frac{C_k}{C_j}.$$

We can see that the Bayes factors are not well defined since there are arbitrary constants C_k/C_j in the equations. If a proper prior is used for each model such that $C_k < \infty$ and $C_j < \infty$, the Bayes factor is well defined as the ratio C_k/C_j is defined.

There are several approaches to dealing with the prior problem. One approach is the use of Bayesian information criteria (5.14), under the assumption that $\log \pi(\boldsymbol{\theta}) = O_p(1)$. As shown in the derivation, under a large sample situation, this prior specification has no effect on the approximation error of the marginal likelihood.

As alternative approaches, a variety of pseudo-Bayes factors have been proposed. The next section covers several types of pseudo-Bayes factors.

5.7.1 Intrinsic Bayes factors

Let the data $\boldsymbol{X}_n$ be divided into N subsets, denoted by $\{\boldsymbol{x}_{n(\ell)}\}_{\ell=1}^N$, such that $\sum_{\ell=1}^N n(\ell) = n$, where $n(\ell)$ denotes the number of observations in the ℓth subset. Let $\boldsymbol{X}_{n(\ell)}$ denote the data of the jth subset and $\boldsymbol{X}_{-n(\ell)}$ be the remaining observations.

Even if the prior is improper, the resulting posterior often becomes proper. Thus, a natural alternative to the marginal likelihood of the model M_k is the following quantity

$$P_s(\boldsymbol{X}_n|M) = \prod_{\ell=1}^N f(\boldsymbol{X}_{n(\ell)}|\boldsymbol{X}_{-n(\ell)}), \tag{5.33}$$

where the posterior predictive density $f(\boldsymbol{X}_{n(\ell)}|\boldsymbol{X}_{-n(\ell)})$ is defined as

$$f(\boldsymbol{X}_{n(\ell)}|\boldsymbol{X}_{-n(\ell)}) = \int f(\boldsymbol{X}_{n(\ell)}|\boldsymbol{\theta})\pi(\boldsymbol{\theta}|\boldsymbol{X}_{-n(\ell)})d\boldsymbol{\theta}.$$

Based on the quantity (5.33), Berger and Pericchi (1996) proposed the intrinsic Bayes factor. The quantity (5.33), the pseudo-marginal likelihood, uses part of the data $\boldsymbol{X}_{-n(\ell)}$ to update an improper prior, and using the remainder of the data $\boldsymbol{X}_{n(\ell)}$ to compute the marginal likelihood. The pseudo-marginal likelihood may be interpreted as a predictive measure for a future replication of the given data.

The Bayes factor, conditioned on a particular partition $n(\ell)$, can be expressed as

$$\begin{aligned}
&\text{Intrinsic Bayes factor}\,(M_k, M_j, n(\ell)) \\
&= \frac{\int f_k(\boldsymbol{X}_{-n(\ell)}|\boldsymbol{\theta}_k)\pi_k(\boldsymbol{\theta}_k|\boldsymbol{X}_{n(\ell)})d\boldsymbol{\theta}_k}{\int f_j(\boldsymbol{X}_{-n(\ell)}|\boldsymbol{\theta}_j)\pi_j(\boldsymbol{\theta}_j|\boldsymbol{X}_{n(\ell)})d\boldsymbol{\theta}_j} \\
&= \frac{\int f_k(\boldsymbol{X}_n|\boldsymbol{\theta}_k)\pi_k(\boldsymbol{\theta}_k)d\boldsymbol{\theta}_k}{\int f_j(\boldsymbol{X}_n|\boldsymbol{\theta}_j)\pi_j(\boldsymbol{\theta}_j)d\boldsymbol{\theta}_j} \times \frac{\int f_j(\boldsymbol{X}_{-n(\ell)}|\boldsymbol{\theta}_j)\pi_j(\boldsymbol{\theta}_j)d\boldsymbol{\theta}_j}{\int f_k(\boldsymbol{X}_{-n(\ell)}|\boldsymbol{\theta}_k)\pi_k(\boldsymbol{\theta}_k)d\boldsymbol{\theta}_k} \\
&= \text{Bayes factor}\,(M_k, M_j) \times \text{Bayes factor}\,(M_j, M_k, n(\ell)).
\end{aligned}$$

Note that the arbitrary constant terms C_k and C_j are removed. This is because the arbitrary constant C_k/C_j of the Bayes factor (M_k, M_j) is cancelled by C_j/C_k in the Bayes factor $(M_j, M_k, n(\ell))$ conditioned on a particular partition $n(\ell)$.

However, this Bayes factor $(M_j, M_k, n(\ell))$ conditioned on a particular partition $n(\ell)$ depends on the division of samples. Berger and Pericchi (1996) considered dividing the observations $\boldsymbol{X}_n$ into N minimal subsets and then take average of the corresponding N Bayes factor $(M_j, M_k, n(\ell))$ conditioned on $n(\ell)$. Taking an arithmetic average, Berger and Pericchi (1996) considered the arithmetic intrinsic Bayes factor (AIBF)

$$\text{AIBF}\,(M_k, M_j) = \frac{1}{N}\sum_{\ell=1}^{N} \text{Intrinsic Bayes factor}\,(M_k, M_j, n(\ell)). \qquad (5.34)$$

Also, Berger and Pericchi (1996) considered the geometric average and proposed the geometric intrinsic Bayes factor (GIBF):

$$\text{GIBF}\,(M_k, M_j) = \left[\prod_{\ell=1}^{N} \text{Intrinsic Bayes factor}\,(M_k, M_j, n(\ell))\right]^{1/N}. \qquad (5.35)$$

Berger and Pericchi (1996) recommended using reference noninformative priors to compute the intrinsic Bayes factor. Also, they summarized advantages and disadvantages of the intrinsic Bayes factor approach. Some of them are given as follows. First, the intrinsic Bayes factor can be applied to nonnested, as well as nested, model comparisons. Therefore, this approach can be employed for various types of model selection problems. However, the intrinsic Bayes factor approach might be computationally intensive, especially when an analytical evaluation of the posterior predictive density $f(\boldsymbol{X}_{n(\ell)}|\boldsymbol{X}_{-n(\ell)})$ is not available. Also, it can be unstable if the sample size is small. We refer to Berger and Pericchi (1996) for more details.

Intrinsic priors

Berger and Pericchi (1996) considered a situation that the intrinsic Bayes factor (AIBF, GIBF) (approximately) corresponds with an actual Bayes factor based on some priors. They are called "intrinsic priors." Advantage of the intrinsic prior approach is that it eliminate the need for consideration of all possible combinations of $n(\ell)$, $\ell = 1, ..., N$ and also concerns about stability of the intrinsic Bayes factor. Berger and Pericchi (1996) considered how to obtain intrinsic priors using asymptotic methods.

We illustrate the intrinsic priors with the arithmetic intrinsic Bayes factor, since similar augment applies to the geometric intrinsic Bayes factor. Modifying the arithmetic intrinsic Bayes factor, we have

$$\begin{aligned}\text{AIBF}\,(M_k, M_j) &= \frac{\displaystyle\int f_k(\boldsymbol{X}_n|\boldsymbol{\theta}_k)\pi_k(\boldsymbol{\theta}_k)d\boldsymbol{\theta}_k}{\displaystyle\int f_j(\boldsymbol{X}_n|\boldsymbol{\theta}_j)\pi_j(\boldsymbol{\theta}_j)d\boldsymbol{\theta}_j} \\ &\quad\times\left[\frac{1}{N}\sum_{\ell=1}^{N}\frac{\displaystyle\int f_j(\boldsymbol{X}_{-n(\ell)}|\boldsymbol{\theta}_j)\pi_j(\boldsymbol{\theta}_j)d\boldsymbol{\theta}_j}{\displaystyle\int f_k(\boldsymbol{X}_{-n(\ell)}|\boldsymbol{\theta}_k)\pi_k(\boldsymbol{\theta}_k)d\boldsymbol{\theta}_k}\right].\end{aligned}$$

From the Laplace approximation method, we have the following approximation.

$$\int f_k(\boldsymbol{X}_k|\boldsymbol{\theta}_k)/\pi(\boldsymbol{\theta}_k)d\boldsymbol{\theta}_k \approx f_k(\boldsymbol{X}_n|\hat{\boldsymbol{\theta}}_k)\pi_k(\hat{\boldsymbol{\theta}}_k) \times \frac{(2\pi)^{p_k/2}}{n^{p_k/2}|J_n(\hat{\boldsymbol{\theta}}_k)|^{1/2}},$$

where $\hat{\boldsymbol{\theta}}_k$ is the maximum likelihood estimate, p_k is the dimension of $\boldsymbol{\theta}_k$, and

$$J_n(\hat{\boldsymbol{\theta}}_k) = -\frac{1}{n}\left.\frac{\partial^2 \log f_k(\boldsymbol{X}_n|\boldsymbol{\theta}_k)}{\partial\boldsymbol{\theta}_k\partial\boldsymbol{\theta}_k^T}\right|_{\boldsymbol{\theta}_k=\hat{\boldsymbol{\theta}}_k}.$$

Applying this approximation to both the numerator and denominator of

AIBF (M_k, M_j) can be shown to be approximately equal to

$$\text{AIBF}\,(M_k, M_j) \approx \frac{f_k(\boldsymbol{X}_n|\hat{\boldsymbol{\theta}}_k)\pi_k(\hat{\boldsymbol{\theta}}_k) \times \dfrac{(2\pi)^{p_k/2}}{n^{p_k/2}|J_n(\hat{\boldsymbol{\theta}}_k)|^{1/2}}}{f_j(\boldsymbol{X}_n|\hat{\boldsymbol{\theta}}_j)\pi_j(\hat{\boldsymbol{\theta}}_j) \times \dfrac{(2\pi)^{p_j/2}}{n^{p_j/2}|J_n(\hat{\boldsymbol{\theta}}_j)|^{1/2}}}$$
$$\times \left[\frac{1}{N}\sum_{\ell=1}^{N} \frac{\displaystyle\int f_j(\boldsymbol{X}_{-n(\ell)}|\boldsymbol{\theta}_j)\pi_j(\boldsymbol{\theta}_j)d\boldsymbol{\theta}_j}{\displaystyle\int f_k(\boldsymbol{X}_{-n(\ell)}|\boldsymbol{\theta}_k)\pi_k(\boldsymbol{\theta}_k)d\boldsymbol{\theta}_k}\right]. \quad (5.36)$$

Suppose that, for some priors $q_k(\boldsymbol{\theta}_k)$ and $q_j(\boldsymbol{\theta}_j)$, AIBF is approximately equal to the Bayes factor based on $\pi_k(\boldsymbol{\theta}_k)$ and $\pi_j(\boldsymbol{\theta}_j)$. Applying this approximation to the Bayes factor (M_k, M_j), we also have

$$\frac{\displaystyle\int f_k(\boldsymbol{X}_n|\boldsymbol{\theta}_k)q_k(\boldsymbol{\theta}_k)d\boldsymbol{\theta}_k}{\displaystyle\int f_j(\boldsymbol{X}_n|\boldsymbol{\theta}_j)q_j(\boldsymbol{\theta}_j)d\boldsymbol{\theta}_j} \approx \frac{f_k(\boldsymbol{X}_n|\hat{\boldsymbol{\theta}}_k)q_k(\hat{\boldsymbol{\theta}}_k) \times \dfrac{(2\pi)^{p_k/2}}{n^{p_k/2}|J_n(\hat{\boldsymbol{\theta}}_k)|^{1/2}}}{f_j(\boldsymbol{X}_n|\hat{\boldsymbol{\theta}}_j)q_j(\hat{\boldsymbol{\theta}}_j) \times \dfrac{(2\pi)^{p_j/2}}{n^{p_j/2}|J_n(\hat{\boldsymbol{\theta}}_j)|^{1/2}}}. \quad (5.37)$$

Equating the Equations (5.37) with (5.36) yields

$$\frac{\pi_k(\hat{\boldsymbol{\theta}}_k)q_j(\hat{\boldsymbol{\theta}}_j)}{\pi_j(\hat{\boldsymbol{\theta}}_j)q_k(\hat{\boldsymbol{\theta}}_k)} \approx \left[\frac{1}{N}\sum_{\ell=1}^{N} \frac{\displaystyle\int f_j(\boldsymbol{X}_{-n(\ell)}|\boldsymbol{\theta}_j)\pi_j(\boldsymbol{\theta}_j)d\boldsymbol{\theta}_j}{\displaystyle\int f_k(\boldsymbol{X}_{-n(\ell)}|\boldsymbol{\theta}_k)\pi_k(\boldsymbol{\theta}_k)d\boldsymbol{\theta}_k}\right] \equiv \bar{B}_{kj}.$$

Assume that, under M_k,

$$\hat{\boldsymbol{\theta}}_k \to \boldsymbol{\theta}_{k0}, \quad \hat{\boldsymbol{\theta}}_j \to a(\boldsymbol{\theta}_{k0}), \quad \text{and} \quad \bar{B}_{kj} \to B^*_{kj}(\boldsymbol{\theta}_{k0}),$$

and under M_j,

$$\hat{\boldsymbol{\theta}}_j \to \boldsymbol{\theta}_{j0}, \quad \hat{\boldsymbol{\theta}}_k \to a(\boldsymbol{\theta}_{j0}), \quad \text{and} \quad \bar{B}_{kj} \to B^*_{kj}(\boldsymbol{\theta}_{j0}),$$

Berger and Pericchi (1996a) obtained

$$\frac{\pi_k(\boldsymbol{\theta}_{k0})q_j(\boldsymbol{\theta}_{k0})}{\pi_j(\boldsymbol{\theta}_{k0})q_k(\boldsymbol{\theta}_{k0})} = B^*_{kj}(\boldsymbol{\theta}_{k0}) \quad \text{and} \quad \frac{\pi_k(\boldsymbol{\theta}_{j0})q_j(\boldsymbol{\theta}_{j0})}{\pi_j(\boldsymbol{\theta}_{j0})q_k(\boldsymbol{\theta}_{j0})} = B^*_{kj}(\boldsymbol{\theta}_{j0}).$$

In the nested model scenario, solutions are given by

$$\pi_k(\boldsymbol{\theta}_{k0}) = q_k(\boldsymbol{\theta}_{k0}) \quad \text{and} \quad \pi_j(\boldsymbol{\theta}_{j0}) = q_j(\boldsymbol{\theta}_{j0})B^*_{kj}(\boldsymbol{\theta}_{j0}).$$

Berger and Pericchi (1996) pointed out this may not be the unique solution. An excellent review of the intrinsic Bayes factor and intrinsic priors is provided in Berger and Pericchi (1998b).

5.7.2 Partial Bayes factor and fractional Bayes factor

When we compare the partial predictive likelihood

$$f(\boldsymbol{X}_{n(1)}|\boldsymbol{X}_{-n(1)}) = \int f(\boldsymbol{X}_{n(1)}|\boldsymbol{\theta})\pi(\boldsymbol{\theta}|\boldsymbol{X}_{-n(1)})d\boldsymbol{\theta},$$

it reduces to the partial Bayes factor (O'Hagan, 1995). Therefore, a subset $\boldsymbol{X}_{-n(1)}$, of the data is used as a estimation sample for updating the priors into proper posterior distributions and the models are compared using the remainder of data $\boldsymbol{X}_{n(1)}$. The computational amount will be smaller than that of $P_s(\boldsymbol{X}_n|M)$ in (5.33). However, it obviously leads to inferior accuracy from the predictive perspective. Also, there is arbitrariness in the choice of a specific estimation sample $\boldsymbol{X}_{n(j)}$.

To solve arbitrariness in the choice of a specific estimation sample $\boldsymbol{X}_{n(1)}$ in the partial Bayes factors, O'Hagan (1995) proposed using the whole likelihood raised to the power $b = n(1)/n$ for the estimation. Motivating by the following approximation:

$$f(\boldsymbol{X}_{n(1)}|\boldsymbol{\theta}) \approx \{f(\boldsymbol{X}_n|\boldsymbol{\theta})\}^b.$$

O'Hagan (1995) proposed the fractional Bayes factor, which is defined as the ratio of the following pseudo-marginal likelihood:

$$P_s(\boldsymbol{X}_n|b) = \int f(\boldsymbol{X}_n|\boldsymbol{\theta})^{1-b}\pi(\boldsymbol{\theta}|\boldsymbol{X}_n, b)d\boldsymbol{\theta},$$

with

$$\pi(\boldsymbol{\theta}|\boldsymbol{X}_n, b) = \frac{f(\boldsymbol{X}_n|\boldsymbol{\theta})^b\pi(\boldsymbol{\theta})}{\int f(\boldsymbol{X}_n|\boldsymbol{\theta})^b\pi(\boldsymbol{\theta})d\boldsymbol{\theta}},$$

called a fractional posterior.

Thus the fractional Bayes factor, given value of b, can be expressed as

$$\begin{aligned}
&\text{Fractional Bayes factor}\,(M_k, M_j, b)\\
&= \frac{\int f_k(\boldsymbol{X}_n|\boldsymbol{\theta}_k)^{1-b}\pi_k(\boldsymbol{\theta}_k|\boldsymbol{X}_n, b)d\boldsymbol{\theta}_k}{\int f_j(\boldsymbol{X}_n|\boldsymbol{\theta}_j)^{1-b}\pi_j(\boldsymbol{\theta}_j|\boldsymbol{X}_n, b)d\boldsymbol{\theta}_j}\\
&= \frac{\int f_k(\boldsymbol{X}_n|\boldsymbol{\theta}_k)\pi_k(\boldsymbol{\theta}_k)d\boldsymbol{\theta}_k}{\int f_k(\boldsymbol{X}_n|\boldsymbol{\theta}_j)\pi_j(\boldsymbol{\theta}_j)d\boldsymbol{\theta}_j}\frac{\int f_j(\boldsymbol{X}_n|\boldsymbol{\theta}_j)^b\pi_j(\boldsymbol{\theta}_j)d\boldsymbol{\theta}_j}{\int f_k(\boldsymbol{X}_n|\boldsymbol{\theta}_k)^b\pi_k(\boldsymbol{\theta}_k)d\boldsymbol{\theta}_k}\\
&= \text{Bayes factor}\,(M_k, M_j) \times \frac{\int f_j(\boldsymbol{X}_n|\boldsymbol{\theta}_j)^b\pi_j(\boldsymbol{\theta}_j)d\boldsymbol{\theta}_j}{\int f_k(\boldsymbol{X}_n|\boldsymbol{\theta}_k)^b\pi_k(\boldsymbol{\theta}_k)d\boldsymbol{\theta}_k}. \qquad (5.38)
\end{aligned}$$

Note that the arbitrary constant terms C_k and C_j are removed. This is because the arbitrary constant C_k/C_j are cancelled.

Example: Linear regression model with improper prior

O'Hagan (1995) applied the fractional Bayes factor to the linear regression model with improper prior. Consider the linear regression model

$$\boldsymbol{y}_n = X_n\boldsymbol{\beta} + \boldsymbol{\varepsilon}_n, \quad \boldsymbol{\varepsilon}_n \sim N(0, \sigma^2 I),$$

where $\boldsymbol{\theta} = (\boldsymbol{\beta}, \sigma^2)$ is the $(p+1)$ dimensional model parameters. The use of improper prior $\pi(\boldsymbol{\beta}, \sigma^2) = 1/\sigma^{2t}$ leads to a situation where the Bayes factor is not well-defined. Now,

$$\begin{aligned} &\int f(\boldsymbol{y}_n|X_n, \boldsymbol{\theta})^b \pi(\boldsymbol{\theta}) d\boldsymbol{\theta} \\ &= \pi^{-\frac{nb}{2}} |X_n^T X_n|^{-\frac{1}{2}} \times 2^{-\frac{r}{2}} b^{-\frac{nb+p+1-r}{2}} S_n^{-(nb-r)} \Gamma\left(\frac{(nb-r)}{2}\right), \end{aligned}$$

where $r = p - 2t + 2$ and S_n^2 is the residual sum of squares

$$S_n^2 = \boldsymbol{y}_n \left(I - X_n (X_n^T X_n)^{-1} X_n^T\right) \boldsymbol{y}_n.$$

Thus, we have

$$\frac{\displaystyle\int f(\boldsymbol{y}_n|X_n, \boldsymbol{\theta}) \pi(\boldsymbol{\theta}) d\boldsymbol{\theta}}{\displaystyle\int f(\boldsymbol{y}_n|X_n, \boldsymbol{\theta})^b \pi(\boldsymbol{\theta}) d\boldsymbol{\theta}} = \pi^{-\frac{n(1-b)}{2}} b^{\frac{nb+p+1-r}{2}} S_n^{-n(1-b)} \frac{\Gamma\left(\frac{(b-r)}{2}\right)}{\Gamma\left(\frac{(nb-r)}{2}\right)},$$

and hence

$$\text{Bayes factor}\,(M_2, M_1, b) = \frac{\Gamma\left(\frac{(b-r_2)}{2}\right)\Gamma\left(\frac{(nb-r_1)}{2}\right)}{\Gamma\left(\frac{(b-r_1)}{2}\right)\Gamma\left(\frac{(nb-r_2)}{2}\right)} b^{t_2-t_1} \left(\frac{S_2^2}{S_1^2}\right)^{n(1-b)/2},$$

where $r_j = r = p_j - 2t + 2$ with p_j is the number of predictors in the regression model M_j. Smith and Spiegelhalter (1980) also proposed other methods as a solution to the regression selection problem with noninformative priors. They discussed the global Bayes factor and the local Bayes factor and showed that the local Bayes factor has a close relationship with the Akaike information criterion.

5.7.3 Posterior Bayes factors

To overcome the difficulties of the Bayes factor, Aitkin (1991) proposed a posterior Bayes factor, defined as the ratio of the following pseudo marginal likelihood:

$$\text{PBF} = \int f(\boldsymbol{X}_n|\boldsymbol{\theta}) \pi(\boldsymbol{\theta}|\boldsymbol{X}_n) d\boldsymbol{\theta}. \tag{5.39}$$

When one wants to check the posterior predictive replications to the observed data $\boldsymbol{X}_n$, this quantity would be useful. Note that, however, all observations $\boldsymbol{X}_n$ are used to construct the posterior and to compute the posterior mean of the likelihood. It is easily presumed that the posterior Bayes factors favor over-fitted models. Therefore, the posterior predictive densities may not generally be used for model comparison.

5.7.4 Pseudo Bayes factors based on cross validation

In the prediction problem, it is natural to assess the predictive ability of the model by using the cross validation. Gelfand et al. (1992) proposed the use of cross validation predictive densities

$$\mathrm{CVPD} = \prod_{\alpha=1}^{n} \int f(\boldsymbol{x}_\alpha|\boldsymbol{\theta})\pi(\boldsymbol{\theta}|\boldsymbol{X}_{-\alpha})d\boldsymbol{\theta}, \tag{5.40}$$

where $\boldsymbol{X}_{-\alpha}$ is all elements of $\boldsymbol{X}_n$ except for $\boldsymbol{x}_\alpha$. An advantage of the cross validation method (Stone (1974)) is that it can be applied in an automatic way to various practical situations. The computational time is, however, enormous for a large sample size.

5.7.4.1 Example: Bayesian linear regression model with improper prior

Consider the linear regression model

$$\boldsymbol{y}_n = X_n\boldsymbol{\beta} + \boldsymbol{\varepsilon}_n, \quad \boldsymbol{\varepsilon}_n \sim N(0, \sigma^2 I),$$

with improper prior $\pi(\boldsymbol{\beta}, \sigma^2) = 1/\sigma^2$. It is known that the predictive distribution for the future observation z, given point $\boldsymbol{x}_0$, is the Student-t distribution

$$\begin{aligned} f(z|\boldsymbol{x}_0, \boldsymbol{y}_n, X_n) &= \int f\left(z|\boldsymbol{x}_0, \boldsymbol{\beta}, \sigma^2\right) \pi\left(\boldsymbol{\beta}, \sigma^2|\boldsymbol{y}_n, X_n\right) d\boldsymbol{\beta} d\sigma^2 \\ &= \frac{\Gamma\left(\frac{\nu+1}{2}\right)}{\Gamma\left(\frac{\nu}{2}\right)(\pi\nu)^{\frac{1}{2}}\sigma^{2*}} \left\{1 + \frac{1}{\nu\sigma^{2*}}\left(z_n - \boldsymbol{x}_0^T\hat{\boldsymbol{\beta}}_{\mathrm{MLE}}\right)^2\right\}^{-\frac{\nu+1}{2}}, \end{aligned}$$

where $\nu = n - p$,

$$\hat{\boldsymbol{\beta}}_{\mathrm{MLE}} = \left(X_n X_n^{-1}\right)^{-1} X_n^T \boldsymbol{y}_n, \quad \text{and} \quad \sigma^{2*} = s^2\left[1 + \boldsymbol{x}_0^T\left(X_n X_n^{-1}\right)^{-1}\boldsymbol{x}_0\right],$$

with $s^2 = (\boldsymbol{y}_n - X_n\hat{\boldsymbol{\beta}}_{\mathrm{MLE}})^T(\boldsymbol{y}_n - X_n\hat{\boldsymbol{\beta}}_{\mathrm{MLE}})$.

Thus, the cross validation predictive density (5.40) is given as

$$CVPD = \prod_{\alpha=1}^{n} f(y_\alpha|\boldsymbol{x}_\alpha, \boldsymbol{y}_{-\alpha}, X_{-\alpha}).$$

5.8 Expected predictive likelihood approach for Bayesian model selection

To overcome the main weakness of the Bayes factor, its sensitivity to the prior distribution, we reviewed a variety of pseudo-Bayes factors to evaluate the goodness of fit of Bayesian models. Dividing the dataset $\boldsymbol{X}_n = \{\boldsymbol{x}_1, ..., \boldsymbol{x}_n\}$ into N subsets, $\{\boldsymbol{X}_{n(k)}\}_{k=1}^N$, Berger and Pericchi (1996, 1998a) proposed the intrinsic Bayes factor as a proxy of the pseudo-marginal likelihood (5.33). Also, Gelfand et al. (1992) and Gelfand (1996) proposed the use of cross validation predictive densities (5.40). Under the usual model assumption, an analytical expression of the posterior predictive density is infrequently available, and the pseudo likelihood is often estimated by using Markov chain Monte Carlo (MCMC) methods with Metropolis-Hasting algorithm and Gibbs sampler.

The pseudo-marginal likelihood can be interpreted as a predictive measure for a future replication of the given data; see e.g., Mukhopadhyaya et al. (2003), Ando and Tsay (2009). However, as pointed out by Eklund and Karlsson (2005) and Ando and Tsay (2009), the use of pseudo-marginal likelihood encounters some practical limitations. First, the division of the data into subsets may affect the results, yet there exist no clear guidelines for the division. Second, the approach is hard to apply when the data are dependent, e.g., time series data. Finally, when the number of observations is large, the approach consumes a substantial amount of computational time. To overcome the difficulties of Bayes factor and to reduce the computation intensity associated with cross-validation, Ando and Tsay (2009) considered an alternative measure for assessing the predictive distributions.

Following the model selection literature (e.g., Konishi and Kitagawa (1996)), Ando and Tsay (2009) proposed to evaluate the predictive ability of a given model M by using the Kullback-Leibler information of Kullback and Leibler (1951)

$$\int \left[\log \frac{g(\boldsymbol{Z}_n)}{f(\boldsymbol{Z}_n|\boldsymbol{X}_n, M)}\right] g(\boldsymbol{Z}_n)d\boldsymbol{Z}_n$$
$$= \int \log g(\boldsymbol{Z}_n)g(\boldsymbol{Z}_n)d\boldsymbol{Z}_n - \int \log f(\boldsymbol{Z}_n|\boldsymbol{X}_n, M)g(\boldsymbol{Z}_n)d\boldsymbol{Z}_n$$

where $\boldsymbol{Z}_n = \{\boldsymbol{z}_1, ..., \boldsymbol{z}_n\}$ is a set of unseen future observations. The density $f(\boldsymbol{Z}_n|\boldsymbol{X}_n, M)$ is the Bayesian predictive distributions

$$f(\boldsymbol{Z}_n|\boldsymbol{X}_n, M) = \int f(\boldsymbol{Z}_n|\boldsymbol{\theta})\pi(\boldsymbol{\theta}|\boldsymbol{X}_n)d\boldsymbol{\theta}.$$

The first term is not relevant to the model $p(\boldsymbol{Z}_n|\boldsymbol{X}_n, M)$, but the second term, which is the expected log-predictive likelihood

$$\eta(M) \equiv \int \log f(\boldsymbol{Z}_n|\boldsymbol{X}_n, M)g(\boldsymbol{Z}_n)d\boldsymbol{Z}_n, \qquad (5.41)$$

is highly relevant. Indeed, an information criterion is obtained as an estimator of the Kullback-Leibler information or equivalently the expected log-predictive likelihood. See, for instance, Akaike (1974) and Konishi and Kitagawa (1996). Note that, similar to the common practice, we use MCMC methods in this paper so that our approach does not require the availability of an analytical expression of the posterior predictive density $f(\boldsymbol{Z}_n|\boldsymbol{X}_n, M)$.

The measure in (5.41) has many advantages. First, from an information theoretic point of view (Akaike, 1974), the measure is a well-known statistic for model evaluation involving log-predictive likelihood. Indeed, it is the negative Kullback-Leibler divergence of the predictive distribution against the true density $g(\boldsymbol{Z}_n)$. Second, Konishi and Kitagawa (1996) employed the measure to evaluate the predictive power of a Bayesian model, but these authors use the maximum likelihood estimates in the evaluation. Third, as pointed out by Zellner (2006), the quantity can be regarded as the expected height of the density $f(\boldsymbol{Z}_n|\boldsymbol{X}_n, M)$ relative to the measure $g(\boldsymbol{Z}_n)$. Finally, when we replace the true density $g(\boldsymbol{Z}_n)$ with the empirical distribution $\hat{g}(\boldsymbol{Z}_n)$ constructed by a set of n future observations $\boldsymbol{Z}_n$, this quantity reduces to the predictive likelihood. Therefore, information theoretic augment is a general approach for evaluating the goodness of fit of statistical models.

5.8.1 Predictive likelihood for model selection

Ando and Tsay (2009) introduced the expected log-predictive likelihood $\eta(M)$ to measure the predictive ability of model M. However, this measure depends on the specified model M and the unknown true model $g(\boldsymbol{Z}_n)$. The problem then is how to estimate $\eta(M)$.

A natural estimator of (5.41) is

$$\hat{\eta}(M) = \frac{1}{n} \log f(\boldsymbol{X}_n|\boldsymbol{X}_n, M), \tag{5.42}$$

which is obtained by replacing the unknown true model $g(\boldsymbol{Z}_n)$ with the empirical distribution $\hat{g}(\boldsymbol{X}_n)$ of the data. This quantity is known as the posterior Bayes factor proposed by Aitkin (1991). However, the quantity generally has a positive bias as an estimator of $\eta(M)$. Employing an information theoretic argument (Akaike, 1974), we define the bias $b(M)$ of $\hat{\eta}(M)$ as an estimator of $\eta(M)$ by

$$b(M) = \int [\hat{\eta}(M) - \eta(M)] g(\boldsymbol{X}_n) d\boldsymbol{X}_n, \tag{5.43}$$

where the expectation is taken over the joint distribution of $\boldsymbol{X}_n$. Once an estimator of the bias is obtained, we can employ a bias corrected version of $\hat{\eta}(M)$, say $\hat{\eta}(M) - \hat{b}(M)$, where $\hat{b}(M)$ denotes the bias.

Under some regularity conditions, Ando and Tsay (2009) evaluated the asymptotic bias (8.11) under model misspecification. Let $\eta(M)$ and $\hat{\eta}(M)$ be defined in (5.41) and (5.42), respectively. Suppose that the specified family of probability distributions does not necessarily contain the true model. Then, under some regularity conditions, the asymptotic bias $\hat{b}(M)$ is

$$\hat{b}(M) \approx \frac{1}{2n}\mathrm{tr}\left[J_n^{-1}\{\hat{\boldsymbol{\theta}}_n\}I_n\{\hat{\boldsymbol{\theta}}_n\}\right], \tag{5.44}$$

where $\hat{\boldsymbol{\theta}}_n$ is the mode of $\log f(\boldsymbol{X}_n|\boldsymbol{\theta}, M) + \log \pi(\boldsymbol{\theta})/2$, $\pi(\boldsymbol{\theta})$ denotes the prior distribution for the parameter vector $\boldsymbol{\theta}$ of model M, $p = \dim\{\boldsymbol{\theta}\}$, and the $p \times p$ matrices $I_n(\boldsymbol{\theta})$ and $J_n(\boldsymbol{\theta})$ are given by

$$\begin{aligned} I_n(\boldsymbol{\theta}) &= \frac{1}{n}\sum_{\alpha=1}^{n}\left\{\frac{\partial \log \zeta(x_\alpha|\boldsymbol{\theta})}{\partial \boldsymbol{\theta}}\frac{\partial \log \zeta(x_\alpha|\boldsymbol{\theta})}{\partial \boldsymbol{\theta}'}\right\}, \\ J_n(\boldsymbol{\theta}) &= -\frac{1}{n}\sum_{\alpha=1}^{n}\left\{\frac{\partial^2 \log \zeta(x_\alpha|\boldsymbol{\theta})}{\partial \boldsymbol{\theta}\partial \boldsymbol{\theta}'}\right\}, \end{aligned}$$

with $\log \zeta(x_\alpha|\boldsymbol{\theta}) = \log f(x_\alpha|\boldsymbol{\theta}, M) + \log \pi(\boldsymbol{\theta})/(2n)$.

Correcting the asymptotic bias of $\hat{\eta}(M)$, Ando and Tsay (2009) estimated the expected log-predictive likelihood of model M by

$$\mathrm{PL}(M) = \frac{1}{n}\left\{\log f(\boldsymbol{X}_n|\boldsymbol{X}_n, M) - \frac{1}{2}\mathrm{tr}\left[J_n^{-1}\{\hat{\boldsymbol{\theta}}_n\}I_n\{\hat{\boldsymbol{\theta}}_n\}\right]\right\}. \tag{5.45}$$

We can choose the model that maximizes this PL score. Since predictive likelihood depends on the prior and the sampling density, the PL score can be used to select the prior distribution if the sampling model is fixed. In general, the PL score selects the best combination between prior and sampling distributions.

If we impose some further assumptions, the bias term (5.44) reduces to a simple form. For instance, if we assume that (a) the prior is dominated by the likelihood as the sample size n increases, say, $\log \pi(\boldsymbol{\theta}) = O_p(1)$, and (b) the specified parametric models contain the true model, then the estimated bias $\hat{b}(M)$ in (5.44) reduces to $\hat{b}(M) \approx p/(2n)$, where p is the dimension of $\boldsymbol{\theta}$. In this situation, the estimate of log-predictive likelihood (5.45) becomes

$$\mathrm{PL}_2(M) = \frac{1}{n}\left\{\log f(\boldsymbol{X}_n|\boldsymbol{X}_n, M) - \frac{p}{2}\right\}. \tag{5.46}$$

Another notable fact is that under the assumption (a) in Remark 1, the proposed approach assumes the expression of TIC asymptotically. If we further assume that the specified parametric family of probability distributions encompasses the true model, then the proposed criterion reduces to the AIC. These are natural results since the Bayesian method reduces to the frequentist method under weak prior assumptions. Therefore, the proposed method is a natural extension of the traditional information criteria.

Although the theoretical setup of the paper is only for iid data, many applications employ observations that are serially correlated. Suppose we have a time-series sequence $\boldsymbol{x}_t$, $t = 1, ..., n$. Ando and Tsay (2009) pointed out that these criteria are still applicable with minor modification. For time-series data, $\boldsymbol{x}_t$ is dependent on the previous observed values $\boldsymbol{X}_{t-1} = \{\boldsymbol{x}_1, ..., \boldsymbol{x}_{t-1}\}$. Consequently, instead of using the joint likelihood $f(\boldsymbol{X}_n|\boldsymbol{\theta}, M) = \prod_{t=1}^n f(\boldsymbol{x}_t|\boldsymbol{\theta}, M)$, we use the following likelihood decomposition

$$f(\boldsymbol{X}_n|\boldsymbol{\theta}) = \prod_{t=1}^{n} \log f(\boldsymbol{x}_t|\boldsymbol{X}_{t-1}; \boldsymbol{\theta}, M),$$

where $\boldsymbol{X}_{t-1} = (\boldsymbol{x}_1, ..., \boldsymbol{x}_{t-1})^T$. In other words, the joint density is decomposed into the product of the conditional densities of $\boldsymbol{x}_t$. For more details, we refer to Ando and Tsay (2009).

5.8.2 Example: Normal model with conjugate prior

To appreciate the effects of bias correction, we consider a simple example. Specifically, we show that the quantity $\hat{\eta}(M)$ provides a positive bias as an estimator of $\eta(M)$ using a simple normal model with known variance.

Suppose that a set of n independent observations $\boldsymbol{x}_n = (x_1, ..., x_n)^T$ are generated from a normal distribution with true mean μ_t and known variance σ^2, i.e. $g(z|\mu_t) = N(\mu_t, \sigma^2)$. Suppose also that we hypothesize that the data are generated from a normal distribution $f(z|\mu) = N(\mu, \sigma^2)$. The use of a normal prior $\mu \sim N(\mu_0, \tau_0^2)$ leads to the posterior distribution of μ being normal with mean $\hat{\mu}_n = (\mu_0/\tau_0^2 + \sum_{\alpha=1}^n x_\alpha/\sigma^2)/(1/\tau_0^2 + n/\sigma^2)$ and variance $\sigma_n^2 = 1/(1/\tau_0^2 + n/\sigma^2)$. In this particular case, the true bias $b(M)$ can be obtained analytically (details omitted to save space), and its estimate $\hat{b}(M) = \mathrm{tr}[J_n^{-1}(\hat{\mu}_n)I_n(\hat{\mu}_n)]/(2n)$, where $\hat{\mu}_n$ is the mode, $J_n(\mu) = 1/\sigma^2 + 1/(2n\tau_0^2)$, and $I_n(\mu) = n^{-1}\sum_{\alpha=1}^n\{(x_\alpha - \mu)/\sigma^2 + (\mu_0 - \mu)/(2n\tau_0^2)\}^2$.

Figure 5.14 shows the true bias and the estimated bias for various prior variances τ_0^2. The quantity $\hat{b}(M)$ is evaluated by a Monte Carlo simulation with 10,000 repetitions. In the simulation, we arbitrarily set the true mean, true variance and the prior mean as $\mu_t = 0$, $\sigma = 0.5$ and $\mu_0 = 0$, respectively. The numbers of observations used are $n = 10$ and 100. The range of prior variance is from $\log(\tau_0) = -3$ to $\log(\tau_0) = 7$. Figure 5.14 shows that $\hat{\eta}(M)$ has a significant bias as an estimator of $\eta(M)$. Also, the true bias converges to half of the dimension of the parameter vector as the amount of the prior information becomes weak. Overall, this simple example shows that, as expected, the estimated bias converges to the true bias as the sample size increases.

5.8.3 Example: Bayesian spatial modeling

There is a substantial amount of study focusing on Bayesian spatial data analysis, including Besag, York and Mollie (1991), Besag and Higdon (1999), Gelfand, Banerjee and Gamerman (2005), Gelfand, Kim, Sirmans and Baner-

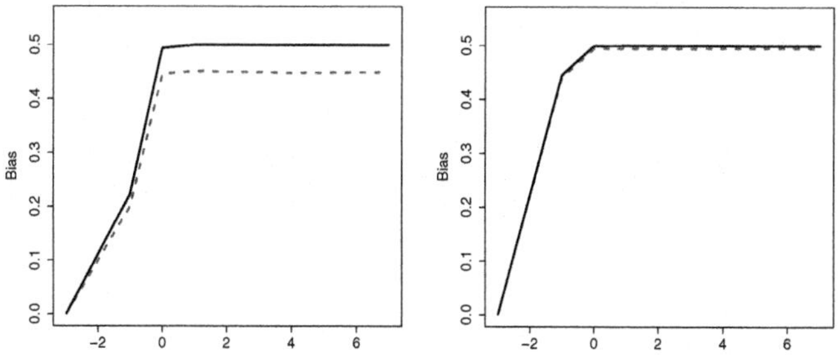

(a) $n = 10$ (b) $n = 100$

FIGURE 5.14: (From Ando, T. and Tsay, S., *Int. J. Forecasting*, in press. With permission.) Ando and Tsay (2009). A simple normal example: Comparison of the true bias b (—) and the estimated asymptotic bias (- - -) for various prior variances $\log(\tau_0)$, where n denotes the sample size.

jee (2003), and Knorr-Held and Rue (2002). See also Banerjee, Carlin and Gelfand (2004) for more references.

A basic spatial model is

$$y(\boldsymbol{s}) = \boldsymbol{\beta}^T \boldsymbol{x}(\boldsymbol{s}) + w(\boldsymbol{s}) + \varepsilon(\boldsymbol{s}),$$

where $\boldsymbol{s}$ denotes a location, $\boldsymbol{\beta} = (\beta_1, ..., \beta_p)^T$ is the p-dimensional unknown parameters, $\boldsymbol{x}(\boldsymbol{s})$ is a location specific predictors, $w(\boldsymbol{s})$ is a location specific noise assumed to be realizations from a zero-centered stationary Gaussian spatial process, $\varepsilon(\boldsymbol{s})$ are the uncorrelated error terms with variance τ^2.

For $w(\boldsymbol{s})$, we especially specify the dependence through the distance $||\boldsymbol{s}_k - \boldsymbol{s}_j||^2$, and use the exponential isotropic form:

$$\text{Cov}\left(w(\boldsymbol{s}_k), w(\boldsymbol{s}_j)\right) = \sigma^2 \exp\left(-\phi||\boldsymbol{s}_k - \boldsymbol{s}_j||^2\right).$$

Let us have a set of n data $\boldsymbol{y}_n = \{y(\boldsymbol{s}_1), ..., y(\boldsymbol{s}_n)\}$, the covariance matrix of the data is then given as

$$\Sigma(\sigma^2, \phi, \tau^2) = \sigma^2 H(\phi) + \tau^2 I,$$

with

$$\begin{aligned} H(\phi) &= \begin{pmatrix} \exp\left(-\phi||\boldsymbol{s}_1 - \boldsymbol{s}_1||^2\right) & \cdots & \exp\left(-\phi||\boldsymbol{s}_1 - \boldsymbol{s}_n||^2\right) \\ \vdots & \ddots & \vdots \\ \exp\left(-\phi||\boldsymbol{s}_n - \boldsymbol{s}_1||^2\right) & \cdots & \exp\left(-\phi||\boldsymbol{s}_n - \boldsymbol{s}_n||^2\right) \end{pmatrix} \\ &= \begin{pmatrix} 1 & \cdots & \exp\left(-\phi||\boldsymbol{s}_1 - \boldsymbol{s}_n||^2\right) \\ \vdots & \ddots & \vdots \\ \exp\left(-\phi||\boldsymbol{s}_n - \boldsymbol{s}_1||^2\right) & \cdots & 1 \end{pmatrix}. \end{aligned}$$

Assuming the normality, the likelihood function of $\boldsymbol{y}_n$ given X_n is

$$\begin{aligned} & f\left(\boldsymbol{Y}_n|X_n,\boldsymbol{\beta},\sigma^2,\tau^2,\phi\right) \\ &= \frac{1}{(2\pi)^{n/2}|\Sigma(\sigma^2,\phi,\tau^2)|^{1/2}} \\ &\quad\times \exp\left[-\frac{1}{2}(\boldsymbol{y}_n - X_n\boldsymbol{\beta})^T\Sigma(\sigma^2,\phi,\tau^2)^{-1}(\boldsymbol{y}_n - X_n\boldsymbol{\beta})\right], \end{aligned}$$

with $\boldsymbol{X}_n = (\boldsymbol{x}(\boldsymbol{s}_1),...,\boldsymbol{x}(\boldsymbol{s}_n))^T$.

To complete the Bayesian model, we shall specify the prior as follows. First, we shall assume the independent priors for different parameters:

$$\pi(\boldsymbol{\beta},\sigma^2,\phi,\tau^2) = \pi(\boldsymbol{\beta})\pi(\sigma^2)\pi(\phi)\pi(\tau^2),$$

with $\pi(\boldsymbol{\beta}) \propto$ Const., $\sigma^2 \sim IG(\nu_\sigma/2,\lambda_\sigma/2,)$, $\tau^2 \sim IG(\nu_\tau/2,\lambda_\tau/2,)$. For the prior density of ϕ, a uniform prior $\phi \sim U[a,b]$ is usually used. These prior specifications are also explained in Banerjee, Carlin and Gelfand (2004). An implementation of the MCMC sampler can be done by the **R** package **spBayes**.

To illustrate the expected predictive likelihood approach, we will use the Bartlett experimental forest inventory data. This dataset holds 1991 and 2002 forest inventory information, including species specific basal area and biomass (BIO), inventory plot coordinates ($\boldsymbol{s}$), elevation (ELEV), slope (SLOPE), and tasseled cap brightness (TC1), greenness (TC2), and wetness (TC3), respectively. The dataset and more information can be obtained through the **R** package **spBayes**. See also (Finley, Banerjee and Carlin (2007)).

First we consider the following mean structure of the log(BIO) $y(\boldsymbol{s})$:

$$\mu(\boldsymbol{s}) = \beta_0 + \beta_1\text{ELEV}(\boldsymbol{s}) + \beta_2\text{SLOPE}(\boldsymbol{s}).$$

Specifying the starting values as $\sigma^2 = 0.005$, $\tau^2 = 0.005$, $\phi = 0.01$, $a = 0.001$, $b = 0.05$, $\nu_\sigma = 10^{-5}$, $\lambda_\sigma = 10^{-5}$, $\nu_\tau = 10^{-5}$ and $\lambda_\tau = 10^{-5}$, we generated a set of 1,000 posterior samples. The first 1,000 samples are discarded as burn-in samples and we then stored every 100th iteration.

Clearly, the Bayes factor is not well defined because of the use of the improper prior for $\boldsymbol{\beta}$. In such a case, the predictive likelihood score (5.46) can be used. The score is calculated as

$$\begin{aligned} & n \times \text{PL}_2 \\ &= \log\int f\left(\boldsymbol{Y}_n|X_n,\boldsymbol{\beta},\sigma^2,\tau^2,\phi\right)\pi\left(\boldsymbol{\beta},\sigma^2,\tau^2,\phi|\boldsymbol{Y}_n,X_n\right)d\boldsymbol{\beta}d\sigma^2 d\tau^2 d\phi - \frac{p}{2}. \end{aligned}$$

Putting the number of observations $n = 415$, the posterior mean of the log-likelihood 253.96, the number of parameters $p = 6$, the corresponding predictive likelihood score (5.46) is $n \times \text{PL}_2 = 250.961$.

As an alternative model, we consider the following mean structure

$$\mu(\boldsymbol{s}) = \beta_0 + \beta_1\text{ELEV}(\boldsymbol{s}) + \beta_2\text{SLOPE}(\boldsymbol{s}) + \beta_3\text{TC1}(\boldsymbol{s}) + \beta_4\text{TC2}(\boldsymbol{s}) + \beta_5\text{TC3}(\boldsymbol{s}).$$

The same estimation procedure results in the predictive likelihood score (5.46) as $PL_2 = 273.724$. Thus, we can consider that the alternative model is preferred.

5.9 Other related topics

5.9.1 Bayes factors when model dimension grows

In Section 5.5, we discussed the consistency of the Bayesian information criterion, or equivalently, that of the Bayes factor. Unfortunately, Stone (1979) pointed out that BIC can be inconsistent when the dimension of the parameter goes to infinity. Berger et al. (2003) showed that BIC may be a poor approximation to the logarithm of Bayes Factor using a normal example of Stone (1979). The problem is highlighted below.

Consider a simple ANOVA model. The set of independent observations y_{ij} follows the following linear model:

$$y_{ij} = \delta + \mu_i + \varepsilon_{ij} \qquad \varepsilon_{ij} \sim N(0, \sigma^2),$$

with $i = 1, ..., p$, $j = 1, ..., r$ and $n = rp$. We consider M_1 and M_2 are two nested linear models for a set of n independent normal random variables with known variance σ^2. Under the model M_1, all n random variables have the same mean (i.e., $\mu_i = 0$ for all i), while, under the model M_2, each block of r random variables has a different mean $\boldsymbol{\mu} = (\mu_1, ..., \mu_p)^T$.

The difference of the values of Bayesian information criteria is

$$-\frac{1}{2}\Delta\text{BIC} = \frac{r}{2\sigma^2}\sum_{i=1}^{p}(\bar{y}_i - \bar{y})^2 - \frac{(p-1)}{2}\log n,$$

where $\bar{y}_i = r^{-1}\sum_{j=1}^{r} y_{ij}$ and $\bar{y} = (rp)^{-1}\sum_{i=1}^{p}\sum_{j=1}^{r} y_{ij}$. The Bayesian information criteria select M_1, if the above score is negative.

It is obvious that the score is negative, if

$$r\sum_{i=1}^{p}(\bar{y}_i - \bar{y})^2/((p-1)\sigma^2) < \log n.$$

Under the Stone (1979)'s assumption that, as $n \to \infty$, $(p/n)\log n \to \infty$ and

$$\sum_{i=1}^{p}\frac{(\mu_i - \mu)^2}{(p-1)} \to \tau^2 > 0,$$

which implies that $r\sum_{i=1}^{p}(\bar{y}_i - \bar{y})^2/[(p-1)(\sigma^2 + r\tau^2)] \to 1$ in probability as $n \to \infty$ $(r/\log n \to 0)$, BIC selects the model M_1 if $(\sigma^2 + r\tau^2) < \log n$, even if the model M_2 is true. Thus, BIC selects the wrong model asymptotically, demonstrating its inconsistency.

Berger et al. (2003) pointed out that the problem lies in the inappropriateness of BIC as an approximation to Bayes factors and developed some new approximations to Bayes factors that are valid for the above situation. As an extension, Chakrabarti and Ghosh (2006) considered a fairly general case where one has p groups of observations coming from an arbitrary general exponential family with each group having a different parameter and r observations.

5.9.2 Bayesian p-values

When one wants to assess the model fitness, the predictive distribution $f(\boldsymbol{z}|\boldsymbol{X}_n)$ can be compared to the observed data $\boldsymbol{X}_n$. We usually consider that the observed data $\boldsymbol{X}_n$ would be likely under the predictive distribution if the model fits the data well. If there is a large discrepancy between the observed data $\boldsymbol{X}_n$ and the predictive distribution $f(\boldsymbol{z}|\boldsymbol{X}_n)$, the model might not capture the data structure.

Let $\boldsymbol{X}_n^{\text{rep}}$ denote replicated values of $\boldsymbol{X}_n$ generated from the predictive distribution $f(\boldsymbol{z}|\boldsymbol{X}_n)$. Then model fit statistics can be assessed by comparing the test statistic based on observed data $T(\boldsymbol{X}_n, \boldsymbol{\theta})$ and the test statistic based on replicates $T(\boldsymbol{X}_n^{\text{rep}}, \boldsymbol{\theta})$. Formally, a Bayesian p value can be defined as

$$\text{Bayesian } p\text{−value} = \Pr\left(T(\boldsymbol{X}_n^{\text{rep}}, \boldsymbol{\theta}) \geq T(\boldsymbol{X}_n, \boldsymbol{\theta})|\boldsymbol{X}_n\right),$$

where the probability is taken over the posterior distribution of $\boldsymbol{\theta}$ and also the predictive distribution of $\boldsymbol{X}_n^{\text{rep}}$. Or equivalently,

$$\begin{aligned}&\text{Bayesian } p\text{−value}\\&= \int\int I\left(T(\boldsymbol{X}_n^{\text{rep}}, \boldsymbol{\theta}) \geq T(\boldsymbol{X}_n, \boldsymbol{\theta})\right) \pi(\boldsymbol{\theta}|\boldsymbol{X}_n) f(\boldsymbol{X}_n^{rep}|\boldsymbol{\theta}) d\boldsymbol{\theta} d\boldsymbol{X}_n^{rep},\end{aligned}$$

where $I(\cdot)$ is the indicator function. Even if the analytical expressions of the predictive distribution and the posterior distribution of $\boldsymbol{\theta}$ are unavailable, one can simulate samples $\boldsymbol{\theta}^{(k)}$, $\boldsymbol{X}_n^{\text{rep}(k)}$, $k = 1, ..., L$ from these densities based on some approaches, including MCMC methods. The classical p value that treats the data $\boldsymbol{X}_n$ as random and the value of parameters $\boldsymbol{\theta}$ is fixed given a null hypothesis. In contrast to the classical p value, under the Bayesian p-value context, the observed data $\boldsymbol{X}_n$ are treated as fixed, and the distribution on $T(\boldsymbol{X}_n^{\text{rep}}, \boldsymbol{\theta})$ depends on randomness of model parameters $\boldsymbol{\theta}$.

An extreme value for Bayesian p-value indicates that the observed data are unlikely from the model. In classical hypothesis testing, p-value seems to have meaning to provide evidence against a null hypothesis. However, we emphasize that small Bayesian p-values just reflect the lack of fit of the model to the observed data.

Gelman et al. (1995) suggested that test quantity $T(\boldsymbol{X}_n, \boldsymbol{\theta})$ is usually chosen to measure characteristics of the observed data $\boldsymbol{X}_n$ that are not directly addressed by the probability model. One of omnibus measures of goodness of fit is

$$T(\boldsymbol{X}_n, \boldsymbol{\theta}) = \sum_{\alpha=1}^{n} \frac{[y_\alpha - \mathrm{E}(x_\alpha|\boldsymbol{\theta})]^2}{\mathrm{Var}(\mathrm{x}_\alpha|\boldsymbol{\theta})}.$$

Using simulated samples $\boldsymbol{\theta}^{(k)}$, $\boldsymbol{X}_n^{\mathrm{rep}(k)}$, $k = 1, ..., L$, the Bayesian p value can be estimated by

$$\text{Bayesian } p\text{−value} = \frac{1}{L}\sum_{k=1}^{L} I\left(T(\boldsymbol{X}_n^{\mathrm{rep}(k)}, \boldsymbol{\theta}^{(k)}) \geq T(\boldsymbol{X}_n, \boldsymbol{\theta}^{(k)})\right).$$

For more details on the Bayesian p value, we refer to Gelman et al. (1995).

Example: Checking the tail behaviors

We have generated a set of $n = 20$ observations $\boldsymbol{X}_{20}$ from the Student-t density with the mean $\mu = 0$, the scale $\sigma = 1$ and the degrees of freedom $\nu = 4$. The following observations are generated.

$$\begin{array}{rrrrr} -0.94, & 0.68, & 1.75, & 1.25, & -0.25, \\ -0.14, & 0.38, & 2.48, & -2.68, & 0.27, \\ -0.43, & -1.07, & 0.87, & 0.11, & -0.66, \\ -0.10, & -0.25, & 0.57, & 0.69, & 4.03. \end{array}$$

We shall fit the normal model $N(\mu, 1)$ with known variance 1. For the prior density, we shall use the conjugate prior $\mu \sim N(\mu_0, \sigma_0^2)$. The posterior of the mean parameter μ is again the normal with the mean $\mu_n = (\mu_0/\sigma_0^2 + n\bar{y}_n)/(1/\sigma_0^2 + n)$ and the variance $\tau^2 = 1/(1/\sigma_0^2 + n)$. Here $\bar{y}_n = n^{-1}\sum_{\alpha=1}^{n} y_\alpha$ is the sample mean. Setting $\mu_0 = 0$ and $\sigma_0^2 = 1$, we generated a set of 10,000 draws $\{\mu^{(1)}, ..., \mu^{(10,000)}\}$ from the posterior of μ and then make a set of 10,000 draws $\boldsymbol{X}_{20}^{\mathrm{rep}(k)}$ from the normal model $N(\mu^{(k)}, 1)$.

To check the fitness, we shall consider the maximum value as a test statistic $T(\boldsymbol{X}_n, \boldsymbol{\theta}) = \max_\alpha x_\alpha$. For the observation $T(\boldsymbol{X}_{20}, \boldsymbol{\theta}) = 4.03$. Figure 5.15 shows the histgram of the observed maximum value of $T(\boldsymbol{X}_n^{\mathrm{rep}(k)}, \mu^{(k)})$, $k = 1, ..., 10,000$. The Bayesian p-value is

$$\text{Bayesian } p\text{−value} = \frac{1}{10,000}\sum_{k=1}^{10,000} I\left(T(\boldsymbol{X}_n^{\mathrm{rep}(k)}, \mu^{(k)}) \geq 4.03\right) = 208/10,000.$$

in this case.

5.9.3 Bayesian sensitivity analysis

We have seen that the Bayesian model consists of the sampling density for the observations, the likelihood $f(\boldsymbol{X}_n|\boldsymbol{\theta})$, and the prior density $\pi(\boldsymbol{\theta})$. The

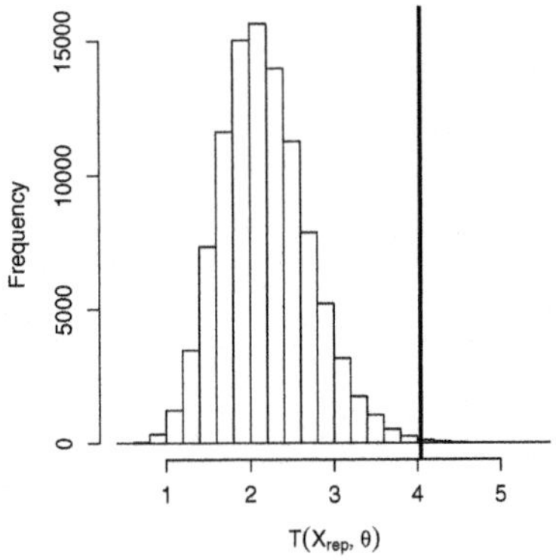

FIGURE 5.15: The histgram of the observed maximum value of $T(\boldsymbol{X}_n^{\text{rep}(k)}, \mu^{(k)})$, $k = 1, ..., 10,000$. The observed maximum value $T(\boldsymbol{X}_{20}, \boldsymbol{\theta}) = 4.03$ is also indicated by the bold line.

results of the posterior inference on parameters $\boldsymbol{\theta}$ are used for the decision problems. When one wants to check the sensitivity of the decisions, a Bayesian sensitivity analysis is useful. The standard approach is to investigate possible decisions from several sampling density and priors. If the decisions are robust to these changes, then one can feel confident in the decisions.

5.9.3.1 Example: Sensitivity analysis of Value at Risk

Value at Risk (VaR), a measure of market risk in finance, is one of the most commonly used tools in financial risk management. In statistical terms, VaR is an extreme conditional quantile of an asset return distribution. We consider daily log returns of Nikkei 225 stock index x_t from August 28, 2001 to September 22, 2005 with 1,000 observations. Bayesian sensitivity analysis is conducted by studying properties of VaR estimates of the following two estimation methods.

In modeling the volatility of financial time series, the generalized autoregressive conditional heteroskedasticity (GARCH) model is one of the most common methods. Here we first consider the Bayesian estimation of the GARCH(1,1) model with Student-t innovations. The GARCH(1,1) model with Student-t innovations may be written as

$$x_t = \varepsilon_t \sqrt{h_t(\nu - 2)/\nu}, \quad t = 1, ..., 1000,$$

where ε_t follows a Student-t distribution with ν degrees of freedom, and the term $(\nu - 2)/\nu$ is a scaling factor which ensures the conditional variance of x_t to be h_t. An innovation of the volatility h_t is

$$h_t = \alpha_0 + \alpha_1 x_{t-1}^2 + \beta h_{t-1},$$

where $\alpha_0 > 0$ and $\alpha_1, \beta \geq 0$ ensures a positive conditional variance.

For the prior density on $\boldsymbol{\alpha} = (\alpha_0, \alpha_1)$, a bivariate truncated Normal distribution with the prior mean $\boldsymbol{\mu}_{\alpha 0}$ and the prior covariance matrix Σ_α is used.

$$\pi(\boldsymbol{\alpha}) \propto N(\boldsymbol{\alpha}_0|\boldsymbol{\mu}_{\alpha 0}, \Sigma_\alpha)I(\alpha_0 > 0, \alpha_1 \geq 0),$$

where $I(\cdot)$ is the indicator function. Due to the restriction on the support for β, the prior distribution on β is again a univariate truncated Normal distribution with the prior mean $\mu_{\beta 0}$ and the prior variance matrix σ_β

$$\pi(\beta) \propto N(\beta|\mu_{\beta 0}, \sigma_\beta)I(\beta > 0).$$

For the prior distribution on the degrees of freedom parameter ν, a translated Exponential distribution is used

$$\pi(\nu) = \lambda \exp\{-\lambda(\nu - \delta)\} I(\nu > \delta),$$

where $\lambda > 0$ and $\delta \geq 2$. The prior mean for ν is thus $\delta + 1/\lambda$. Assuming prior independence of the parameters, the joint prior on parameter $\boldsymbol{\theta} = (\boldsymbol{\alpha}, \beta, \nu)$ is obtained

$$\pi(\boldsymbol{\theta}) = \pi(\boldsymbol{\alpha})\pi(\beta)\pi(\nu).$$

The values of hyperparameters are set to be $\boldsymbol{\mu}_\alpha = \mathbf{0}$, $\Sigma_\alpha = 10,000I$, $\mu_{\beta 0} = 0$, $\sigma_\beta = 10,000$, $\lambda = 0.01$, and $\delta = 2$, which leads to a rather vague prior. In posterior sampling, 110,000 samples are generated from MCMC algorithms. The first 10,000 samples are discarded as a burn in sample and then every 10-th sample is stored. The convergence of MCMC simulation was checked by calculating the Geweke's (1992) convergence test at a significance level of 5% for all parameters. For more details on the Bayesian analysis of the GARCH model with Student-t innovations, we refer to Nakatsuma (1998, 2000), Geweke (1993) and Deschamps (2006) and Ardia (2009).

Figure 5.16 shows the fluctuations of the forecasted 99% VaR point for the next day, September 22, 2005. Generally, 99% VaR (VaR_{99}) for x_{t+1} is defined as

$$P(\text{VaR}_{99} \leq x_{t+1}) = 0.99.$$

As an alternative model, we consider the GARCH(1,1) model with Normal innovation

$$x_t = \varepsilon_t \sqrt{h_t}, \quad t = 1, ..., 1000,$$

where ε_t follows the standard normal distribution. Setting $\lambda = 100$ and $\delta = 50$ in the GARCH(1,1) model with Student-t innovations, we shall implement the Bayesian estimation of the GARCH(1,1) model with Normal innovations. Since the degrees of freedom ν is above 50, we can consider the resulting model ensures an approximate Normality for the innovations. Figure 5.17 shows the

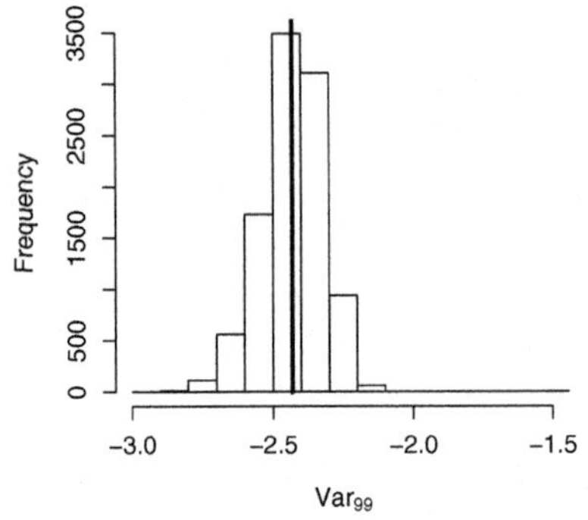

FIGURE 5.16: The fluctuations of the forecasted 99% VaR point for the next day, September 22, 2005. The mean value of the forecasted 99% VaR is also indicated by the bold line. The results are from the Bayesian estimation of the GARCH(1,1) model with Student-t innovations.

fluctuations of the forecasted 99% VaR point for the next day, September 22, 2005.

The fluctuation of the forecasted 99% VaR point shows different characteristics. First the distributional range has moved to the positive side slightly. The mean values of the forecasted 99% VaR are the GARCH(1,1) model with Student-t innovations $\mathrm{Va\bar{R}}_{99} = -2.43$ the GARCH(1,1) model with (approximated) normal innovation. $\mathrm{Va\bar{R}}_{99} = -2.29$, respectively. Thus the 99% VaR from the GARCH(1,1) model with Student-t innovations is more conservative than that based on that with (approximated) normal innovations. If the decision is sensitive to this change, the distributional assumption is an important factor for the risk management decision.

5.9.3.2 Example: Bayesian change point analysis

Cobb (1978) introduced the change point detection problem through the analysis of the annual volume of discharge from the Nile River at Aswan y_t for the years $t = 1871, ..., 1970$. Here we consider a convenient method that detects a change point automatically. We consider a regression model for y_t

$$y_t = \beta_1 I(t \leq \gamma) + \beta_2 I(\gamma < t) + \beta_3 I(t \leq \gamma) \times t + \beta_4 I(\gamma < t) \times t + \varepsilon_t, \quad (5.47)$$

where the variable γ denotes the change point and $I(\cdot)$ is an indicator function, the errors ε_t are independently, normally distributed with mean zero and variance σ^2. In matrix notation,

$$\boldsymbol{y}_n = X_n(\gamma)\boldsymbol{\beta} + \boldsymbol{\varepsilon}_n, \quad \boldsymbol{\varepsilon}_n \sim N(0, \sigma^2 I),$$

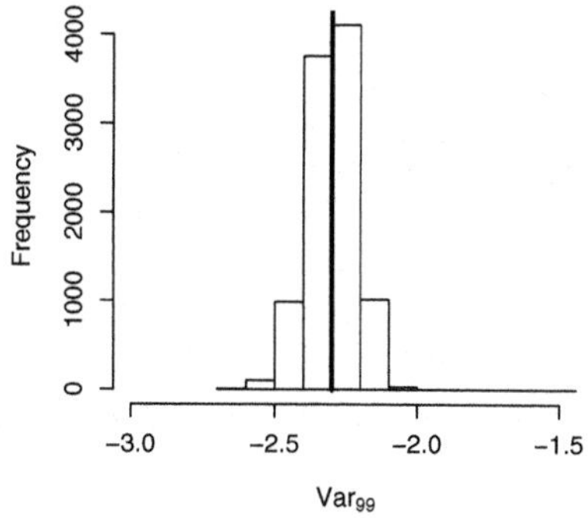

FIGURE 5.17: The fluctuations of the forecasted 99% VaR point for the next day, September 22, 2005. The mean value of the forecasted 99% VaR is also indicated by the bold line. The results are from the Bayesian estimation of the GARCH(1,1) model with approximately Normal innovation.

where

$$X_n(\gamma) = \begin{pmatrix} I(1871 \le \gamma) & I(\gamma < 1871) & I(1871 \le \gamma)1871 & I(\gamma < 1871)1871 \\ I(1872 \le \gamma) & I(\gamma < 1872) & I(1872 \le \gamma)1872 & I(\gamma < 1872)1982 \\ \vdots & \ddots & \vdots & \\ I(1970 \le \gamma) & I(\gamma < 1970) & I(1970 \le \gamma)1970 & I(\gamma < 1970)1970 \end{pmatrix}.$$

We denote the design matrix $X_n(\gamma)$ as a function of γ because different change points will yield a different design matrix.

For the prior of $\boldsymbol{\beta}$ and σ^2, we shall use a conjugate normal inverse-gamma prior $\pi(\boldsymbol{\beta}, \sigma^2) = \pi(\boldsymbol{\beta}|\sigma^2)\pi(\sigma^2)$,

$$\pi(\boldsymbol{\beta}|\sigma^2) = N\left(\mathbf{0}, \sigma^2 A^{-1}\right) \quad \text{and} \quad \pi(\sigma^2) = IG\left(\frac{\nu_0}{2}, \frac{\lambda_0}{2}\right).$$

Given the value of a change point γ, the marginal likelihood of $\boldsymbol{y}_n$ is given as follows:

$$P\left(\boldsymbol{y}_n \middle| X_n(\gamma)\right) = \frac{\left|\hat{A}_n\right|^{1/2} |A|^{1/2} \left(\frac{\lambda_0}{2}\right)^{\frac{\nu_0}{2}} \Gamma\left(\frac{\hat{\nu}_n}{2}\right)}{\pi^{\frac{n}{2}} \Gamma\left(\frac{\nu_0}{2}\right)} \left(\frac{\hat{\lambda}_n}{2}\right)^{-\frac{\hat{\nu}_n}{2}}, \tag{5.48}$$

where $\hat{\nu}_n = \nu_0 + n$, $\hat{A}_n(\gamma) = (X_n^T(\gamma)X_n(\gamma) + A)^{-1}$ and

$$\begin{aligned} \hat{\lambda}_n &= \lambda_0 + \left(\boldsymbol{y}_n - X_n(\gamma)\hat{\boldsymbol{\beta}}\right)^T \left(\boldsymbol{y}_n - X_n(\gamma)\hat{\boldsymbol{\beta}}\right) \\ &\quad + \hat{\boldsymbol{\beta}}^T \left((X_n^T(\gamma)X_n(\gamma))^{-1} + A^{-1}\right)^{-1} \hat{\boldsymbol{\beta}}^T \end{aligned}$$

with $\hat{\boldsymbol{\beta}} = \left(X_n^T(\gamma)X_n(\gamma)\right)^{-1} X_n^T(\gamma)\boldsymbol{y}_n$.

Setting the uniform prior density for γ between the range $\gamma \in \{1872, ..., 1969\}$ we can calculate the posterior distribution of γ as

$$\pi(\gamma|\boldsymbol{y}_n) = \frac{P\left(\boldsymbol{y}_n \middle| X_n(\gamma)\right)}{\sum_{t=1872}^{1969} P\left(\boldsymbol{y}_n \middle| X_n(\gamma)\right)},$$

where $P\left(\boldsymbol{y}_n \middle| X_n(\gamma)\right)$ is given in (5.48).

Figure 5.18 plots the posterior distribution of γ. We set the hyperparameter values as $A^{-1} = I_p$ and $a = b = 10^{-10}$, which makes the prior diffused. As shown in Figure 5.18, the mode of the posterior is at $t = 1899$. The corresponding fitted curve $X_n(1899)\hat{\boldsymbol{\beta}}_n$ is shown in Figure 5.19. Here

$$\hat{\boldsymbol{\beta}} = \left(X_n^T(\gamma)X_n(\gamma) + A^{-1}\right)^{-1} X_n^T(\gamma)\boldsymbol{y}_n.$$

We conduct the sensitivity analysis and simplify the mean structure as

$$y_t = \beta_1 I(t \le \gamma) + \beta_2 I(\gamma < t). + \varepsilon_t, \tag{5.49}$$

Figure 5.19 plots the posterior distribution of γ. The same prior settings are used. As shown in Figure 5.19, the mode of the posterior is again at $t = 1899$. However, the posterior distribution of γ is concentrated around the posterior mode.

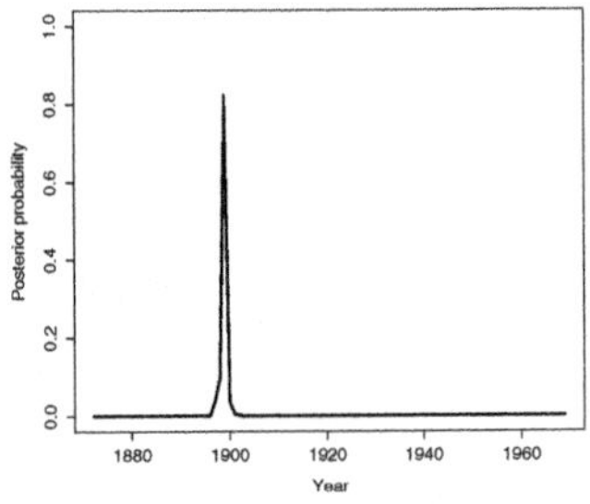

FIGURE 5.18: The posterior distribution of γ based on the model (5.47).

Exercises

1. *In example 5.2.3, show that the posterior distribution of λ is a Gamma distribution with parameter $(n\bar{x}_n + \alpha, n + \beta)$. Also, find the pair of (α, β) that maximizes the marginal likelihood.*

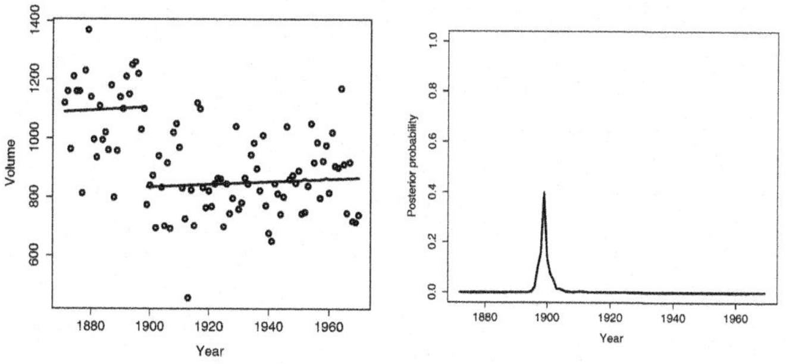

FIGURE 5.19: The fitted curve $\boldsymbol{y}_n = X_n(1899)\hat{\boldsymbol{\beta}}_n$ based on the model (5.47). The posterior distribution of γ based on the model (5.49).

2. *In example 5.3.1, show that the posterior distribution of p is the Beta density with parameter $(y_n + \alpha, n - y_n + \beta)$. Also, find the pair of (α, β) that maximizes the marginal likelihood.*

3. *In example 5.3.2, under the Zellner's g-prior, show that the conditional posterior distribution of $\boldsymbol{\beta}$ given σ^2 and the posterior distribution of σ^2 becomes*

$$\pi\left(\boldsymbol{\beta}, \sigma^2 \middle| \boldsymbol{y}_n, X_n\right) = \pi\left(\boldsymbol{\beta} \middle| \sigma^2, \boldsymbol{y}_n, X_n\right)\pi\left(\sigma^2 \middle| \boldsymbol{y}_n, X_n\right)$$

with

$$\pi\left(\boldsymbol{\beta} \middle| \sigma^2, \boldsymbol{y}_n, X_n\right) = N\left(\frac{g}{g+1}\hat{\boldsymbol{\beta}}_{\mathrm{MLE}} + \frac{1}{g+1}\boldsymbol{\beta}_0, \frac{\sigma^2 g}{g+1}\left(X_n^T X_n\right)^{-1}\right),$$

$$\pi\left(\sigma^2 \middle| \boldsymbol{y}_n, X_n\right)$$

$$= IG\left(\frac{n}{2}, \frac{R^2}{2} + \frac{1}{2(g+1)}\left(\boldsymbol{\beta}_0 - \hat{\boldsymbol{\beta}}_{\mathrm{MLE}}^T\right) X_n^T X_n \left(\boldsymbol{\beta}_0 - \hat{\boldsymbol{\beta}}_{\mathrm{MLE}}^T\right)\right),$$

where $\hat{\boldsymbol{\beta}}_{\mathrm{MLE}} = \left(X_n^T X_n\right)^{-1} X_n^T \boldsymbol{y}_n$ is the maximum likelihood estimate and

$$R^2 = \left(\boldsymbol{y}_n - X_n\hat{\boldsymbol{\beta}}_{\mathrm{MLE}}\right)^T\left(\boldsymbol{y}_n - X_n\hat{\boldsymbol{\beta}}_{\mathrm{MLE}}\right)$$

is the sum of squared errors.

4. *Example 5.5.1 investigated the approximation error of BIC as an estimator of the marginal likelihood $f(\boldsymbol{X}_n)$ through the conjugate prior analysis of the binomial distribution. Using the result of Example 5.2.3, implement the same evaluation. The exact marginal likelihood is in Example*

5.5.1. Also, the BIC score is given as

$$-\frac{1}{2}\mathrm{BIC} = -n\hat{\lambda}_{\mathrm{MLE}} \sum_{\alpha=1}^{n} x_\alpha \log(\hat{\lambda}_{\mathrm{MLE}}) - \sum_{\alpha=1}^{n} \log(x_\alpha!),$$

where $\hat{\lambda}_{\mathrm{MLE}} = \sum_{\alpha=1}^{n} x_\alpha/n$ *is the maximum likelihood estimator.*

5. *Example 5.5.2 used the BIC score for selecting the link function through the analysis of O-ring data. As an alternative, we can also consider the complementary log-log function.*

$$p(t;\boldsymbol{\beta}) = 1 - \exp\left(-\exp\left(\beta_0 + \beta_1 t\right)\right).$$

Calculate the BIC score for this model. Then compare the fitted probability curse with those from the logistic function, the probit function in Figure 5.3. The **R** *function* **glm** *might be helpful.*

6. *Example 5.5.3 applied BIC for selecting the number of factors in factor analysis model. Ando (2009a) applied the Bayesian factor analysis model with fat-tailed factors to the customer satisfaction data (Rossi et al. (2001)). The data contain responses to a satisfaction survey for a Yellow Pages advertising product. The number of observations and the dimension of the datasets are* $(n, p) = (1811, 10)$. *Table 5.7 summarizes the variables. The scores of data sets are recorded on the 10 point measure. Using the BIC, select the number of factors in the factor analysis model. The dataset can be obtained through the* **R** *package* **bayesm**.

TABLE 5.7: Variable descriptions for customer satisfaction data.

1	Overall Satisfaction
2	Setting Competitive Prices
3	Holding Price Increase to a Minimum
4	Appropriate Pricing given Volume
5	Demonstrating Effectiveness of Purchase
6	Reach a Large Number of Customers
7	Reach of Advertising
8	Long-term Exposure
9	Distribution
10	Distribution to Right Geographic Areas

7. *The DNA microarray measures the activities of several thousand genes simultaneously and the gene expression profiles are increasingly being performed in biological and medical studies. Since transcriptional changes accurately reflect the status of cancers, the expression level of genes contains the keys to address fundamental problems relating to the prevention and cure of tumors, biological evolution mechanisms and drug*

discovery. The gene expression data has very unique characteristics. First, it has very high-dimensionality and usually contains up to tens of thousands of genes. Second, the publicly available data size is very small; some have sizes below 100.

To get an idea, obtain the diffuse large B-cell lymphoma data (Alizadeh et al. (2000)) through the supplemental website. The data set consists of gene expression levels from cDNA experiments involving three prevalent adult lymphoid malignancies: diffuse large B-cell lymphoma, B-cell chronic lymphocytic leukemia, and follicular lymphoma. Using a hierarchical clustering method, Alizadeh et al. (2000) identified two molecularly distinct types of B-cell chronic lymphocytic leukemia, which had gene expression patterns indicative of different stages of B-cell differentiation: germinal center B-like (GCB) and activated B-like (AB). In Section 5.5.4, we considered the survival analysis model. Using a stepwise procedure, try to find the identification of "marker" genes that characterize the survival time for GCB. In the same way, try to identify "marker" genes that characterize the survival time for AB.

8. *Suppose that we have n independent observations y_α, each from a Poisson distribution with conditional expectation $E[Y_\alpha|\boldsymbol{x}_\alpha] = \gamma(\boldsymbol{x}_\alpha)$, where $\boldsymbol{x}_\alpha$ is a vector of p covariates. Assume that the conditional expectation is of the form:*

$$\log\left(\gamma(\boldsymbol{x}_\alpha)\right) = \sum_{k=1}^{m} w_k b_k(\boldsymbol{x}_\alpha) = \boldsymbol{w}^T \boldsymbol{b}(\boldsymbol{x}_\alpha), \qquad \alpha = 1, ..., n.$$

Show that the log-likelihood function is expressed as

$$\log f(\boldsymbol{y}_n|X_n, \boldsymbol{w}) = \sum_{\alpha=1}^{n} \left[y_\alpha \boldsymbol{w}^T \boldsymbol{b}(\boldsymbol{x}_\alpha) - \exp\{\boldsymbol{w}^T \boldsymbol{b}(\boldsymbol{x}_\alpha)\} - \log y_\alpha! \right].$$

9. *(Continued). Generate a set of n independent observations $\boldsymbol{y}_n = \{y_1, ..., y_n\}$ from a Poisson distribution with conditional expectation*

$$\log\{\gamma(x_\alpha)\} = \sin(3\pi x^2),$$

where the design points x_α are uniformly distributed in $[0, 1]$. **R** *function* **rpois** *generates random samples. Then using a singular multivariate normal prior density,*

$$\pi(\boldsymbol{w}) = \left(\frac{n\lambda}{2\pi}\right)^{(m-2)/2} |R|_+^{1/2} \exp\left\{-\frac{n\lambda}{2} \boldsymbol{w}^T D_2^T D_2 \boldsymbol{w}\right\},$$

find the posterior mode $\hat{\boldsymbol{w}}_n$. The mode can be estimated by Fisher scoring iterations:

$$\boldsymbol{w}^{new} = \left(B^T W B + n\lambda D_2^T D_2\right)^{-1} B^T W \boldsymbol{\zeta},$$

where W is an $n \times n$ diagonal matrix, and $\boldsymbol{\zeta}$ is an n-dimensional vector:

$$\begin{aligned} W_{\alpha\alpha} &= \gamma(\boldsymbol{x}_\alpha), \\ \zeta_\alpha &= \{y_\alpha - \gamma(\boldsymbol{x}_\alpha; \boldsymbol{w})\}/\gamma(\boldsymbol{x}_\alpha; \boldsymbol{w}) + \boldsymbol{w}^T \boldsymbol{b}(\boldsymbol{x}_\alpha). \end{aligned}$$

Update the parameter vector $\boldsymbol{w}$ until a suitable convergence criterion is satisfied.

10. *(Continued). Show that GBIC score for evaluating the Poisson regression model is*

$$\begin{aligned} \text{GBIC} = & \; 2\sum_{\alpha=1}^{n} \left[y_\alpha \hat{\boldsymbol{w}}_n^T \boldsymbol{b}(x_\alpha) - \exp\{\hat{\boldsymbol{w}}_n^T \boldsymbol{b}(x_\alpha)\} - \log y_\alpha! \right] + n\lambda \hat{\boldsymbol{w}}_n^T D_2^T D_2 \hat{\boldsymbol{w}}_n \\ & -2\log(2\pi/n) + \log|S_n(\hat{\boldsymbol{w}}_n)| - (m-2)\log\lambda, \end{aligned}$$

where $S_n(\hat{\boldsymbol{w}}_n)$ is the $m \times m$ matrix given as

$$S_n(\hat{\boldsymbol{w}}_n) = \frac{1}{n}\left(B^T \Gamma B + n\lambda D_2^T D_2\right),$$

with $\Gamma = \text{diag}\left[\gamma(x_1; \hat{\boldsymbol{w}}_n), \cdots, \gamma(x_n; \hat{\boldsymbol{w}}_n)\right]$. Search an optimal combination of λ and m which minimizes GBIC.

11. *The vowel recognition data (Hastie et al. (1994)) is a popular benchmark for neural network algorithms, and consists of training and test data with 10 measurements and 11 classes. An ascii approximation to the International Phonetic Association symbol and the words in which the eleven vowel sounds were recorded, are given in Table 5.8. The word was uttered once by each of the fifteen speakers and for each utterance, ten floating-point input values were measured. In detail, the speech signals were low-filtered at 4.7kHz and then digitized to 12 bits with a 10kHz sampling rate.*

TABLE 5.8: Words used in recording the vowels.

vowel	word	vowel	word
i	heed	I	hid
E	head	A	had
a:	hard	Y	hud
O	hod	Ɔ:	hoard
U	hood	u:	who'd
ɜ:	heard		

In Section 5.6.2, the GBIC was applied for multinomial logistic model with basis expansion predictors. Using 528 training data from eight speakers (4 male and 4 female), construct the model using GBIC. Then test the prediction capability of a model constructed on 462 data from seven speakers (4 male and 3 female).

12. *Berger and Pericchi (1996). Suppose that we have a set of* n *independent observations* $\boldsymbol{X}_n$. *Consider the following models.* M_1: $N(0, \sigma_1^2)$ *with* $\pi(\sigma_1) \propto 1/\sigma_1$, *and* M_2: $N(\mu, \sigma_2^2)$ *with* $\pi(\mu, \sigma_2) \propto 1/\sigma_2^2$. *It is obvious that the priors are improper. Show that*

$$\text{AIBF}\,(M_2, M_1) = \sqrt{\frac{2\pi}{n}} \left(1 + \frac{n\bar{x}^2}{s^2}\right)^{n/2} \left[\frac{1}{N} \sum_{\ell=1}^{N} \frac{(x_{1(\ell)} - x_{2(\ell)})^2}{2\sqrt{\pi}\left(x_{1(\ell)}^2 + x_{2(\ell)}^2\right)}\right],$$

$$\text{GIBF}\,(M_2, M_1) = \sqrt{\frac{2\pi}{n}} \left(1 + \frac{n\bar{x}^2}{s^2}\right)^{n/2} \left[\prod_{\ell=1}^{N} \frac{(x_{1(\ell)} - x_{2(\ell)})^2}{2\sqrt{\pi}\left(x_{1(\ell)}^2 + x_{2(\ell)}^2\right)}\right]^{1/N},$$

respectively. Here $s^2 = \sum_{\alpha=1}^{n}(x_\alpha - \bar{x}_n)$ *with* $\bar{x}_n = \sum_{\alpha=1}^{n} x_\alpha / n$.

13. *Berger and Pericchi (1996). Consider the linear regression model*

$$M_j \quad : \quad \boldsymbol{y}_n = X_{jn}\boldsymbol{\beta}_j + \boldsymbol{\varepsilon}_{jn}, \quad \boldsymbol{\varepsilon}_{jn} \sim N(0, \sigma_j^2 I),$$

with diffuse prior

$$\pi_j(\boldsymbol{\beta}_j, \sigma_j^2) \propto 1/\sigma_j^{1+q_j}, \quad q_j > -1.$$

Here $\boldsymbol{\beta}_j$ *is the* p_j *dimensional parameter vector. Taking* $q_j = 0$ *corresponds to the reference prior and* $q_j = p_j$ *reduces to Jeffreys prior. The minimal training sample of* $X_n(\ell)$ *is* $m = \max_j p_j + 1$ *such that all* $X_{jn}^T X_{jn}$ *are non-singular.*

Let C_k *and* C_j *be the arbitrary normalizing constant term of the priors. Then, show that*

$$\begin{aligned} &\text{Bayes factor}(M_k, M_j) \\ &= \frac{\displaystyle\int f_k(\boldsymbol{y}_n|X_n, \boldsymbol{\theta}_k)\pi_k(\boldsymbol{\theta}_k)d\boldsymbol{\theta}_k}{\displaystyle\int f_j(\boldsymbol{y}_n|X_n, \boldsymbol{\theta}_j)\pi_j(\boldsymbol{\theta}_j)d\boldsymbol{\theta}_j} \times \frac{C_k}{C_j} \\ &= \frac{\pi^{-\frac{p_k - p_j}{2}}}{2^{-\frac{q_j - q_k}{2}}} \frac{\Gamma\left(\frac{(n-p_k+q_k)}{2}\right)}{\Gamma\left(\frac{(n-p_j+q_j)}{2}\right)} \frac{|X_{jn}^T X_{jn}|^{\frac{1}{2}}}{|X_{kn}^T X_{kn}|^{\frac{1}{2}}} \frac{S_j^{n-p_j-q_j}}{S_k^{n-p_k-q_k}} \times \frac{C_k}{C_j}, \end{aligned}$$

where S_j^2 *is the residual sum of squares*

$$S_j^2 = \boldsymbol{y}_n \left(I - X_{jn}(X_{jn}^T X_{jn})^{-1} X_{jn}^T\right) \boldsymbol{y}_n.$$

Similarly, show that the Bayes factor conditioned on a particular partition $n(\ell)$

$$\text{Bayes factor}\,(M_j, M_k, n(\ell)) = \frac{\displaystyle\int f_j(\boldsymbol{y}_{-n(\ell)}|\boldsymbol{\theta}_j)\pi_j(\boldsymbol{\theta}_j)d\boldsymbol{\theta}_j}{\displaystyle\int f_k(\boldsymbol{y}_{-n(\ell)}|\boldsymbol{\theta}_k)\pi_k(\boldsymbol{\theta}_k)d\boldsymbol{\theta}_k} \times \frac{C_k}{C_j}$$

is given by the inverse of the above expression with n, X_{jn}, X_{kn}, S_j^2 *and* S_k^2 *replaced by* m, $X_{j,n(\ell)}$, $X_{k,n(\ell)}$, $S_j^2(\ell)$ *and* $S_k^2(\ell)$, *respectively. Here* $S_j^2(\ell)$ *and* $S_k^2(\ell)$ *are the residual sum of squares for the training sample* $\boldsymbol{y}_{n(\ell)}$. *Noting that the arbitrary constant terms* C_k *and* C_j *can be cancelled, obtain the arithmetic intrinsic Bayes factor.*

14. *In Section 5.8.3, the predictive likelihood approach with* PL_2 *score was used for the analysis of forest inventory data. Zurichberg forest inventory data holds the coordinates for all trees in the Zurichberg Forest. Species (SPP), basal area (BAREA) diameter at breast height (DBH), and volume (VOL) are recorded for each tree. The dataset is obtained in the* **R** *package* **spBayes**. *Develop the Bayesian linear spatial model with the help of* PL_2 *score.*

15. *Section 5.9.2 applied the Bayesian p-value to check the tail beheavior of the data. To check the fitness, consider the minimum value as a test statistic* $T(\boldsymbol{X}_n, \boldsymbol{\theta}) = \min_\alpha x_\alpha$.

    Generate the a set of $n = 100$ *observations from the normal distribution* $N(0, 1)$, *and consider fitting the normal model* $N(\mu, 1)$ *with the conjugate prior* $\mu \sim N(\mu_0, \sigma_0^2)$. *Setting* $\mu_0 = 0$ *and* $\sigma_0^2 = 10$, *generate a set of 10,000 draws* $\{\mu^{(1)}, ..., \mu^{(10{,}000)}\}$ *from the posterior of* μ *and then make a set of 10,000 draws* $\boldsymbol{X}_n^{\mathrm{rep}(k)}$ *from the normal model* $N(\mu^{(k)}, 1)$. *Calculate the observed minimum value of* $T(\boldsymbol{X}_n^{\mathrm{rep}(k)}, \mu^{(k)})$, $k = 1, ..., 10,000$. *Then calculate the Bayesian p-value as*

$$\text{Bayesian } p\text{−value} = \frac{1}{10,000} \sum_{k=1}^{10,000} I\left(T(\boldsymbol{X}_n^{\mathrm{rep}(k)}, \mu^{(k)}) \leq T(\boldsymbol{X}_n, \mu)\right).$$

16. *In Section 5.9.3, we have conducted the sensitivity analysis of Value at Risk. Obtain time series data and consider the Bayesian estimation of the GARCH(1,1) model with Student-t innovations used in Section 5.9.3. Also, setting* $\lambda = 100$ *and* $\delta = 50$ *in the GARCH(1,1) model with Student-t innovations, implement the (approximately) Bayesian estimation of the GARCH(1,1) model with Normal innovations. Compare the fluctuations of the forecasted 99% VaR.*

17. *In Section 5.9.3.2, we have conducted the sensitivity analysis of the Bayesian change point analysis model. Generate the data* y_t *from*

$$y_t = \beta_1 \times I(t \leq 33) + \varepsilon_t,$$

    with $\beta_1 = 5.2$ *and* $\varepsilon_t \sim N(0, 1.3)$. *Then fit the model (5.49) with hyperparameter values* $A^{-1} = 0.000I$ *and* $A^{-1} = 10,000I$. *Check the posterior distribution of the change point* γ *from these two models.*

Chapter 6

Simulation approach for computing the marginal likelihood

The calculation of the posterior probabilities for a set of competing models is essential in the Bayesian approach for model selection. However, the marginal likelihood is generally not easily computed.

An easy way to use simulation to estimate the marginal likelihood is to sample from the prior $\pi(\boldsymbol{\theta})$. Thus, generating a set of samples $\{\boldsymbol{\theta}^{(1)}, ..., \boldsymbol{\theta}^{(L)}\}$ from the prior distribution, then

$$P(\boldsymbol{X}_n|M) = \frac{1}{L}\sum_{j=1}^{L} f\left(\boldsymbol{X}_n|\boldsymbol{\theta}^{(j)}\right)$$

estimates the marginal likelihood consistently. Unfortunately, this estimate is usually quite a poor approximation (McCulloch and Rossi 1992).

Many studies that take advantage of modern Markov chain Monte Carlo computing methods are available to estimate the marginal likelihood. This book covers some of these studies, including the Laplace-Metropolis estimator (Lewis and Raftery, 1997), the so-called candidate formula (Chib, 1995), the harmonic mean estimator (Newton and Raftery, 1994), Gelfand and Dey's estimator (Gelfand and Dey, 1994), the bridge sampling estimator (Meng and Wong, 1996) and so on. Readers also might want to refer to DiCiccio et al. (1997), Gelman and Meng (1998), Verdinelli and Wasserman (1995), and Lopes and West (2004). Recent textbooks reviewing modern computational approaches for marginal likelihood include Carlin and Louis (2000), Chen et al. (2000), and Gamerman and Lopes (2006). In this chapter, we delete the notation M from $P(\boldsymbol{X}_n|M)$ and express the marginal likelihood simply as $P(\boldsymbol{X}_n)$ except for differenciating several models.

6.1 Laplace-Metropolis approximation

In the previous section, we provided approximation of the marginal likelihood based on the normal approximation, which is based on the posterior mode $\hat{\boldsymbol{\theta}}_n$ and the inverse Hessian of the penalized log-likelihood function evaluated at the posterior mode.

If we have a set of L sample values $\{\boldsymbol{\theta}^{(1)}, ..., \boldsymbol{\theta}^{(L)}\}$ from the posterior distribution, we can estimate the posterior mode by

$$\begin{aligned}\hat{\boldsymbol{\theta}} &\approx \max_j \pi(\boldsymbol{\theta}^{(j)}|\boldsymbol{X}_n)\\ &= \max_j f(\boldsymbol{X}_n|\boldsymbol{\theta}^{(j)})\pi(\boldsymbol{\theta}^{(j)}).\end{aligned}$$

Similarly, under the i.i.d assumption on $\boldsymbol{X}_n$, we estimate the posterior covariance matrix by

$$\hat{V}_n \approx \frac{1}{L}\sum_{j=1}^{n}\left\{(\boldsymbol{\theta}^{(j)} - \bar{\boldsymbol{\theta}})^T(\boldsymbol{\theta}^{(j)} - \bar{\boldsymbol{\theta}})\right\}, \qquad (6.1)$$

where $\bar{\boldsymbol{\theta}}$ is the posterior mean.

Putting these quantities into (5.8), we obtain the Laplace approximation to the marginal likelihood in the form

$$P(\boldsymbol{X}_n) \approx f(\boldsymbol{X}_n|\hat{\boldsymbol{\theta}})\pi(\hat{\boldsymbol{\theta}}) \times (2\pi)^{p/2}|\hat{V}_n|^{1/2}. \qquad (6.2)$$

6.1.1 Example: Multinomial probit models

The multinomial probit model is often used to analyze the discrete choices made by respondents. Let $\boldsymbol{y}_\alpha = (y_{1\alpha}, .., y_{J\alpha})^T$ be a multinomial vector, with $y_{j\alpha} = 1$ if an individual α chooses alternative j, and $y_{j\alpha} = 0$ otherwise. Thus likelihood function for the multinomial probit model is then

$$f(\boldsymbol{y}_n|X_n, \boldsymbol{\beta}, \Sigma) = \prod_{\alpha=1}^{n}\left[\prod_{j=1}^{J}\Pr(y_{j\alpha} = 1|\boldsymbol{X}_n, \boldsymbol{\beta}, \Sigma)^{y_{j\alpha}}\right],$$

where

$$\Pr(y_{j\alpha} = 1|\boldsymbol{X}_n, \boldsymbol{\beta}, \Sigma) = \int_{S_j}\frac{1}{(2\pi)^{(J-1)/2}|\Sigma|^{-1/2}}\exp\left[-\frac{1}{2}\boldsymbol{\varepsilon}_\alpha^T\Sigma^{-1}\boldsymbol{\varepsilon}_\alpha\right]d\boldsymbol{\varepsilon}_\alpha.$$

The sets S_j are given by

$$S_j = \cap_{k\neq j}\left\{\varepsilon_{j\alpha} - \varepsilon_{k\alpha} > (\boldsymbol{x}_{k\alpha} - \boldsymbol{x}_{j\alpha})^T\boldsymbol{\beta}\right\} \cap \left\{\varepsilon_{j\alpha} > -\boldsymbol{x}_{j\alpha}^T\boldsymbol{\beta}\right\}.$$

To complete the specification of a Bayesian model, we assign a prior distribution for the parameter. A convenient prior specification for the multinomial probit model is used in Imai and van Dyk (2005):

$$\pi(\boldsymbol{\beta}) = N(\boldsymbol{0}, A^{-1}) \quad \text{and} \quad \pi(\Sigma) = IW(\Lambda_0, \nu_0),$$

where A is the prior precision matrix of $\boldsymbol{\beta}$, ν_0 is the prior degrees of freedom parameter, and Λ_0 is $(J-1) \times (J-1)$ positive definite matrix. For alternate prior specifications, see McCulloch and Rossi (1994), McCulloch et al. (2000). An implementation of the sampler proposed by Imai and van Dyk (2005) can be done by the **R** package **MNP**.

We shall analyze the Dutch voting behavior data in 1989, a subset of multiply imputed datasets used in Quinn and Martin (2002). A set of $n = 1754$ individuals gives the self-reported vote choice. The choices ($J = 4$) are CDA (Christen Democratisch Appel), D66 (Democraten 66), Pvda (Partij van de Arbeid), and VVD (Volkspartij voor Vrijheid en Democratie), respectively. For covariates $\boldsymbol{x}$, we shall use the following 4 predictors; x_1 (distPvdA), giving the squared ideological distance between the respondent and the distPvdA, x_2 (distVVD), giving the squared ideological distance between the respondent and the VVD x_3 (distCDA), giving the squared ideological distance between the respondent and the CDA. Larger values indicate ideological dissimilarity between the respondent and the party. The reaming 2 predictors are x_4 (income: 0 is lowest and 6 is highest) and x_5 (age: 0 is lowest and 12 is highest).

Setting $A^{-1} = 10,000I$, $\nu = 3$ and $\Lambda_0 = I$, we use a vague prior distribution. We generated a set of 1,000 samples from the posterior distribution, saving every 100th sample after discarding the first 5,000 samples. Convergence diagnostics can be checked by for e.g., Geweke (1992)'s approach.

After we get a set of posterior samples $\{\boldsymbol{\beta}^{(k)}, \Sigma^{(k)}\}$, the marginal likelihood of this model M is given as

$$P(\boldsymbol{X}_n|M) \approx f(\boldsymbol{y}_n|X_n, \hat{\boldsymbol{\beta}}, \hat{\Sigma})\pi(\hat{\boldsymbol{\beta}})\pi(\hat{\Sigma}) \times (2\pi)^{q/2}|\hat{V}_n|^{1/2},$$

where q is the number of parameters, $\hat{\boldsymbol{\beta}}$ and $\hat{\Sigma}$ are their posterior modes, and $\hat{V}_n$ is the posterior covariance matrix given in (6.1)

As a result, the estimated marginal likelihood is $\log P(\boldsymbol{X}_n|M) = -117.6487$. As an alternative model, we can consider the multinomial probit model based only on the first 3 predictors: x_1 distPvdA, x_2 distVVD, and x_3 distCDA, respectively. In the same way, the marginal likelihood for this alternative model M_a is calculated. The estimated marginal likelihood $\log P(\boldsymbol{X}_n|M_a) = -98.3482$ indicates that the alternative model specification is preferred.

6.2 Gelfand-Day's approximation and the harmonic mean estimator

For any probability density function $h(\boldsymbol{\theta})$ with H normalizing constant, we have

$$\begin{aligned}\int \frac{h(\boldsymbol{\theta})}{f(\boldsymbol{X}_n|\boldsymbol{\theta})\pi(\boldsymbol{\theta})}\pi(\boldsymbol{\theta}|\boldsymbol{X}_n)d\boldsymbol{\theta} &= \int \frac{h(\boldsymbol{\theta})}{f(\boldsymbol{X}_n|\boldsymbol{\theta})\pi(\boldsymbol{\theta})}\frac{f(\boldsymbol{X}_n|\boldsymbol{\theta})\pi(\boldsymbol{\theta})}{P(\boldsymbol{X}_n)}d\boldsymbol{\theta} \\ &= \int h(\boldsymbol{\theta})d\boldsymbol{\theta} \times \frac{1}{P(\boldsymbol{X}_n)} \\ &= \frac{1}{P(\boldsymbol{X}_n)}.\end{aligned}$$

Therefore, specifying a density function $h(\boldsymbol{\theta})$, the Gelfand-Day's approximation can be used to approximate the marginal likelihood. One of the specifications of $h(\boldsymbol{\theta})$ is the use of the prior density function $h(\boldsymbol{\theta}) = \pi(\boldsymbol{\theta})$. This specification leads the Gelfand-Day's approximation formula to

$$\begin{aligned}P(\boldsymbol{X}_n) &= \frac{1}{\displaystyle\int \frac{\pi(\boldsymbol{\theta})}{f(\boldsymbol{X}_n|\boldsymbol{\theta})\pi(\boldsymbol{\theta})}\pi(\boldsymbol{\theta}|\boldsymbol{X}_n)d\boldsymbol{\theta}} \\ &\approx \frac{1}{\displaystyle\frac{1}{L}\sum_{j=1}^{L}\frac{1}{f(\boldsymbol{X}_n|\boldsymbol{\theta}^{(j)})}},\end{aligned} \tag{6.3}$$

where the $\boldsymbol{\theta}^{(j)}$, $j = 1, ..., L$ are posterior samples.

This estimator is called the harmonic mean estimator (Newton and Raftery (1994)). As shown in the equation, this estimator is based on a harmonic mean of the likelihood values. Though it has been quite widely used, it might be noticed that this estimator can be unstable in some applications because a few outlying values with small likelihood values can have a large effect on this estimate. This is because the inverse likelihood does not possess a finite variance (Chib (1995)).

6.2.1 Example: Bayesian analysis of the ordered probit model

In the context of the ordered probit model with J possible choices, the probability that the α-th observation y_α, given predictors $\boldsymbol{x}$, is assigned to be the category k is given as

$$P(y_\alpha = k|\boldsymbol{x}_\alpha) = \Phi\left(\gamma_k - \boldsymbol{x}_\alpha^T\boldsymbol{\beta}\right) - \Phi\left(\gamma_{k-1} - \boldsymbol{x}_\alpha^T\boldsymbol{\beta}\right),$$

where $-\infty = \gamma_0, \gamma_1, ..., \gamma_{J-1}, \gamma_J = \infty$ are the cut-off points, and $\Phi(\cdot)$ is the cumulative probability distribution function of the standard normal distribution.

Using the algorithm Albert and Chib (2001), we can generate a set of posterior samples $\{\boldsymbol{\beta}^{(j)}, \boldsymbol{\gamma}^{(j)}\}$, $j = 1, ..., L$. In a practical situation, the **R** packge **MCMCpack** will implement MCMC sampling. Then the harmonic mean estimator is

$$P(\boldsymbol{X}_n|M) \approx \frac{1}{\displaystyle\frac{1}{L}\sum_{j=1}^{L}\frac{1}{f(\boldsymbol{y}_n|X_n, \boldsymbol{\beta}^{(j)}, \boldsymbol{\gamma}^{(j)})}},$$

with

$$f(\boldsymbol{y}_n|X_n, \boldsymbol{\beta}^{(j)}, \boldsymbol{\gamma}^{(j)}) = \prod_{\alpha=1}^{n}\left[\sum_{k=1}^{J} I(y_\alpha = k) \times \left\{\Phi\left(\gamma_k^{(j)} - \boldsymbol{x}_\alpha^T\boldsymbol{\beta}^{(j)}\right) - \Phi\left(\gamma_{k-1}^{(j)} - \boldsymbol{x}_\alpha^T\boldsymbol{\beta}^{(j)}\right)\right\}\right].$$

According to the original Basel Accord issued in 1988, internationally active banks are required to hold their percentage of capital divided by the risk-weighted asset ratio to at least 8%. In the new Basel Accord, banks are provided a range of options to calculate their capital charges, reflecting risk differences across individual credit exposures. Beginning with the standardized approach, where banks are required to differentiate their credit exposures into broad categories, the range of options was expanded to include two approaches based on internal credit ratings. The internal rating approach permits banks to perform internal credit risk assessments, while the standardized approach relies on externally provided risk assessments.

Fundamentally, a firm's credit rating is a measure of financial strength, in the sense that the firm meets its interest commitments and honors its payments promptly. When banks employ the standardized approach, the credit rating provided by external rating agencies may play an important role in calculating the risk weights for asset classes. Credit ratings also supply useful information to bond issuers, investors, financial intermediaries, brokers, regulation authorities, and the government. For example, investors generally use credit ratings as one source of information when dealing with debenture. For example, portfolio managers such as pension fund managers might refrain from purchasing bonds attached to a low credit rating. The term, long-term ratings, usually represent the degree of certainty that interest and principal payments will be made against the firm's long-term obligations.

In this section, we analyze a set of $n = 147$ synthetic credit rating data, which is obtained by adding a certain portion of noise to the real dataset. The credit ratings are categorized into 4 groups AAA ($y_\alpha = 1$), AA and A ($y_\alpha = 2$), BBB ($y_\alpha = 3$), below BB ($y_\alpha = 4$). AAA is the highest credit rating. For the predictors, accounting variables related to profitability, stability, scale, efficiency, cash flow, are usually used. We shall use the three predictors, return on asset (ROA), log(sales), and cash flow-sales ratio. Figure 6.1 shows a matrix of scatterplots. As shown in Figure 6.1, three variables seem to be related to the credit ratings.

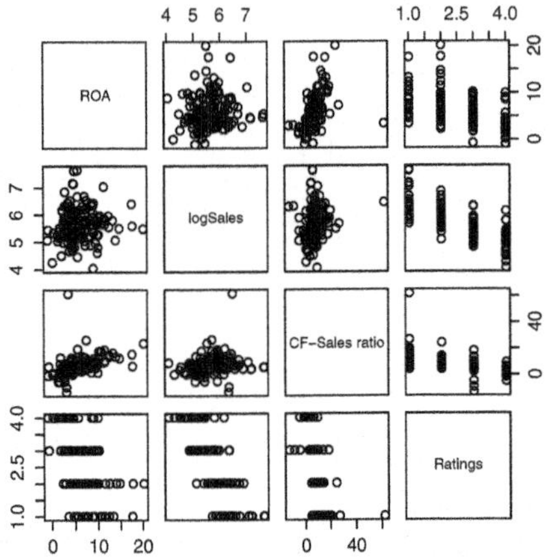

FIGURE 6.1: A matrix of scatterplots of credit rating data.

Using the **R** function, **MCMCoprobit**, we generated a set of 1,000 posterior samples. For the prior density of $\boldsymbol{\beta}$, we use the normal distribution with prior mean $\boldsymbol{\beta}_0 = \mathbf{0}$, and the prior covariance matrix $10,000I$. Note that if we use the improper prior, the marginal likelihood is not well-defined. For more details, we refer to Albert and Chib (2001). As a result, the estimated log-marginal likelihood value under this model M is $P(\boldsymbol{X}_n|M) = -2.91$. It is clear that we can identify an optimal subset of accounting variables by investigating the marginal likelihood score for possible combinations. For example, we consider the model with 2 predictors; return on asset (ROA), log(sales) as an alternative model M_a. The corresponding log-marginal likelihood score is $P(\boldsymbol{X}_n|M_a) = -1.97$. This implies that the alternative model M_a is favored. One of the reasons might be that information coming from all three predictors is redundant for predicting the credit ratings.

6.3 Chib's estimator from Gibb's sampling

To compute the marginal likelihood, Chib (1995) provided a method to estimate the posterior ordinate in the context of Gibb's sampling. Modifying the definition of the posterior density, Chib (1995) noted that

$$\log P(\boldsymbol{X}_n) = \log f(\boldsymbol{X}_n|\boldsymbol{\theta}) + \log \pi(\boldsymbol{\theta}) - \log \pi(\boldsymbol{\theta}|\boldsymbol{X}_n), \tag{6.4}$$

for any value of $\boldsymbol{\theta}$. When the three terms on the right hand side of Equation (6.4) are analytically available, the marginal likelihood can be evaluated easily. The value of $\boldsymbol{\theta}$ is usually chosen as a point of high posterior density, to maximize the accuracy of this approximation.

Even when the joint posterior distribution of $\boldsymbol{\theta}$ is not available, we can deal with this situation. Chib (1995) dealt with the case where the parameter vector $\boldsymbol{\theta}$ can be partitioned into several blocks so that the full conditional for each block is available in closed form. To illustrate the idea, let us consider the case of two blocks, $\boldsymbol{\theta} = (\boldsymbol{\theta}_1^T, \boldsymbol{\theta}_2^T)^T$ where $\pi(\boldsymbol{\theta}_1|\boldsymbol{X}_n, \boldsymbol{\theta}_2)$, and $\pi(\boldsymbol{\theta}_2|\boldsymbol{X}_n, \boldsymbol{\theta}_1)$ are available in closed form. Note that the joint posterior distribution of $\boldsymbol{\theta}$ in (6.4) can be estimated as follows:

$$\pi(\boldsymbol{\theta}|\boldsymbol{X}_n) = \pi(\boldsymbol{\theta}_1|\boldsymbol{X}_n, \boldsymbol{\theta}_2)\pi(\boldsymbol{\theta}_2|\boldsymbol{X}_n), \tag{6.5}$$

where an appropriate Monte Carlo estimate of $\pi(\boldsymbol{\theta}_2|\boldsymbol{X}_n)$ is given as

$$\hat{\pi}(\boldsymbol{\theta}_2|\boldsymbol{X}_n) = \frac{1}{L}\sum_{j=1}^{L} \pi(\boldsymbol{\theta}_2|\boldsymbol{X}_n, \boldsymbol{\theta}_1^{(j)}).$$

Here $\{\boldsymbol{\theta}_1^{(j)}; j = 1, ..., L\}$ is a set of posterior samples. Under regularity conditions, this estimate is simulation consistent, i.e., $\hat{\pi}(\boldsymbol{\theta}_2|\boldsymbol{X}_n) \to \pi(\boldsymbol{\theta}_2|\boldsymbol{X}_n)$ as L becomes large, as a consequence of the ergodic theorem (Tierney (1994)).

Therefore, we have an estimator of the marginal likelihood as

$$\begin{aligned}&\log P(\boldsymbol{X}_n)\\ &\approx \log f(\boldsymbol{X}_n|\boldsymbol{\theta}_1^*, \boldsymbol{\theta}_2^*) + \log \pi(\boldsymbol{\theta}_1^*, \boldsymbol{\theta}_2^*) - \log \pi(\boldsymbol{\theta}_1^*|\boldsymbol{X}_n, \boldsymbol{\theta}_2^*) - \log \hat{\pi}(\boldsymbol{\theta}_2^*|\boldsymbol{X}_n),\end{aligned}$$

where the first three terms on the right side are available in closed form. The value of $\boldsymbol{\theta}^*$ may be chosen as the posterior mode, posterior mean, or posterior median and other point estimates so as to maximize the accuracy of this approximation.

The extension from two to B parameter blocks $\boldsymbol{\theta} = (\boldsymbol{\theta}_1^T, ..., \boldsymbol{\theta}_B^T)^T$ replaces the Equation (6.5) with a factoring of the joint posterior into B components. As well as the two blocks case, suppose that the conditional posterior densities $\pi(\boldsymbol{\theta}_1|\boldsymbol{X}_n, \boldsymbol{\theta}_2, ..., \boldsymbol{\theta}_B)$, $\pi(\boldsymbol{\theta}_2|\boldsymbol{X}_n, \boldsymbol{\theta}_1, \boldsymbol{\theta}_3, ..., \boldsymbol{\theta}_B)$, and $\pi(\boldsymbol{\theta}_B|\boldsymbol{X}_n, \boldsymbol{\theta}_1, ..., \boldsymbol{\theta}_{B-1})$ are available in closed form.

Begin by writing the joint posterior density as

$$\pi(\boldsymbol{\theta}|\boldsymbol{X}_n) = \pi(\boldsymbol{\theta}_1|\boldsymbol{X}_n, \boldsymbol{\theta}_2, ..., \boldsymbol{\theta}_B)\pi(\boldsymbol{\theta}_2|\boldsymbol{X}_n, \boldsymbol{\theta}_3, ..., \boldsymbol{\theta}_B) \times \cdots \times \pi(\boldsymbol{\theta}_B|\boldsymbol{X}_n), \tag{6.6}$$

where the first term is the known conditional posterior density form. The typical terms are the reduced conditional ordinate $\pi(\boldsymbol{\theta}_k|\boldsymbol{X}_n, \boldsymbol{\theta}_{k+1}, ..., \boldsymbol{\theta}_B)$. This is given by

$$\int \pi(\boldsymbol{\theta}_k|\boldsymbol{X}_n, \boldsymbol{\theta}_1, ..., \boldsymbol{\theta}_{k-1}, \boldsymbol{\theta}_{k+1}, ..., \boldsymbol{\theta}_B) d\pi(\boldsymbol{\theta}_1, ..., \boldsymbol{\theta}_{k-1}|\boldsymbol{X}_n, \boldsymbol{\theta}_{k+1}, ..., \boldsymbol{\theta}_B).$$

If the draws from the conditional density $\pi(\boldsymbol{\theta}_1, ..., \boldsymbol{\theta}_{k-1}|\boldsymbol{X}_n, \boldsymbol{\theta}_{k+1}, ..., \boldsymbol{\theta}_B)$ are denoted by $\{\boldsymbol{\theta}_1^{(j)}, ..., \boldsymbol{\theta}_{k-1}^{(j)}\}, j = 1, ..., L$, then an estimate of the above quantity is

$$\hat{\pi}(\boldsymbol{\theta}_k|\boldsymbol{X}_n, \boldsymbol{\theta}_{k+1}, ..., \boldsymbol{\theta}_B) = \frac{1}{L}\sum_{j=1}^{L} \pi(\boldsymbol{\theta}_k|\boldsymbol{X}_n, \boldsymbol{\theta}_1^{(j)}, ..., \boldsymbol{\theta}_{k-1}^{(j)}, \boldsymbol{\theta}_{k+1}, ..., \boldsymbol{\theta}_B), \quad (6.7)$$

which is simulation consistent.

Noting that an estimate of the joint posterior density (6.6) is

$$\hat{\pi}(\boldsymbol{\theta}|\boldsymbol{X}_n) = \pi(\boldsymbol{\theta}_1|\boldsymbol{X}_n, \boldsymbol{\theta}_2, ..., \boldsymbol{\theta}_B)\hat{\pi}(\boldsymbol{\theta}_2|\boldsymbol{X}_n, \boldsymbol{\theta}_3, ..., \boldsymbol{\theta}_B) \times \cdots \times \hat{\pi}(\boldsymbol{\theta}_B|\boldsymbol{X}_n),$$

the log of the marginal likelihood (6.4) is estimated by

$$\log P(\boldsymbol{X}_n) \approx \log f(\boldsymbol{X}_n|\boldsymbol{\theta}^*) + \log \pi(\boldsymbol{\theta}^*) - \sum_{k=1}^{B} \log \hat{\pi}(\boldsymbol{\theta}_k^*|\boldsymbol{X}_n, \boldsymbol{\theta}_{k+1}^*, ..., , \boldsymbol{\theta}_B^*). (6.8)$$

Again, the posterior mode, posterior mean, or posterior median and other point estimates can be used for $\boldsymbol{\theta}^*$.

This approach requires us to know the normalizing constant for the full conditional distributions. However, it often happens that the full conditional distributions are not analytically available. In such a case, rather than Gibb's sampling, Metropolis-Hastings sampling steps are used. The next section covers Chib and Jeliazkov (2001)'s approach that overcomes the problems associated with the presence of intractable full conditional densities.

6.3.1 Example: Seemingly unrelated regression model with informative prior

As shown in Section 4.2.4, the SUR model can be expressed as the following from: $\boldsymbol{y}_n = X_n\boldsymbol{\beta} + \boldsymbol{\varepsilon}$, $\boldsymbol{\varepsilon} \sim N(\boldsymbol{0}, \Sigma \otimes I)$. Therefore, another expression of the likelihood function of the SUR model is given in the following form:

$$f(\boldsymbol{Y}_n|X_n, \boldsymbol{\beta}, \Sigma) = \frac{1}{(2\pi)^{nm/2}|\Sigma|^{n/2}} \exp\left[-\frac{1}{2}(\boldsymbol{y}_n - X_n\boldsymbol{\beta})(\Sigma \otimes I)^{-1}(\boldsymbol{y}_n - X_n\boldsymbol{\beta})\right].$$

A conventional Bayesian analysis is used to conjugate priors. Since the use of conjugate priors results in the posterior distribution of the parameters having the same form as the prior density, we can calculate the marginal likelihood easily. However, there is no conjugate prior for SUR model (Richard and Steel (1988)).

Instead we use the normal and the inverse Wishart priors for $\boldsymbol{\beta}$ and Σ, $\pi_2(\boldsymbol{\beta}, \Sigma) = \pi_2(\boldsymbol{\beta})\pi_2(\Sigma)$, with

$$\pi_2(\boldsymbol{\beta}) = N(\boldsymbol{0}, A^{-1}) \quad \text{and} \quad \pi_2(\Sigma) = IW(\Lambda_0, \nu_0).$$

The joint posterior density function for the parameters is then:

$$\pi_1(\boldsymbol{\beta}, \Sigma|\boldsymbol{Y}_n, X_n) \propto |\Sigma|^{-(n+\nu_0+m+1)/2} \exp\left[-\frac{1}{2}\mathrm{tr}\left\{\Sigma^{-1}\Lambda_0\right\}\right]$$
$$\times \exp\left[-\frac{1}{2}\boldsymbol{\beta}^T A\boldsymbol{\beta} + (\boldsymbol{y}_n - X_n\boldsymbol{\beta})(\Sigma\otimes I)^{-1}(\boldsymbol{y}_n - X_n\boldsymbol{\beta})\right].$$

Again, the informative prior distribution just allows us to get the analytical conditional posterior densities of $\boldsymbol{\beta}$ and Σ, but the analytical joint posterior density. Holmes et al. (2002) showed that the conditional posterior of $\boldsymbol{\beta}$ is expressed as

$$\pi_2(\boldsymbol{\beta}|\boldsymbol{Y}_n, X_n, \Sigma) \propto \exp\left[-\frac{1}{2}\left\{(\boldsymbol{\beta}-\bar{\boldsymbol{\beta}})^T\bar{\Omega}^{-1}(\boldsymbol{\beta}-\bar{\boldsymbol{\beta}}) + b\right\}\right],$$

with

$$\bar{\boldsymbol{\beta}} = \left\{X_n^T\left(\Sigma^{-1}\otimes I\right)X_n + A\right\}^{-1}X_n^T\left(\Sigma^{-1}\otimes I\right)\boldsymbol{y}_n,$$
$$\bar{\Omega} = \left(X_n^T\left(\Sigma^{-1}\otimes I\right)X_n + A\right)^{-1},$$
$$b = \mathrm{tr}\left\{\Sigma^{-1}\Lambda_0\right\} + \boldsymbol{y}_n^T\left(\Sigma^{-1}\otimes I\right)\boldsymbol{y}_n - \bar{\boldsymbol{\beta}}^T\bar{\Omega}^{-1}\bar{\boldsymbol{\beta}}.$$

Therefore, the conditional posterior of $\boldsymbol{\beta}$ given Σ, is normal with mean $\bar{\boldsymbol{\beta}}$ and covariance matrix $\bar{\Omega}$.

Using the other likelihood form given in Section 4.2.4, the use of the inverse Wishart prior to leave the posterior probability density

$$\pi_2(\Sigma|\boldsymbol{Y}_n, X_n, \boldsymbol{\beta}) \propto |\Sigma|^{(n+\nu_0+m+1)/2} \exp\left[-\frac{1}{2}\mathrm{tr}\left\{\Sigma^{-1}(R+\Lambda_0)\right\}\right],$$

which is $IW(R+\Lambda_0, n+\nu_0)$. Here R is given in Section 4.2.4. Although the posteriors of $\boldsymbol{\beta}$ and Σ are depending upon each other, we can use the Gibb's sampler. Replacing the conditional posteriors of $\boldsymbol{\beta}$ and Σ in the Gibb's sampling by $\pi_2(\boldsymbol{\beta}|\boldsymbol{Y}_n, X_n, \Sigma)$ and $\pi_2(\Sigma|\boldsymbol{Y}_n, X_n, \boldsymbol{\beta})$, we can use the Gibb's sampling approach.

6.3.1.1 Calculation of the marginal likelihood

Using the generated posterior samples $\{(\boldsymbol{\beta}^{(j)}, \Sigma^{(j)}),\ j = 1, ..., L\}$, we can estimate the posterior mode by

$$\left\{\hat{\boldsymbol{\beta}}_n, \hat{\Sigma}_n\right\} \approx \max_j f\left(\boldsymbol{Y}_n|X_n, \boldsymbol{\beta}^{(j)}, \Sigma^{(j)}\right)\pi_2\left(\boldsymbol{\beta}^{(j)}, \Sigma^{(j)}\right).$$

Calculating the posterior covariance matrix $\hat{V}_n$ by (6.1), we can obtain the Laplace-Metropol estimator (6.2):

$$P(\boldsymbol{Y}_n) \approx f\left(\boldsymbol{Y}_n|X_n, \hat{\boldsymbol{\beta}}_n, \hat{\Sigma}_n\right)\pi_2\left(\hat{\boldsymbol{\beta}}_n, \hat{\Sigma}_n\right) \times (2\pi)^{p/2}\left|\hat{V}_n\right|^{1/2}, \quad (6.9)$$

where $p = \sum_{k=1}^{m} \text{dim}\boldsymbol{\beta}_k + \text{dim}\Sigma$ is the number of free parameters included in the coefficient vector and the covariance matrix. Also, the Harmonic mean estimator is

$$P(\boldsymbol{Y}_n) \approx \frac{1}{\dfrac{1}{L}\displaystyle\sum_{j=1}^{L} \frac{1}{f\left(\boldsymbol{Y}_n|X_n, \boldsymbol{\beta}^{(j)}, \Sigma^{(j)}\right)}}.$$

Note that the conditional posterior distributions $\pi_2(\boldsymbol{\beta}|\boldsymbol{Y}_n, X_n, \Sigma)$, and $\pi_2(\Sigma|\boldsymbol{Y}_n, X_n, \boldsymbol{\beta})$ are available in closed form; Chib's estimator from Gibb's sampling can be calculated. Noting that the joint posterior distribution of $\boldsymbol{\theta}$ in (6.4) can be estimated as

$$\pi_2(\boldsymbol{\beta}, \Sigma|\boldsymbol{Y}_n, X_n) = \pi_2(\boldsymbol{\beta}|\boldsymbol{Y}_n, X_n, \Sigma)\, \pi_2(\Sigma|\boldsymbol{Y}_n, X_n)$$

with an appropriate Monte Carlo estimate of $\pi_2(\Sigma|\boldsymbol{Y}_n, X_n)$ given as

$$\hat{\pi}_2(\Sigma|\boldsymbol{Y}_n, X_n) = \frac{1}{L}\sum_{j=1}^{L} \pi_2\left(\Sigma|\boldsymbol{Y}_n, X_n, \boldsymbol{\beta}^{(j)}\right),$$

we have an estimator of the marginal likelihood as

$$\begin{aligned}\log P(\boldsymbol{Y}_n) \approx\ & \log f(\boldsymbol{Y}_n|X_n, \boldsymbol{\beta}^*, \Sigma^*) + \log \pi_2(\boldsymbol{\beta}^*, \Sigma^*) \\ & + \log \pi_2(\boldsymbol{\beta}^*|\boldsymbol{Y}_n, X_n, \Sigma^*) + \log \hat{\pi}_2(\Sigma^*|\boldsymbol{Y}_n, X_n).\end{aligned}$$

The values of $\boldsymbol{\beta}^*$ and Σ^* can be chosen as their posterior mode $\hat{\boldsymbol{\beta}}_n$ and $\hat{\Sigma}_n$.

Practical implementation of MCMC

To implement the Gibb's sampling procedure with the informative prior, we simulate data sets from the $m = 2$ dimensional SUR model

$$\begin{pmatrix} \boldsymbol{y}_1 \\ \boldsymbol{y}_2 \end{pmatrix} = \begin{pmatrix} X_{n1} & O \\ O & X_{n2} \end{pmatrix}\begin{pmatrix} \boldsymbol{\beta}_1 \\ \boldsymbol{\beta}_2 \end{pmatrix} + \begin{pmatrix} \boldsymbol{\varepsilon}_1 \\ \boldsymbol{\varepsilon}_2 \end{pmatrix}, \quad i = 1, ..., n,$$

where $\boldsymbol{y}_j$ and $\boldsymbol{\varepsilon}_j$ are $n \times 1$ vectors, X_j is the $n \times 3$ matrix and $\boldsymbol{\beta}_j$ is the 3-dimensional vector. Each element of Σ is set to be

$$\Sigma = \begin{pmatrix} \sigma_1^2 & \sigma_{12} \\ \sigma_{21} & \sigma_2^2 \end{pmatrix} = \begin{pmatrix} 0.35 & -0.15 \\ -0.15 & 0.43 \end{pmatrix}.$$

The design matrices X_{nj} $j = 1, 2$ were generated from a uniform density over the interval $[-2, 2]$. The coefficient vectors were set to be $\boldsymbol{\beta}_1 = (-2, 0, 1)^T$ and $\boldsymbol{\beta}_2 = (0, 3, 1)^T$, respectively. Therefore, this true model just contains the $\boldsymbol{x}_1 = (x_{11}, x_{13})^T$ and $\boldsymbol{x}_2 = (x_{22}, x_{23})^T$ as the predictors. In this simulation we set the number of observations to be $n = 100$.

When one wants to select a set of variables that contribute to the prediction, the marginal likelihood can be used. To the generated data, we fit the following models:

$$\begin{aligned} M_1 &: \boldsymbol{x}_1 = (x_{11}, x_{13}) \quad \text{and} \quad \boldsymbol{x}_2 = (x_{22}, x_{23}), \\ M_2 &: \boldsymbol{x}_1 = (x_{12}, x_{13}) \quad \text{and} \quad \boldsymbol{x}_2 = (x_{21}, x_{23}), \\ M_3 &: \boldsymbol{x}_1 = (x_{11}, x_{12}, x_{13}) \quad \text{and} \quad \boldsymbol{x}_2 = (x_{21}, x_{22}, x_{23}). \end{aligned}$$

Thus, the model M_1 is the true model specification.

Setting the values of hyperparameters in the prior to be diffuse, $\nu_0 = 5$, $\Lambda_0 = I$, $A = 10^5 I$, we generated 6,000 Markov chain Monte Carlo samples, of which the first 1,000 iterations are discarded. To check whether the posterior sample is taken from the stationary distribution, the convergence diagnostic (CD) test statistics (Geweke (1992)) were calculated. All the results we report in this paper are based on samples that have passed the Geweke's (1992) convergence test at a significance level of 5% for all parameters.

Table 6.1 compares the calculated marginal likelihood values based on the Laplace-Metropolis approximation, the harmonic mean estimator, and Chib's estimator from Gibb's sampling. Although the estimated values from three methods are different, all estimators selected the true model.

TABLE 6.1: Comparison of the calculated marginal likelihood values based on the Laplace-Metropolis approximation (LM), the harmonic mean estimator (HM), and Chib's estimator from Gibb's sampling (Chib).

Model	Predictors	LM	HM	Chib
True	$\boldsymbol{x}_1 = (x_{11}, x_{13})$, $\boldsymbol{x}_2 = (x_{22}, x_{23})$	–	–	–
M_1	$\boldsymbol{x}_1 = (x_{11}, x_{13})$, $\boldsymbol{x}_2 = (x_{22}, x_{23})$	-195.583	-194.285	-189.495
M_2	$\boldsymbol{x}_1 = (x_{12}, x_{13})$, $\boldsymbol{x}_2 = (x_{21}, x_{23})$	-553.981	-550.566	-544.715
M_3	$\boldsymbol{x}_1 = (x_{11}, x_{12}, x_{13})$			
	$\boldsymbol{x}_2 = (x_{21}, x_{22}, x_{23})$	-202.726	-203.031	-198.714

6.4 Chib's estimator from MH sampling

Chib and Jeliazkov (2001) extended Chib's approach in the context of MCMC chains produced by the Metropolis-Hastings algorithm, whose building blocks are used both for sampling and marginal likelihood estimation.

Let $\boldsymbol{\theta}$ be updated in a single block and define $p(\boldsymbol{\theta}, \boldsymbol{\theta}^*)$ as the proposal density for the transition from $\boldsymbol{\theta}$ to $\boldsymbol{\theta}^*$, and the acceptance probability as

$$\alpha(\boldsymbol{\theta}, \boldsymbol{\theta}^*) = \min\left\{1, \frac{f(\boldsymbol{X}_n|\boldsymbol{\theta}^*)\pi(\boldsymbol{\theta}^*)p(\boldsymbol{\theta}^*, \boldsymbol{\theta})}{f(\boldsymbol{X}_n|\boldsymbol{\theta})\pi(\boldsymbol{\theta})p(\boldsymbol{\theta}, \boldsymbol{\theta}^*)}\right\}.$$

Also, let $q(\boldsymbol{\theta}, \boldsymbol{\theta}^*) = \alpha(\boldsymbol{\theta}, \boldsymbol{\theta}^*)p(\boldsymbol{\theta}^*, \boldsymbol{\theta})$ denote the sub-kernel of the Metropolis-Hastings algorithm.

From the reversibility condition of the sub-kernel, we have

$$q(\boldsymbol{\theta}, \boldsymbol{\theta}^*)\pi(\boldsymbol{\theta}|\boldsymbol{X}_n) = q(\boldsymbol{\theta}^*, \boldsymbol{\theta})\pi(\boldsymbol{\theta}^*|\boldsymbol{X}_n).$$

Upon integrating both sides of this equation with respect to $\boldsymbol{\theta}$, the posterior ordinate is given as follows:

$$\pi(\boldsymbol{\theta}^*|\boldsymbol{X}_n) = \frac{\int \alpha(\boldsymbol{\theta}, \boldsymbol{\theta}^*)p(\boldsymbol{\theta}, \boldsymbol{\theta}^*)\pi(\boldsymbol{\theta}|\boldsymbol{X}_n)d\boldsymbol{\theta}}{\int \alpha(\boldsymbol{\theta}^*, \boldsymbol{\theta})p(\boldsymbol{\theta}^*, \boldsymbol{\theta})d\boldsymbol{\theta}}$$

for any value of $\boldsymbol{\theta}^*$. Note that the expectation in the denominator is with respect to the posterior density $\pi(\boldsymbol{\theta}|\boldsymbol{X}_n)$ and the expectation in the numerator with respect to the candidate density $p(\boldsymbol{\theta}^*, \boldsymbol{\theta})$.

Thus the numerator is then estimated by averaging the product in braces with respect to draws from the posterior, and the denominator is estimated by averaging the acceptance probability with respect to draws from the proposal density. A simulation-consistent estimate of the posterior ordinate is

$$\pi(\boldsymbol{\theta}^*|\boldsymbol{X}_n) \approx \frac{L^{-1}\sum_{j=1}^{L} \alpha(\boldsymbol{\theta}^{(j)}, \boldsymbol{\theta}^*)p(\boldsymbol{\theta}^{(j)}, \boldsymbol{\theta}^*)}{M^{-1}\sum_{j'=1}^{M} \alpha(\boldsymbol{\theta}^*, \boldsymbol{\theta}^{(j')})}, \tag{6.10}$$

where $\boldsymbol{\theta}^{(j)}$, $j = 1, ..., L$ are the sampled draws from the posterior distribution and $\boldsymbol{\theta}^{(j')}$, $j' = 1, ..., M$ are draws from the proposal density $p(\boldsymbol{\theta}^{(j)}, \boldsymbol{\theta}^*)$ given the fixed value $\boldsymbol{\theta}^*$.

Substituting this estimate in the log of the marginal likelihood (6.4), we obtain

$$\log P(\boldsymbol{X}_n) \approx \log f(\boldsymbol{X}_n|\boldsymbol{\theta}^*) + \log \pi(\boldsymbol{\theta}^*) - \log \hat{\pi}(\boldsymbol{\theta}^*|\boldsymbol{X}_n), \tag{6.11}$$

where the third term is given in (6.10). Though the choice of point $\boldsymbol{\theta}^*$ is arbitrary, it is usually chosen as a point that has high posterior density.

Even there are two or more parameter blocks, Chib and Jeliazkov (2001) illustrated an extended version of this algorithm for estimating the marginal likelihood. It is similar to the Chib (1995)'s approach for the Gibb's sampler outlined earlier.

6.5 Bridge sampling methods

Meng and Wong (1996) studied an innovative method based on bridge sampling. The method starts from an identity

$$1 = \frac{\int \alpha(\boldsymbol{\theta})\pi(\boldsymbol{\theta}|\boldsymbol{X}_n)p(\boldsymbol{\theta})d\boldsymbol{\theta}}{\int \alpha(\boldsymbol{\theta})p(\boldsymbol{\theta})\pi(\boldsymbol{\theta}|\boldsymbol{X}_n)d\boldsymbol{\theta}}$$

with a pair of functions $\alpha(\boldsymbol{\theta})$ and $p(\boldsymbol{\theta})$ such that $\int \alpha(\boldsymbol{\theta})p(\boldsymbol{\theta})\pi(\boldsymbol{\theta}|\boldsymbol{X}_n)d\boldsymbol{\theta} > 0$. Noting the relationship between the posterior $\pi(\boldsymbol{\theta}|\boldsymbol{X}_n)$ and the marginal likelihood $P(\boldsymbol{X}_n)$, we obtain the following identity from the above identity

$$P(\boldsymbol{X}_n) = \frac{\int \alpha(\boldsymbol{\theta})f(\boldsymbol{X}_n|\boldsymbol{\theta})\pi(\boldsymbol{\theta})p(\boldsymbol{\theta})d\boldsymbol{\theta}}{\int \alpha(\boldsymbol{\theta})p(\boldsymbol{\theta})\pi(\boldsymbol{\theta}|\boldsymbol{X}_n)d\boldsymbol{\theta}}.$$

Both expectations in the numerator and denominator can be estimated by the draws from $p(\boldsymbol{\theta})$ and the MCMC sample values from the posterior distribution $\pi(\boldsymbol{\theta}|\boldsymbol{X}_n)$

$$P(\boldsymbol{X}_n) = \frac{M^{-1}\sum_{j'=1}^{M} \alpha\left(\boldsymbol{\theta}^{(j')}\right) f\left(\boldsymbol{X}_n|\boldsymbol{\theta}^{(j')}\right) \pi\left(\boldsymbol{\theta}^{(j')}\right)}{L^{-1}\sum_{j=1}^{L} \alpha\left(\boldsymbol{\theta}^{(j)}\right) p\left(\boldsymbol{\theta}^{(j)}\right)},$$

where $\boldsymbol{\theta}^{(j)}$ $(j = 1, ..., L)$ are the sampled draws from the posterior distribution, and $\boldsymbol{\theta}^{(j')}$ $(j' = 1, ..., M)$ are the draws from $p(\boldsymbol{\theta})$.

Meng and Wong (1996) provided some discussions on the relationship between their estimator and other marginal likelihood estimators. Suppose that we take $\alpha(\boldsymbol{\theta})^{-1} = f(\boldsymbol{X}_n|\boldsymbol{\theta})\pi(\boldsymbol{\theta})p(\boldsymbol{\theta})$ the corresponding estimator resembles the harmonic mean estimator. It is obtained by

$$P(\boldsymbol{X}_n) = \frac{M^{-1}\sum_{j'=1}^{M} \left[p\left(\boldsymbol{\theta}^{(j')}\right)\right]^{-1}}{L^{-1}\sum_{j=1}^{L} \left[f\left(\boldsymbol{X}_n|\boldsymbol{\theta}^{(j)}\right) \pi\left(\boldsymbol{\theta}^{(j)}\right)\right]^{-1}}.$$

Lopes and West (2004) investigated the performance of the bridge sampling estimator and various marginal likelihood evaluation methods in the context of factor analysis.

6.6 The Savage-Dickey density ratio approach

When we compare the nested models, the Savage-Dickey density ratio approach is a convenient tool for calculating the marginal likelihood. Suppose that we want to compare the two Bayesian model specifications: M_1: the model $f(\boldsymbol{X}_n|\boldsymbol{\theta}, \boldsymbol{\psi})$ with the prior $\pi(\boldsymbol{\theta}, \boldsymbol{\psi})$, and M_2: in the model M_1, the parameter value $\boldsymbol{\theta} = \boldsymbol{\theta}_0$ is a fixed value both in the model $f(\boldsymbol{X}_n|\boldsymbol{\theta}_0, \boldsymbol{\psi})$ and the prior $\pi(\boldsymbol{\theta}_0, \boldsymbol{\psi})$.

Under these model specifications, Dickey (1971) showed that the Bayes factor B_{21} for two models is expressed as

$$B_{21} = \frac{\displaystyle\int f(\boldsymbol{X}_n|\boldsymbol{\theta}_0, \boldsymbol{\psi})\pi(\boldsymbol{\theta}_0, \boldsymbol{\psi})d\boldsymbol{\psi}}{\displaystyle\int f(\boldsymbol{X}_n|\boldsymbol{\theta}, \boldsymbol{\psi})\pi(\boldsymbol{\theta}, \boldsymbol{\psi})d\boldsymbol{\theta}d\boldsymbol{\psi}}$$

$$= \frac{\displaystyle\int \pi(\boldsymbol{\theta}_0, \boldsymbol{\psi}|\boldsymbol{X}_n)d\boldsymbol{\psi}}{\displaystyle\int \pi(\boldsymbol{\theta}_0, \boldsymbol{\psi})d\boldsymbol{\psi}},$$

where $\pi(\boldsymbol{\theta}_0, \boldsymbol{\psi}|\boldsymbol{X}_n)$ and $\pi(\boldsymbol{\theta}_0, \boldsymbol{\psi})$ are the posterior and prior densities of the parameters under the model M_1. Thus computing the Bayes factor reduces to the problem of estimating the marginal posterior density $\pi(\boldsymbol{\theta}_0|\boldsymbol{X}_n)$ at the point $\boldsymbol{\theta}_0$.

6.6.1 Example: Bayesian linear regression model

Consider the Bayesian linear regression model $\boldsymbol{y}_n = X_n\boldsymbol{\beta} + \boldsymbol{\varepsilon}_n$, $\boldsymbol{\varepsilon}_n \sim N(0, \sigma^2 I)$ in (2.3) with the following prior

$$\begin{aligned}
\pi(\boldsymbol{\beta}, \sigma^2) &= \pi(\boldsymbol{\beta})\pi(\sigma^2), \\
\pi(\boldsymbol{\beta}) &= N\left(\boldsymbol{\beta}_0, A^{-1}\right) = \frac{1}{(2\pi)^{p/2}}|A|^{1/2}\exp\left[-\frac{(\boldsymbol{\beta}-\boldsymbol{\beta}_0)^T A(\boldsymbol{\beta}-\boldsymbol{\beta}_0)}{2}\right], \\
\pi(\sigma^2) &= IG\left(\frac{\nu_0}{2}, \frac{\lambda_0}{2}\right) = \frac{\left(\frac{\lambda_0}{2}\right)^{\nu_0/2}}{\Gamma\left(\frac{\nu_0}{2}\right)}(\sigma^2)^{-\left(\frac{\nu_0}{2}+1\right)}\exp\left[-\frac{\lambda_0}{2\sigma^2}\right].
\end{aligned} \tag{6.12}$$

The joint posterior distribution is then

$$\begin{aligned}
\pi\left(\boldsymbol{\beta},\sigma^2\middle|\boldsymbol{y}_n,X_n\right) &\propto f\left(\boldsymbol{y}_n|X_n,\boldsymbol{\beta},\sigma^2\right)\pi(\boldsymbol{\beta},\sigma^2)\\
&\propto \frac{1}{(\sigma^2)^{n/2}}\exp\left[-\frac{(\boldsymbol{y}_n-X_n\boldsymbol{\beta})^T(\boldsymbol{y}_n-X_n\boldsymbol{\beta})}{2\sigma^2}\right]\\
&\quad\times\exp\left[-\frac{(\boldsymbol{\beta}-\boldsymbol{\beta}_0)^TA(\boldsymbol{\beta}-\boldsymbol{\beta}_0)}{2}\right]\\
&\quad\times\frac{1}{(\sigma^2)^{\nu_0/2+1}}\exp\left[-\frac{\lambda_0}{2\sigma^2}\right].
\end{aligned}$$

Using the follwoing identities,

$$\begin{aligned}
&\sigma^{-2}\left(\boldsymbol{y}_n-X_n\boldsymbol{\beta}\right)^T\left(\boldsymbol{y}_n-X_n\boldsymbol{\beta}\right)+(\boldsymbol{\beta}-\boldsymbol{\beta}_0)^TA(\boldsymbol{\beta}-\boldsymbol{\beta}_0)\\
&=\left(\boldsymbol{\beta}-\hat{\boldsymbol{\beta}}_n\right)^T\hat{A}_n^{-1}\left(\boldsymbol{\beta}-\hat{\boldsymbol{\beta}}_n\right)+R,
\end{aligned}$$

with

$$\begin{aligned}
\hat{\boldsymbol{\beta}}_n &= (\sigma^{-2}X_n^TX_n+A)^{-1}\left(\sigma^{-2}X_n^T\boldsymbol{y}_n+A\boldsymbol{\beta}_0\right),\\
\hat{A}_n &= (\sigma^{-2}X_n^TX_n+A)^{-1},
\end{aligned}$$

and

$$R=\sigma^{-2}\boldsymbol{y}_n^T\boldsymbol{y}_n+\boldsymbol{\beta}_0^TA\boldsymbol{\beta}_0-\hat{\boldsymbol{\beta}}_n^T\hat{A}_n^{-1}\hat{\boldsymbol{\beta}}_n$$

proves the term R does not contain $\boldsymbol{\beta}$. We thus obtain

$$\pi\left(\boldsymbol{\beta}\middle|\sigma^2,\boldsymbol{y}_n,X_n\right)\propto\exp\left[-\frac{\left(\boldsymbol{\beta}-\hat{\boldsymbol{\beta}}_n\right)^T\hat{A}_n^{-1}\left(\boldsymbol{\beta}-\hat{\boldsymbol{\beta}}_n\right)}{2}\right].$$

and find that the conditional posterior distribution of $\boldsymbol{\beta}$ is

$$\pi\left(\boldsymbol{\beta}\middle|\sigma^2,\boldsymbol{y}_n,X_n\right)=N\left(\hat{\boldsymbol{\beta}}_n,\hat{A}_n\right).$$

Modifying the joint posterior density $\pi\left(\boldsymbol{\beta},\sigma^2\middle|\boldsymbol{y}_n,X_n\right)$ with respect to σ^2, we have

$$\pi\left(\sigma^2\middle|\boldsymbol{\beta},\boldsymbol{y}_n,X_n\right)\propto\frac{1}{(\sigma^2)^{(n+\nu_0)/2+1}}\exp\left[-\frac{(\boldsymbol{y}_n-X_n\boldsymbol{\beta})^T(\boldsymbol{y}_n-X_n\boldsymbol{\beta})+\lambda_0}{2\sigma^2}\right].$$

We then find that the conditional posterior distribution of σ^2 given $\boldsymbol{\beta}$ is inverse-gamma distribution:

$$\pi\left(\sigma^2\middle|\boldsymbol{\beta},\boldsymbol{y}_n,X_n\right)=IG\left(\frac{\hat{\nu}_n}{2},\frac{\hat{\lambda}_n}{2}\right)$$

with

$$\hat{\nu}_n = \nu_0 + n,$$
$$\hat{\lambda}_n = (\boldsymbol{y}_n - X_n\boldsymbol{\beta})^T(\boldsymbol{y}_n - X_n\boldsymbol{\beta}) + \lambda_0.$$

In contrast to the previous normal and inverse-gamma prior case, the posterior distributions $\boldsymbol{\beta}$ and σ^2 are depending on each other:

$$\pi\left(\boldsymbol{\beta}\middle|\sigma^2, \boldsymbol{y}_n, X_n\right) = N\left(\hat{\boldsymbol{\beta}}_n, \hat{A}_n\right), \quad \pi\left(\sigma^2\middle|\boldsymbol{\beta}, \boldsymbol{y}_n, X_n\right) = IG\left(\frac{\hat{\nu}_n}{2}, \frac{\hat{\lambda}_n}{2}\right).$$

The posterior inference can be done by Gibb's sampling approach.

Let us consider the two models. M_1: the model $f(X_n|\boldsymbol{\beta}, \sigma^2)$ with the prior $\pi(\boldsymbol{\beta}, \sigma^2)$ in (6.12), and M_2: In the model M_1, the parameter value $\boldsymbol{\beta}$ is fixed at $\boldsymbol{\beta} = \boldsymbol{\beta}^*$ both in the model $f(\boldsymbol{y}_n|X_n, \boldsymbol{\beta}^*, \sigma^2)$ and the prior $\pi(\boldsymbol{\beta}^*, \sigma^2)$. From the Savage-Dickey density ratio, the Bayes factor B_{21} for two models is expressed as

$$\begin{aligned} B_{21} &= \frac{\int f(\boldsymbol{y}_n|X_n, \boldsymbol{\beta}^*, \sigma^2)\pi(\boldsymbol{\beta}^*, \sigma^2)d\sigma^2}{\int f(\boldsymbol{y}_n|X_n, \boldsymbol{\beta}, \sigma^2)\pi(\boldsymbol{\beta}, \sigma^2)d\boldsymbol{\beta}d\sigma^2} \\ &= \frac{\int \pi(\boldsymbol{\beta}^*, \sigma^2|\boldsymbol{y}_n, X_n)d\sigma^2}{\int \pi(\boldsymbol{\beta}^*, \sigma^2)d\sigma^2}, \end{aligned}$$

where $\pi(\boldsymbol{\beta}, \sigma^2|X_n)$ and $\pi(\boldsymbol{\beta}, \sigma^2)$ are the posterior and prior densities of the parameters under the model M_1.

To compute the Bayes factor, we have to evaluate the numerator and the denominator in B_{12}. We can easily evaluate the numerator.

$$\begin{aligned} \int \pi(\boldsymbol{\beta}^*, \sigma^2)d\sigma^2 &= \pi(\boldsymbol{\beta}^*)\int \pi(\sigma^2)d\sigma^2 \\ &= \frac{1}{(2\pi)^{p/2}}|A|^{1/2}\exp\left[-\frac{(\boldsymbol{\beta}^* - \boldsymbol{\beta}_0)^T A(\boldsymbol{\beta}^* - \boldsymbol{\beta}_0)}{2}\right]. \end{aligned}$$

To calculate the denominator, we can use the generated posterior samples $\sigma^{2(j)}$, $j = 1, ..., L$. The denominator is estimated by

$$\frac{1}{L}\sum_{j=1}^{L}\pi\left(\boldsymbol{\beta}^*\middle|\sigma^{2(j)}, \boldsymbol{y}_n, X_n\right) \to \int \pi(\boldsymbol{\beta}^*, \sigma^2|X_n)d\sigma^2.$$

Noting that the Bayes factor can be written as the product of a quantity called the Savage-Dickey density ratio and a correction factor, Verdinelalni and Wasserman (1995) proposed an alternative method for computing Bayes factors. Verdinelalni and Wasserman (1995) obtained a generalized version of the Savage-Dickey density ratio.

6.7 Kernel density approach

By rearranging the definition of the posterior distribution, Chib (1995) evaluated the log of the marginal likelihood as follows:

$$\log P(\boldsymbol{X}_n) = \log f(\boldsymbol{X}_n|\boldsymbol{\theta}^*) + \log \pi(\boldsymbol{\theta}^*) - \log \pi(\boldsymbol{\theta}^*|\boldsymbol{X}_n), \tag{6.13}$$

for any values of $\boldsymbol{\theta}^*$. The first term and the second term on the right hand side of equation can be evaluated easily. As an alternative of the Chib's estimator from Gibb's and MH sampling for sampling and marginal likelihood estimation, the third term can be evaluated by using a multivariate kernel density estimate based on the posterior sample (Kim et al. (1998), Berg et al. (2004), and Ando (2006)).

Estimating the posterior density by the kernel density estimate

$$\hat{\pi}(\boldsymbol{\theta}|\boldsymbol{X}_n) = \frac{1}{L}\sum_{j=1}^{L} K\left(\frac{||\boldsymbol{\theta} - \boldsymbol{\theta}^{(j)}||^2}{\sigma^2}\right),$$

one can estimate the marginal likelihood as

$$\log P(\boldsymbol{X}_n) = \log f(\boldsymbol{X}_n|\boldsymbol{\theta}^*) + \log \pi(\boldsymbol{\theta}^*) - \log \hat{\pi}(\boldsymbol{\theta}^*|\boldsymbol{X}_n).$$

Here $\boldsymbol{\theta}^{(j)}$, $j = 1, ..., L$ is a set of posterior samples, $||\cdot||^2$ is the norm, $K(\cdot)$ is the standardized kernel function with $\int K(x)dx = 1$. The parameter σ^2 adjusts the variance of $\boldsymbol{\theta}$ in the kernel function. This procedure is employed in the evaluation of the goodness of the SV models (Kim et al. (1998), Berg et al. (2004) and Ando (2006)).

6.7.1 Example: Bayesian analysis of the probit model

A revision of the Basel Accord (Basel II; Basel Committee on Banking Supervision, 2004) to achieve stability in international financial systems has attracted much attention in recent years. Basel II involves three principles to strengthen the soundness of the international banking system: (i) minimum capital requirements; (ii) supervisory review; and (iii) market discipline. Maintaining the original Basel Accord requirement that banks restrict their capital to a risk-weighted asset ratio of at least 8%, the first principle allows banks to allocate their capital by considering credit risk, operational risk and market risk. This paper focuses only on credit risk in this context.

Under approval of bank supervisors, the revision allows banks to determine the credit risk of each borrower. This method is called the internal ratings-based (IRB) system. Under the IRB approach, a bank estimates the creditworthiness of its obligors and translates this into estimates of potential future losses in combination with other credit risk factors (Basel Committee on Banking Supervision, 2004). With the environmental changes that followed revision of the Basel Capital Accord and the introduction of financial internationalization, banks have realized the need to evaluate various types of credit risk factors.

Credit risk is commonly defined as the loss resulting from obligor default on payments. There are various types of credit risk factors, such as probability of default (PD), loss given default, exposure at default, maturity, default correlations, fluctuation of exposure, and so on. Although the first four factors are key parameters in the IRB approach, PD is a central premise of credit risk modeling.

One of the most commonly used approaches is the probit model.

$$f(\boldsymbol{y}_n|X_n, \boldsymbol{\beta}) = \prod_{\alpha=1}^{n} \Phi\left(\boldsymbol{x}_\alpha^T\boldsymbol{\beta}\right)^{y_\alpha}\left[1 - \Phi\left(\boldsymbol{x}_\alpha^T\boldsymbol{\beta}\right)\right]^{1-y_\alpha},$$

where $\boldsymbol{\beta}$ is the p-dimensional parameter vector, $\boldsymbol{y}_n = \{y_1, .., y_n\}$ contains information on default ($y_\alpha = 1$: default, and $y_\alpha = 0$: non-default), $\boldsymbol{X}_n = \{\boldsymbol{x}_1, .., \boldsymbol{x}_n\}$ contains information on accounting variables, changes in economic conditions and stock prices that affect the default probability, $\Phi(\cdot)$ is the distribution function of the standard normal, y_α takes values 0 or 1, and $\boldsymbol{x}_\alpha$ is the p-dimensional predictors. Here we shall use the three predictors, return on asset (ROA), log(sales), and cash flow-sales ratio. A matrix of scatterplots is given in Figure 6.2.

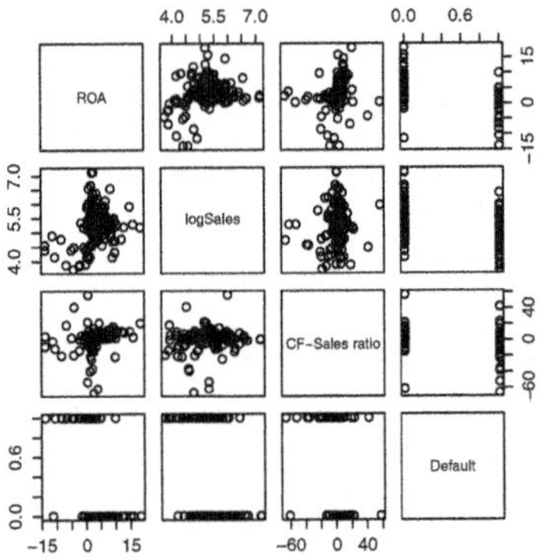

FIGURE 6.2: A matrix of scatterplots of default data.

For the prior density of $\boldsymbol{\beta}$, we use the normal distribution with prior mean $\boldsymbol{\beta}_0$, and the prior covariance matrix A.

$$\pi(\boldsymbol{\beta}) = \frac{1}{(2\pi)^{p/2}} |A|^{1/2} \exp\left[-\frac{1}{2}(\boldsymbol{\beta} - \boldsymbol{\beta}_0)^T A(\boldsymbol{\beta} - \boldsymbol{\beta}_0)\right].$$

Posterior sampling can be done by using the **R** function **MCMCoprobit**. Using the **MCMCoprobit**, we generated a set of 1,000 posterior samples. For the prior density of $\boldsymbol{\beta}$, we use the normal distribution with prior mean $\boldsymbol{\beta}_0 = \mathbf{0}$, and the prior covariance matrix $10,000I$. Using the generated posterior samples $\boldsymbol{\beta}^{(j)}$, $j = 1, ..., L$, we obtain the posterior mean $\bar{\boldsymbol{\beta}}_n = \sum_{j=1}^{L} \boldsymbol{\beta}^{(j)}/L$. Using the posterior mean, the first term and the second term on the right hand side of Equation (6.13) can be evaluated easily. Using the Gaussian kernel, the third term on the right hand side of Equation (6.13) can be estimated by the kernel density estimate

$$\hat{\pi}(\bar{\boldsymbol{\beta}}_n|\boldsymbol{y}_n) = \frac{1}{L}\sum_{j=1}^{L} \frac{1}{(2\pi)^{p/2}|\Sigma_\beta|^{1/2}} \exp\left[-\frac{1}{2}(\bar{\boldsymbol{\beta}}_n - \boldsymbol{\beta}_0)^T \Sigma_\beta^{-1} (\bar{\boldsymbol{\beta}}_n - \boldsymbol{\beta}_0)\right],$$

where Σ_β is the estimated posterior covariance matrix of $\boldsymbol{\beta}$. For the kernel density estimation, we refer to Silverman (1986) and Sheather and Jones (1991).

Thus, the log-marginal likelihood is estimated by

$$\log P(\boldsymbol{X}_n) = f(\boldsymbol{y}_n|X_n, \bar{\boldsymbol{\beta}}_n) + \log \pi(\bar{\boldsymbol{\beta}}_n) - \log \hat{\pi}(\bar{\boldsymbol{\beta}}_n|\boldsymbol{y}_n).$$

Putting the numbers $f(\boldsymbol{y}_n|X_n, \bar{\boldsymbol{\beta}}_n) = -70.67$, $\log \pi(\bar{\boldsymbol{\beta}}_n) = 1.53$ and $\log \hat{\pi}(\bar{\boldsymbol{\beta}}_n|\boldsymbol{y}_n) = 13.67$, the log-marginal likelihood is $\log P(\boldsymbol{X}_n) = -82.81$.

6.8 Direct computation of the posterior model probabilities

In the previous sections, we provided several approaches for computing the marginal likelihood. Once we obtain an estimator of the marginal likelihood, the best model is selected as the maximizer of the posterior model probability. However, the marginal likelihood methods seem impractical if the number of candidate models is very large. For example, consider the variable selection problems with p variables. Having 2^p possible models, corresponding to each of p predictors being either included or excluded, we cannot compute the marginal likelihood for every model when the dimension p is large. This section describes the reversible jump MCMC algorithm as Green (1995) and the product space search (Carlin and Chib (1995)) for computing the posterior model probabilities of each model directly.

6.8.1 Reversible jump MCMC

We overview the reversible jump Markov chain Monte Carlo (Green (1995)) algorithm, which allow the Markov chain to move between different models. Although the reversible jump Markov chain Monte Carlo approaches require appropriate jumping rules, we can directly compute the posterior model probabilities of each model. Thus, we can avoid computing the marginal likelihood.

Suppose we are interested in comparing a set of models $\{M_1, ..., M_r\}$. The reversible jump MCMC method samples over the model and parameter space $(M_k, \boldsymbol{\theta}_k)$, $k = 1, ..., r$ by generating a Markov chain that can jump between models with parameter spaces of different dimensions. It is necessary to design the reversible jump MCMC algorithm to be reversible so as to maintain detailed balance of a irreducible and a periodic chain. Details of the general methodology and ideas can be found in Green (1995).

Let the current state of the Markov chain be $(M_k, \boldsymbol{\theta}_k)$ where $\boldsymbol{\theta}_k$ is of dimension p_k, then one possible version of the reversible jump MCMC algorithm proceeds as follows:

Step 1. Propose a new model M_j with probability $p(k, j) \equiv p(M_k \to M_j)$.
Step 2. Generate $\boldsymbol{u}$ from a proposal density $p_{kj}(\boldsymbol{u}_k|\boldsymbol{\theta}_k)$.
Step 3. Set $(\boldsymbol{\theta}_j, \boldsymbol{u}_j) = g_{kj}(\boldsymbol{\theta}_k, \boldsymbol{u}_k)$, where $g_{kj}(\cdot, \cdot)$ is a predetermined function that makes one to one mapping between $(\boldsymbol{\theta}_j, \boldsymbol{u}_j)$ and $(\boldsymbol{\theta}_k, \boldsymbol{u}_k)$. This function is needed for dimension matching so that $p_k + \dim(\boldsymbol{u}_k) = p_j + \dim(\boldsymbol{u}_j)$.
Step 4. Accept the proposed move (from $M_k \to M_j$) with probability

$$\begin{aligned} \alpha &= \min\left\{1, \text{Likelihood ratio} \times \text{Prior ratio} \times \text{Proposal ratio}\right\} \\ &= \min\left\{1, \frac{f_k(\boldsymbol{X}_n|\boldsymbol{\theta}_k)}{f_j(\boldsymbol{X}_n|\boldsymbol{\theta}_j)} \times \frac{\pi_k(\boldsymbol{\theta}_k)P(M_k)}{\pi_j(\boldsymbol{\theta}_j)P(M_j)} \times \frac{p(k,j)p_{kj}(\boldsymbol{u}_k|\boldsymbol{\theta}_k)}{p(j,k)p_{jk}(\boldsymbol{u}_j|\boldsymbol{\theta}_j)} \left| \frac{\partial g_{jk}(\boldsymbol{\theta}_j, \boldsymbol{u}_j)}{\partial(\boldsymbol{\theta}_j, \boldsymbol{u}_j)} \right| \right\}, \end{aligned}$$

where $P(M_k)$ is the prior model probability of the model M_k.

When the proposed model equals the current model, the reversible jump MCMC algorithm corresponds to the traditional Metropolis-Hastings algorithm. Generating the Markov chain samples for a long time, the posterior model probability is estimated as

$$P(M_k|\boldsymbol{X}_n) \approx \frac{1}{L} \sum_{j=1}^{L} \mathbf{1}(M^{(j)} = M_k),$$

where L is the number of samples, and $\mathbf{1}(M^{(j)} = M_k) = 1$ if $M^{(j)} = M_k$ and zero otherwise.

6.8.2 Example: Reversible jump MCMC for seemingly unrelated regression model with informative prior

In Section 6.3.1, we consider Gibb's algorithm for the m-system SUR model with informative prior, conditioned on the design matrix X_n. With the use of normal and the inverse Wishart priors for $\boldsymbol{\beta}$ and Σ, $\pi_2(\boldsymbol{\beta}, \Sigma)$, we obtain the analytical conditional posterior densities of $\boldsymbol{\beta}$ and Σ as $\pi_2(\boldsymbol{\beta}|\boldsymbol{Y}_n, X_n, \Sigma)$ and $\pi_2(\Sigma|\boldsymbol{Y}_n, X_n, \boldsymbol{\beta})$. In some situations, we want to select a subset of the design matrix X_n, denoted as $X_{n(k)}$, where k is the k-th design matrix specification. When the number of columns of the design X_n is large, it is very difficult to calculate the marginal likelihood for all possible specifications of $X_{n(k)}$. Moreover, when we additionally consider arbitrary transformations of the original explanatory variables as a column of the design matrix, this calculation problem becomes more difficult. In such a case, we can consider a reversible jump MCMC algorithm.

Reversible jump MCMC algorithm

Holmes et al. (2002) proposed a reversible jump MCMC algorithm for Bayesian inference on the SUR models with informative prior, discussed in Section 6.3.1. Initializing the parameter values $\boldsymbol{\beta}^{(0)}$ and $\Sigma^{(0)}$ and the design matrix $X_{n(k)}^{(0)}$, the basic sampling steps are as follows:

Step 1. Update the design matrix $X_{n(k)}^{(j)}$ using a Metropolis-Hastings algorithm described below.

Step 2. Update the coefficient vector $\boldsymbol{\beta}^{(j)}$ by drawing a new value from the conditional posterior density $\pi_2\left(\boldsymbol{\beta}|\boldsymbol{Y}_n, X_{n(k)}^{(j)}, \Sigma\right)$.

Step 3. Update Σ by drawing a new value from the conditional posterior density $\pi_2\left(\Sigma|\boldsymbol{Y}_n, \boldsymbol{\beta}^{(j)}, X_{n(k)}^{(j)}\right)$ given below.

The process is then repeated a large number of times $j = 1, ...,$. An initial number of generates is discarded as being unrepresentative of the posterior distribution. The remaining samples are then used for the posterior inference. Steps 2 and 3 are straightforward. As a sampling step of updating $X_{n(k)}$, we can use a reversible step that attempts one of three move proposal types:

Proposal 1. Add a predictor (column) to $X_{n(k)}$: Select one of the m regression systems at random and then add a potential predictor associated with that regression that is not currently in the selected system.

Proposal 2. Remove a predictor (column) to $X_{n(k)}$: Select one of the m regression systems at random and then choose to remove one of the predictors associated with the selected system.

Proposal 3. Alter a predictor in $X_{n(k)}$: Select one of m regression systems at random and then swap a current predictor in the regression with a potential predictor that is currently not in the selected system. Thus, the dimension of the design matrix $X_{n(k)}$ is unchanged for this move.

Under the equal prior model probabilities $P(M_j)$, and the equal three move proposal probabilities $p(j, k)$, the acceptance probability is given as

$$\begin{aligned}\alpha &= \min\left\{1, \text{Likelihood ratio} \times \text{Prior ratio} \times \text{Proposal ratio}\right\} \\ &= \min\left\{1, \frac{f(\boldsymbol{Y}_n|X_{n(j)}, \Sigma_j, M_j)}{f(\boldsymbol{Y}_n|X_{n(k)}, \Sigma_k, M_k)} \times 1 \times \text{Proposal ratio}\right\}.\end{aligned}$$

with the likelihood ratio is

$$\frac{f(\boldsymbol{Y}_n|X_{n(j)}, \Sigma, M_j)}{f(\boldsymbol{Y}_n|X_{n(k)}, \Sigma, M_k)} = \frac{|A_j|^{1/2}|\bar{\Omega}_j|^{-1/2}\exp(-b_j/2)}{|A_k|^{1/2}|\bar{\Omega}_k|^{-1/2}\exp(-b_k/2)}$$

with

$$\begin{aligned}\bar{\Omega}_j &= \left(X_{n(j)}^T\left(\Sigma^{-1}\otimes I\right)X_{n(j)} + A_j\right)^{-1}, \\ b_j &= \text{tr}\left\{\Sigma^{-1}\Lambda_{j0}\right\} + \boldsymbol{y}_n^T\left(\Sigma^{-1}\otimes I\right)\boldsymbol{y}_n - \bar{\boldsymbol{\beta}}_j^T\bar{\Omega}^{-1}\bar{\boldsymbol{\beta}}_j, \\ \bar{\boldsymbol{\beta}}_j &= \left\{X_{n(j)}^T\left(\Sigma^{-1}\otimes I\right)X_{n(j)} + A_j\right\}^{-1} X_{n(j)}^T\left(\Sigma^{-1}\otimes I\right)^{-1}\boldsymbol{y}_n,\end{aligned}$$

where A_j is the prior variance matrix of the coefficient vector, and Λ_{j0} is hyperparameter to be specified in the prior for Σ. See also Holmes et al. (2002), the discussion about the above proposal steps for updating $X_{n(k)}$.

There is extensive literature for reversible jump MCMC techniques for nonlinear regression models, piecewise polynomials, smoothing splines and so on. With respect to the above two nonparametric regression models, we refer to Hastie and Tibshirani (1990) and Green and Silverman (1994). To select the number and positions of the knots of splines, many studies are carried out by using reversible jump MCMC. Denison et al. (2002) provides a general discussion of the Bayesian curve-fitting with free-knot procedure. Smith and Kohn (1996) developed a reversible jump MCMC for additive regression models. See also Denison et al. (1998), DiMatteo et al. (2001), Holmes and Mallick (2003), Mallick (1998), Liang et al. (2001) that developed reversible jump MCMC for various types of regression models.

6.8.3 Product space search and metropolized product space search

In this section, we describe the product space search method (Carlin and Chib (1995)). Let $\boldsymbol{\theta}^T = (\boldsymbol{\theta}_1^T, ..., \boldsymbol{\theta}_r^T)$ be the vector containing the parameters of all competing models $\{M_1, ..., M_r\}$. In order to use a Gibb's sampler to generate the posterior samples $\{M_k, \boldsymbol{\theta}\}$ from the joint posterior distribution, Carlin and Chib (1995) proposed to use the pseudo priors or linking densities $\pi(\boldsymbol{\theta}_j|M_k, k \neq j)$ for a case that the current model in the MCMC is not M_j. Under the conditional independence assumption, the joint density of the data

$\boldsymbol{x}_n$, $\boldsymbol{\theta}$ and the model M_k is

$$f(\boldsymbol{X}_n, \boldsymbol{\theta}, M_j) = f(\boldsymbol{X}_n|\boldsymbol{\theta}_j, M_j) \prod_{i=1}^{r} \pi(\boldsymbol{\theta}_i|M_j) P(M_j)$$

which implies

$$\begin{aligned} P(\boldsymbol{X}_n|M_j) &= \int f(\boldsymbol{X}_n|\boldsymbol{\theta}, M_j)\pi(\boldsymbol{\theta}|M_j)d\boldsymbol{\theta} \\ &= \int f(\boldsymbol{X}_n|\boldsymbol{\theta}_j, M_j)\pi(\boldsymbol{\theta}_j|M_j)d\boldsymbol{\theta}_j. \end{aligned}$$

Then, a Gibb's sampler is defined over the product space by the full conditional distributions

$$\pi(\boldsymbol{\theta}_j|\boldsymbol{\theta}_{-j}, \boldsymbol{X}_n, M_k) \propto \begin{cases} f(\boldsymbol{X}_n|\boldsymbol{\theta}_j, M_j)\pi_j(\boldsymbol{\theta}_j|M_j), & k = j \\ \pi(\boldsymbol{\theta}_j|M_k, k \neq j), & k \neq j \end{cases},$$

where $\boldsymbol{\theta}_{-j} = (\boldsymbol{\theta}_1^T, ..., \boldsymbol{\theta}_{j-1}^T, \boldsymbol{\theta}_{j+1}^T, ..., \boldsymbol{\theta}_r^T)^T$ and $\pi(\boldsymbol{\theta}_j|M_k, k \neq j)$ is a pseudo prior. When the current model is M_j, we generate the sample $\boldsymbol{\theta}_j$ from the full conditional density. If the current model is M_k $(k \neq j)$, the sample $\boldsymbol{\theta}_k$ is generated from the the pseudo prior $\pi(\boldsymbol{\theta}_j|M_k, k \neq j)$.

For the conditional posterior of the model M, we have

$$P(M_j|\boldsymbol{\theta}, \boldsymbol{X}_n) = \frac{f(\boldsymbol{X}_n|\boldsymbol{\theta}_j, M_j)\left[\prod_{i=1}^{r} \pi(\boldsymbol{\theta}_i|, M_j)\right] P(M_j)}{\sum_{k=1}^{r} f(\boldsymbol{X}_n|\boldsymbol{\theta}_k,, M_k)\left[\prod_{i=1}^{r} \pi(\boldsymbol{\theta}_i|M_k)\right] P(M_k)}.$$

Under the usual regularity conditions, this Gibb's sampling algorithm will produce posterior samples from the joint posterior distribution. Using the estimated posterior probability of model M_k,

$$\hat{P}(M_k|\boldsymbol{X}_n) = \frac{1}{L}\sum_{j=1}^{L} \mathbf{1}\left(M^{(j)} = M_k\right),$$

we can estimate a Bayes factor as

$$\text{Bayes factor}(M_k, M_j) \approx \frac{\hat{P}(M_k|\boldsymbol{X}_n)/\hat{P}(M_j|\boldsymbol{X}_n)}{P(M_k)/P(M_j)}.$$

Since this approach involves generating directly from the pseudo priors, we might need to optimize it so that each pseudo prior density is close to the corresponding conditional posterior density (Dellaportas et al. (2002)). In fact, Carlin and Chib (1995) reported that the sampling efficiency of this approach is optimized when we take the pseudo priors as nearly as possible to the corresponding model specific posteriors.

A main operational drawback of Carlin and Chib (1995)'s Gibb's sampler is the need of evaluating and drawing made from each pseudo prior at every iteration (Han and Carlin 2001). To overcome this problem Dellaportas et al. (2002) proposed "Metropolizing" Carlin and Chib (1995)'s Gibb's sampler. In their approach, the model selection step is based on a proposal for a move from model M_j to M_k, and then accepting/rejecting a move to a new model. If the current state of the Markov chain is at $\{M_k, \boldsymbol{\theta}_k\}$, then the Metropolized Carlin and Chib approach proceeds as follows:

Step 1. Propose a new model M_j with probability $p(k,j) = p(M_k \to M_j)$.
Step 2. Generate $\boldsymbol{\theta}_k$ from a prior density $\pi_k(\boldsymbol{\theta}_k)$.
Step 3. Generate $\boldsymbol{\theta}_j$ from a pseudo prior density $\pi(\boldsymbol{\theta}_j|M_k, k \neq j)$.
Step 4. Accept the proposed move (from $M_k \to M_j$) with probability

$$\alpha = \min\left\{1, \frac{f(\boldsymbol{X}_n|\boldsymbol{\theta}_k, M_k)}{f(\boldsymbol{X}_n|\boldsymbol{\theta}_j, M_j)} \times \frac{\pi(\boldsymbol{\theta}_k|M_k)\pi(\boldsymbol{\theta}_j|M_k, k \neq j)P(M_k)}{\pi(\boldsymbol{\theta}_j|M_j)\pi(\boldsymbol{\theta}_k|M_j, j \neq k)P(M_j)} \times \frac{p(k,j)}{p(j,k)}\right\},$$

where $P(M_k)$ is the prior model probability of the model M_k. The process is then repeated a large number of times. By "Metropolizing" the model selection step, we just need to sample only from the pseudo prior for the proposed model M_j.

Comprehensive review of current approaches that take advantage of modern Bayesian computational approaches is provided by Han and Carlin (2001). They compared several methods in the context of three examples: a simple regression example, a more challenging hierarchical longitudinal model, and a binary data latent variable model. As a result, they found that the joint model-parameter space search approaches perform adequately However, they also reported that the joint model-parameter space search methods might be difficult to program and tune, whereas the marginal likelihood computing methods are more user friendly. Also, Han and Carlin (2001) reported a doubt about the ability of joint model and parameter space search methods to sample effectively over a large space (Clyde et al. (1996)).

6.8.4 Bayesian variable selection for large model space

Bayesian variable selection problem has been attracting many researchers in recent years. In this subsection, the variable selection problem is considered (Smith and Kohn (1996)). When the number of predictors p is very large, it is impossible to calculate the model selection scores over this model space. Therefore, rather than fixing the number of predictors in regression models, prior distributions over each predictor are assigned. As pointed out in Lee et al. (2003), this approach is more efficient than most other existing variable selection algorithms. Let $\boldsymbol{\gamma} = (\gamma_1, ..., \gamma_p)^T$ is p-dimensional vector of indicator variables with j-th element $\gamma_j = 1$ if the variable x_j is included and $\gamma_j = 0$

otherwise. Then we consider the probit model

$$f(\boldsymbol{y}_n|X_{\gamma n}, \boldsymbol{\beta}_\gamma) = \prod_{\alpha=1}^{n} \Phi\left(\boldsymbol{x}_{\gamma\alpha}^T \boldsymbol{\beta}_\gamma\right)^{y_\alpha} \left[1 - \Phi\left(\boldsymbol{x}_{\gamma\alpha}^T \boldsymbol{\beta}_\gamma\right)\right]^{1-y_\alpha},$$

where $X_{\gamma n}$ is the subset of columns of $X_n = (\boldsymbol{x}_1, ..., \boldsymbol{x}_n)^T$ corresponding to those elements of $\boldsymbol{\gamma}$ equal to $\gamma_j = 1$. Following Albert and Chib (1993), we can introduce n independent latent variables $\boldsymbol{z}_n = (z_1, ..., z_n)^T$ with

$$z_\alpha = \boldsymbol{x}_{\gamma\alpha}^T \boldsymbol{\beta}_\gamma + \varepsilon_\alpha, \quad \varepsilon_\alpha \sim N(0, 1),$$

such that

$$y_\alpha = \begin{cases} 1, & (z_\alpha \geq 0) \\ 0, & (z_\alpha < 0) \end{cases}.$$

To complete the hierarchical model, Lee et al. (2003) used the following specifications. Given $\boldsymbol{\gamma}$, the Zellner's g-prior is used for $\boldsymbol{\beta}_\gamma$ with $\boldsymbol{\beta}_\gamma \sim N(\boldsymbol{0}, g(X_{\gamma n}^T X_{\gamma n})^{-1})$. Lee et al. (2003) fixed the value of $g = 100$ so that the prior density contains very little information about $\boldsymbol{\beta}_\gamma$ compared to the likelihood (see also Smith and Kohn (1996)). The indicator variables γ_j are assumed to be a priori independent with $\Pr(\gamma_j = 1) = \pi_j$ for $j = 1, ..., p$. Lee et al. (2003) assigned the small values for π_j, while one can assign a prior on π_j for e.g., uniform distribution. Bayesian inference on the unknown parameters $\boldsymbol{\beta}_\gamma$, $\boldsymbol{\gamma}$ and $\boldsymbol{z}_n$ can be done by implementing a Gibb's sampling.

The conditional posterior distribution of $\boldsymbol{\beta}_\gamma$, given $\boldsymbol{z}_n$ and $\boldsymbol{\gamma}$, is

$$\pi(\boldsymbol{\beta}_\gamma|\boldsymbol{y}_n, X_n, \boldsymbol{z}_n, \boldsymbol{\gamma}) = N\left(\frac{g}{g+1}\left(X_{\gamma n}^T X_{\gamma n}\right)^{-1} X_{\gamma n}^T \boldsymbol{z}_n, \frac{g}{g+1}\left(X_{\gamma n}^T X_{\gamma n}\right)^{-1}\right),$$

and that of z_α, given $\boldsymbol{\beta}_\gamma$ and $\boldsymbol{\gamma}$, is

$$\pi(z_\alpha|y_\alpha, \boldsymbol{x}_{\gamma\alpha}, \boldsymbol{\beta}_\gamma, \boldsymbol{\gamma}) = \begin{cases} TN\left(\boldsymbol{x}_{\gamma\alpha}^T \boldsymbol{\beta}_\gamma, 1, +\right), & (y_\alpha = 1) \\ TN\left(\boldsymbol{x}_{\gamma\alpha}^T \boldsymbol{\beta}_\gamma, 1, -\right), & (y_\alpha = 0) \end{cases},$$

where TN is the truncated normal distribution.

Noting that

$$\begin{aligned} &\pi(\boldsymbol{z}_n|\boldsymbol{y}_n, X_n, \boldsymbol{\gamma}) \\ &\propto \int \pi(\boldsymbol{z}_n|\boldsymbol{\beta}_\gamma, \boldsymbol{\gamma})\pi(\boldsymbol{\beta}_\gamma|\boldsymbol{\gamma})d\boldsymbol{\beta}_\gamma \\ &\propto \exp\left[-\frac{1}{2}\left\{\boldsymbol{z}_n^T \boldsymbol{z}_n - \frac{g}{g+1}\boldsymbol{z}_n^T X_{\gamma n}\left(X_{\gamma n}^T X_{\gamma n}\right)^{-1} X_{\gamma n}^T \boldsymbol{z}_n\right\}\right], \end{aligned}$$

the conditional distribution of $\boldsymbol{\gamma}$, given $\boldsymbol{z}_n$, is

$$\begin{aligned}
&\pi(\boldsymbol{\gamma}|\boldsymbol{z}_n, \boldsymbol{y}_n, X_n) \\
&\propto \pi(\boldsymbol{z}_n|\boldsymbol{\gamma}, \boldsymbol{y}_n, X_n)\pi(\boldsymbol{\gamma}) \\
&\propto \exp\left[-\frac{1}{2}\left\{\boldsymbol{z}_n^T\boldsymbol{z}_n - \frac{g}{g+1}\boldsymbol{z}_n^T X_{\gamma n}\left(X_{\gamma n}^T X_{\gamma n}\right)^{-1} X_{\gamma n}^T \boldsymbol{z}_n\right\}\right] \\
&\quad \times \prod_{j=1}^{p} \pi^{\gamma_j}(1-\pi)^{1-\gamma_j}.
\end{aligned}$$

Thus the conditional posterior of γ_j, given $\boldsymbol{\gamma}_{-j}$ and $\boldsymbol{z}_n$, is

$$\begin{aligned}
&\pi(\gamma_j|\boldsymbol{\gamma}_{-j}, \boldsymbol{z}_n, \boldsymbol{y}_n, X_n) \\
&\propto \pi(\boldsymbol{z}_n|\boldsymbol{\gamma}, \boldsymbol{y}_n, X_n)\pi(\gamma_j) \\
&\propto \exp\left[-\frac{1}{2}\left\{\boldsymbol{z}_n^T\boldsymbol{z}_n - \frac{g}{g+1}\boldsymbol{z}_n^T X_{\gamma n}\left(X_{\gamma n}^T X_{\gamma n}\right)^{-1} X_{\gamma n}^T \boldsymbol{z}_n\right\}\right] \pi^{\gamma_j}(1-\pi)^{1-\gamma_j}.
\end{aligned}$$

Based on the above analysis, Lee et al. (2003) proposed the Gibb's sampling algorithm as follows.

Gibb's sampling

Step 1. Start with initial values $\boldsymbol{\gamma}$, $\boldsymbol{z}_n$, $\boldsymbol{\beta}_\gamma$.

Step 2. Update γ_j by drawing a new value from $\pi(\gamma_j|\boldsymbol{\gamma}_{-j}, \boldsymbol{z}_n, \boldsymbol{y}_n, X_n)$ for $j = 1, ..., p$.

Step 2. Update $\boldsymbol{z}_n$ by drawing a new value from $\pi(\boldsymbol{z}_n|\boldsymbol{y}_n, X_n, \boldsymbol{\gamma})$.

Step 3. Update $\boldsymbol{\beta}_\gamma$ by drawing a new value from the conditional posterior density $\pi(\boldsymbol{\beta}_\gamma|\boldsymbol{y}_n, X_n, \boldsymbol{z}_n, \boldsymbol{\gamma})$.

After suitable burn-in period, we obtain the posterior distributions for posterior inference and prediction. An extension to multi-class classification done by Sha (2004).

There are substantial studies on Bayesian variable selection. For e.g., George and McCulloch (1993) proposed Stochastic Search Variable Selection (SVSS) approach. See also Hall et al. (2001), George and McCulloch (1997), Clyde et al. (1996), Geweke (1996), Mitchell and Beauchamp (1988), Phillips and Smith (1995), and Smith and Kohn (1996).

Exercises

1. *In Section 2.7, Bayesian inference on linear regression models is provided. Then Section 4.6.1 provided the direct Monte Carlo method for generating a set of posterior samples of $\boldsymbol{\beta}$ and σ^2. Setting the number of posterior samples to be $L = \{10, 100, 1000, 10000, 100000\}$, report the*

marginal likelihood estimate based on the harmonic mean estimator. The harmonic mean estimator is given as

$$P(\boldsymbol{y}_n) = \frac{1}{\dfrac{1}{L}\displaystyle\sum_{j=1}^{L} \dfrac{1}{f\left(\boldsymbol{y}_n|X_n, \boldsymbol{\beta}^{(j)}, \sigma^{2(j)}\right)}}.$$

Also, under the setting $L = \{10, 100, 1000, 10000, 100000\}$, *compare the true marginal likelihood value and report the results. Set* $A = 10^{-5} \times I_p$ *and* $a = b = 10^{-10}$, *which make the prior to be diffuse.*

2. *In Section 6.1.1, the marginal likelihood is estimated by Laplace-Metropolis approximation. As an alternative model, we can consider the multinomial probit model based only on the two predictors:* x_4 *income,* x_5 *age, respectively. Using the same procedure, estimate the marginal likelihood for this alternative model.*

3. *Under the Gaussian linear regression model*

$$\boldsymbol{y}_n = X_n\boldsymbol{\beta} + \boldsymbol{\varepsilon}_n, \quad \boldsymbol{\varepsilon}_n \sim N(0, \sigma^2 I),$$

with a conjugate normal inverse-gamma prior $\pi(\boldsymbol{\beta}, \sigma^2) = \pi(\boldsymbol{\beta}|\sigma^2)\pi(\sigma^2)$ *with* $\pi(\boldsymbol{\beta}|\sigma^2) = N\left(\boldsymbol{\beta}_0, \sigma^2 A^{-1}\right)$ *and* $\pi(\sigma^2) = IG(\nu_0/2, \lambda_0/2)$, *the marginal likelihood* $P(\boldsymbol{y}_n \big| X_n, M)$ *was obtained analytically as given in (2.5).*

Generate a set of $n = 100$ *samples from* $y_\alpha = 1.5x_{1\alpha} - 3.1x_{2\alpha} + 0.4x_{3\alpha} + \varepsilon_\alpha$ *where* $x_{j\alpha}$ *are uniformly distributed between* $[-2, 2]$ *and* $\varepsilon_\alpha \sim N(0, 0.5)$. *Setting* $A = 10^5 \times I_p$ *and* $a = b = 10^{-10}$, *which make the prior to be diffuse, generate the posterior samples of* $\boldsymbol{\beta}$ *and* σ^2 *by using the direct Monte Carlo algorithm. Then compare the marginal likelihood value based on the Laplace-Metropolis approximation and its exact value. Try various sample sizes* n, *and investigate the approximation error.*

4. *In Section 6.2.1, the marginal likelihood for the ordered probit model is estimated by the harmonic mean method. Generate a set of* $n = 200$ *observations as follows. The 2 dimensional predictors* $\boldsymbol{x}$ *follow the normal with mean* $\mathbf{0}$, *and unit covariance matrix* I. *Then generate* $z = 0.3 + 0.9x_1 - 1.5x_2 + \varepsilon$, *where* ε *follows the standard normal distribution. The choice variable* y *is then given as* $y = 1$ *if* $z < -2$, $y = 2$ *if* $-2 \le z < 1$, $y = 3$ *if* $1 \le z < 2.5$, $y = 4$ *if* $2.5 \le z$. *Then implement the Bayesian analysis of the ordered probit model and calculate the harmonic mean.* **R** *package* **MCMCpack** *that contains the function* **MCMCoprobit** *might be useful.*

5. *In example 6.6.1, consider the Bayesian linear regression model* $\boldsymbol{y}_n = X_n\boldsymbol{\beta} + \boldsymbol{\varepsilon}_n$, $\boldsymbol{\varepsilon}_n \sim N(0, \sigma^2 I)$ *based on the prior* $\pi(\boldsymbol{\beta}, \sigma^2) = \pi(\boldsymbol{\beta})\pi(\sigma^2)$ *with*

$\pi(\boldsymbol{\beta}) = N\left(\boldsymbol{\beta}_0, A^{-1}\right)$ *and* $\pi(\sigma^2) = IG(\nu_0/2, \lambda_0/2)$. *Since this prior specification does not allow us to evaluate the marginal likelihood analytically, we used the Savage-Dickey density ratio approach.*

Generate the dataset from the simulation model used in the problem set 2. Setting $A = 10^5 \times I_p$ *and* $a = b = 10^{-10}$, *generate the posterior samples of* $\boldsymbol{\beta}$ *and* σ^2 *by using the Gibb's sampling approach. Compute the marginal likelihood value based on Savage-Dickey density ratio approach.*

6. *Consider a generalized linear model*

$$f(y_\alpha|\boldsymbol{x}_\alpha; \xi_\alpha, \phi) = \exp\left\{\frac{y_\alpha \xi_\alpha - u(\xi_\alpha)}{\phi} + v(y_\alpha, \phi)\right\},$$

where the unknown predictors η_α *are approximated by a linear combination of basis functions*

$$\eta_\alpha = \sum_{k=1}^{m} w_k b_k(\boldsymbol{x}_\alpha) = \boldsymbol{w}^T \boldsymbol{b}(\boldsymbol{x}_\alpha), \qquad \alpha = 1, 2, ..., n,$$

where $b_k(\boldsymbol{x}_\alpha)$ *are a set of* m *basis functions. For the basis functions, we can use B-spline basis functions, wavelet basis functions, fouriere basis functions, kernel basis functions, etc.. Depending on the characteristics of basis function, we might need a constant term. In such a case* $b_1(\boldsymbol{x})$ *can be specified as* $b_1(\boldsymbol{x}) = 1$.

Consider a singular multivariate normal prior density

$$\pi(\boldsymbol{\theta}) = \left(\frac{n\lambda}{2\pi}\right)^{(m-d)/2} |R|_+^{1/2} \exp\left\{-n\lambda \frac{\boldsymbol{\theta}^T R \boldsymbol{\theta}}{2}\right\},$$

with $\boldsymbol{\theta} = (\boldsymbol{w}, \phi)$ *and the matrix* R *is given in (5.24).*

Implement the Metropolis-Hastings algorithm to generate a random draw of $\boldsymbol{\theta}$ *from the posterior density* $\pi(\boldsymbol{\theta}|\boldsymbol{Y}_n, X_n) \propto f(\boldsymbol{Y}_n|X_n, \boldsymbol{\theta})\pi(\boldsymbol{\theta})$, *where* $f(\boldsymbol{Y}_n|X_n, \boldsymbol{\theta})$ *is the likelihood function. Using the Metropolis-Hastings algorithm, we can generate a set of posterior samples from the posterior distribution* $\pi(\boldsymbol{\theta}|\boldsymbol{y}_n, X_n)$. *Then calculate the marginal likelihood by using the Laplace-Metropolis estimator (Lewis and Raftery (1997)), the harmonic mean estimator (Newton and Raftery (1994)), and Chib and Jeliazkov (2001) estimator from MH sampling.*

The Metropolis-Hastings algorithm is given as follows:

Step 1. Take an initial value of $\boldsymbol{\theta}^{(0)}$.

Step 2. Given $\boldsymbol{\theta}^{(j-1)}$, *draw a candidate value* $\boldsymbol{\theta}$ *from* $p(\boldsymbol{\theta}|\boldsymbol{\theta}^{(j-1)})$.

Step 3. Set $\boldsymbol{\theta}^{(t)} = \boldsymbol{\theta}$ *with probability* α *and set* $\boldsymbol{\theta}^{(j)} = \boldsymbol{\theta}^{(j-1)}$ *with probability* $1 - \alpha$

Step 4. Repeat Step 2 and Step 3 for $t = 1, 2, \cdots$.

The details follow. Let us define $p(\boldsymbol{\theta}|\boldsymbol{\theta}^*)$ *as the proposal density and the acceptance probability as*

$$\alpha = \min\left\{1, \frac{\pi(\boldsymbol{\theta}|\boldsymbol{Y}_n, X_n)p(\boldsymbol{\theta}^*|\boldsymbol{\theta})}{\pi(\boldsymbol{\theta}^*|\boldsymbol{Y}_n, X_n)p(\boldsymbol{\theta}|\boldsymbol{\theta}^*)}\right\},$$

if $\pi(\boldsymbol{\theta}^*|\boldsymbol{Y}_n, X_n)p(\boldsymbol{\theta}|\boldsymbol{\theta}^*) > 0$ *and* $\alpha = 1$ *otherwise.*

Common choices of the proposal density are the random walk sampler and the independence sampler (Tierney (1994)). For example, for the proposal density $p(\boldsymbol{\theta}|\boldsymbol{\theta}^*)$*, we can use the truncated multivariate Student-t distribution:*

$$p(\boldsymbol{\theta}|\boldsymbol{\mu}, \Sigma, \nu)$$
$$= \frac{\Gamma\left(\frac{\nu+p}{2}\right)}{\Gamma\left(\frac{\nu}{2}\right)(\pi\nu)^{\frac{p}{2}}}|\Sigma|^{-\frac{1}{2}}\left\{1 + \frac{1}{\nu}(\boldsymbol{\theta} - \boldsymbol{\theta}^*)'\Sigma^{-1}(\boldsymbol{\theta} - \boldsymbol{\theta}^*)\right\}^{-\frac{\nu+p}{2}} \times I(\phi > 0),$$

If we specify the mean value of the proposal density as $\boldsymbol{\mu} = \boldsymbol{\theta}^*$*, the current parameter value in the algorithm, the proposal density will become the random walk sampler. When we want to use the independence sampler, we can specify the mean value of the proposal density as the posterior mode* $\boldsymbol{\mu} = \hat{\boldsymbol{\theta}}_n$*, for example. We often specify the covariance matrix as* $\Sigma = n^{-1}S_n^{-1}(\hat{\boldsymbol{\theta}}_n)$ *with*

$$S_n(\hat{\boldsymbol{\theta}}_n) = -\frac{1}{n}\left.\frac{\partial^2 \log\{f(\boldsymbol{Y}_n|X_n, \boldsymbol{\theta})\pi(\boldsymbol{\theta})\}}{\partial\boldsymbol{\theta}\partial\boldsymbol{\theta}^T}\right|_{\boldsymbol{\theta}=\hat{\boldsymbol{\theta}}_n}$$

User can specify the value of the degrees of freedom parameter ν*, for e.g.,* $\nu = 10$*. The multivariate Student-t distribution have "fatter tail" than the multivariate normal distribution. As the number of degrees of freedom grows, the multivariate Student-t distribution approaches the multivariate normal distribution.*

7. *Generate a set of data from nonlinear logistic regression model. You can generate* $n = 100$ *binary observations according to the model:*

$$P(y = 1|\boldsymbol{x}) = \frac{1}{1 + \exp\{-0.3\exp(x_1 + x_2) + 0.8\}},$$

where the design points $\boldsymbol{x} = (x_1, x_2)$ *are uniformly distributed in* $[-1, 1] \times [-1, 1]$*. Then calculate the marginal likelihood by using the Laplace-Metropolis estimator (Lewis and Raftery, 1997), the harmonic mean estimator (Newton and Raftery, 1994), and Chib and Jeliazkov (2001) estimator from MH sampling.*

8. *In problem 6, we considered the generalized linear models with informative prior. Taking the logarithm of the posterior density, we have*

$$\log \pi(\boldsymbol{\theta}|\boldsymbol{Y}_n, X_n) \propto \log f(\boldsymbol{Y}_n|X_n, \boldsymbol{\theta}) + \log \pi(\boldsymbol{\theta}).$$

Show that finding the posterior mode corresponds to the penalized maximum likelihood method.

9. *To obtain the posterior mode $\hat{\boldsymbol{w}}_n$, we can use Fisher scoring algorithm. It is the standard method and is a Newton-Raphson algorithm using the expected rather than the observed information matrix (Green and Silverman (1994), Hastie and Tibshirani (1990), Nelder and Wedderburn (1972)). Show that the Fisher scoring iterations are expressed by*

$$\boldsymbol{w}^{new} = \left(B^T W B + n\beta D_2^T D_2\right)^{-1} B^T W \boldsymbol{\zeta},$$

where $\beta = \phi\lambda$ and

$$B = \begin{pmatrix} \boldsymbol{b}(\boldsymbol{x}_1)^T \\ \vdots \\ \boldsymbol{b}(\boldsymbol{x}_n)^T \end{pmatrix} = \begin{pmatrix} b_1(\boldsymbol{x}_1) & \cdots & b_m(\boldsymbol{x}_1) \\ \vdots & \ddots & \vdots \\ b_1(\boldsymbol{x}_n) & \cdots & b_m(\boldsymbol{x}_n) \end{pmatrix},$$

W is an $n \times n$ diagonal matrix and $\boldsymbol{\zeta}$ is an n-dimensional vector with α-th elements

$$\begin{aligned} W_{\alpha\alpha} &= \frac{1}{u''(\xi_\alpha)h'(\mu_\alpha)^2}, \\ \zeta_\alpha &= (y_\alpha - \mu_\alpha)h'(\mu_\alpha) + \eta_\alpha, \end{aligned}$$

respectively.

Chapter 7

Various Bayesian model selection criteria

7.1 Bayesian predictive information criterion

The basic idea behind Bayesian models is to specify both a data sampling density and a prior distribution of all unknowns. A joint probability distribution then expresses the correct relationships between the unknowns and the data. Any inference on a specific Bayesian model is based on the posterior distribution, i.e., the conditional probability distribution of the unknowns given the data. The results of such posterior inferences can be used for decision making, forecasting, stochastic structure exploration and many other problems. However, the quality of these solutions depends heavily on the underlying Bayesian model. This crucial issue had been recognized by researchers and practitioners, leading to extensive investigations on the subject of Bayesian model evaluation.

The Bayesian predictive information criterion (BPIC) proposed by Ando (2007) is a useful tool for evaluating the quality of Bayesian models from a predictive point of view. When the specified family of probability distributions $f(x|\boldsymbol{\theta})$ does not contain the true distribution $g(x)$, the BPIC is an estimator for the posterior mean of the expected log-likelihood of the Bayesian predictive distribution. The BPIC is derived by correcting the asymptotic bias in the posterior mean of the log-likelihood. This chapter begins by explaining the general framework of the BPIC.

7.1.1 The posterior mean of the log-likelihood and the expected log-likelihood

Akaike (1974) proposed an information criterion (the AIC) under two assumptions: (a) that a specified parametric family of probability distributions encompasses the true model, and (b) that the model can be estimated by the maximum likelihood method. The divergence of the fitted model from the true model is measured by the Kullback-Leibler information number or, equivalently, by the expected log-likelihood $\int \log f(z|\hat{\boldsymbol{\theta}}_{\rm MLE})dG(z)$, where $\hat{\boldsymbol{\theta}}_{\rm MLE}$ is the maximum likelihood estimator. Subsequent generalizations of AIC include the TIC (Takeuchi (1976)), which relaxes assumption (a), and the GIC (Konishi and Kitagawa (1996)), which relaxes both (a) and (b).

As a Bayesian version of these fitness criteria, Ando (2007) proposed that the posterior mean of the expected log-likelihood

$$\eta(G) = \int \left\{ \int \log f(z|\boldsymbol{\theta})\pi(\boldsymbol{\theta}|\boldsymbol{X}_n)d\boldsymbol{\theta} \right\} dG(z) \tag{7.1}$$

could measure the deviation of the predictive distribution from the true model $g(z)$. When selecting among various Bayesian models, the best one is chosen by maximizing the posterior mean of the expected log-likelihood.

One might consider the plug-in version of the utility function for evaluating the Bayesian models, e.g.,

$$\ell(G) = \int \log f(z|\hat{\boldsymbol{\theta}}_n)dG(z) \ ,$$

where $\hat{\boldsymbol{\theta}}_n$ is the posterior mode. (For some applications, this term can be replaced with the posterior mean $\bar{\boldsymbol{\theta}}_n$.) If we replace the posterior mode $\hat{\boldsymbol{\theta}}_n$ by the maximum likelihood estimator $\hat{\boldsymbol{\theta}}_{\mathrm{MLE}}$, this quantity corresponds to the expected log-likelihood (Akaike, 1973).

Although the posterior mean of the expected log-likelihood, $\eta(G)$ in (8.6), and the plug-in version of the utility function are both based on the log-likelihood function, there are some important differences between these two utility functions. As pointed out by Plummer (2008), the log-likelihood is sensitive to re-parameterization. Changing the parameterization of the model parameter $\boldsymbol{\theta}$ might change the definition of the posterior expectation of the transformed parameter, and hence the value of the utility function. On the other hand, the posterior mean of the expected log-likelihood is independent of the parameterization. Also, Plummer (2008) noted that the plug-in log-likelihood gives equal utility to all models that yield the same posterior expectation of $\boldsymbol{\theta}$, regardless of the covariance of the parameter estimates. In contrast with the plug-in log-likelihood, the posterior mean of the expected log-likelihood takes the covariance of the parameter estimates into account. Thus, the use of the posterior mean of the expected log-likelihood has several advantages.

However, note that the posterior mean of the expected log-likelihood depends on the unknown true distribution $G(z)$, on the observed data $\boldsymbol{X}_n$ taken from the joint distribution of $\boldsymbol{X}_n$ and $G(\boldsymbol{X}_n)$, and on the Bayesian model being fitted. Therefore, the crucial issue is how to construct an estimator for the posterior mean of the expected log-likelihood. Once we construct an estimator, it can be used to select the best model. A natural estimator of η is the posterior mean of the log-likelihood itself:

$$\eta(\hat{G}) = \frac{1}{n} \int \log f(\boldsymbol{X}_n|\boldsymbol{\theta})\pi(\boldsymbol{\theta}|\boldsymbol{X}_n)d\boldsymbol{\theta}. \tag{7.2}$$

The integrand is formally obtained by replacing the unknown distribution $G(z)$ in (8.6) with the empirical distribution, $\hat{G}$, putting a probability weight of $1/n$ on each observation of $\boldsymbol{X}_n$.

7.1.2 Bias correction for the posterior mean of the log-likelihood

The posterior mean of the log-likelihood $\eta(\hat{G})$ is generally positively biased with respect to the posterior mean of the expected log-likelihood $\eta(G)$. This occurs because the same data are used to estimate the parameters of the model and to evaluate the posterior mean of the expected log-likelihood. We should therefore consider how to correct this bias.

The bias is defined as

$$\begin{aligned}
b(G) &= \int \left\{ \eta(\hat{G}) - \eta(G) \right\} dG(\boldsymbol{X}_n) \\
&= \int \left[\frac{1}{n} \int \log f(\boldsymbol{X}_n|\boldsymbol{\theta})\pi(\boldsymbol{\theta}|\boldsymbol{X}_n)d\boldsymbol{\theta} \right. \\
&\qquad \left. - \int \left\{ \int \log f(z|\boldsymbol{\theta})\pi(\boldsymbol{\theta}|\boldsymbol{X}_n)d\boldsymbol{\theta} \right\} dG(z) \right] dG(\boldsymbol{X}_n),
\end{aligned}$$

where $G(\boldsymbol{X}_n)$ is the joint density of $\boldsymbol{X}_n$.

Assuming the bias $b(G)$ has been estimated by appropriate procedures, the bias-corrected posterior mean of the log-likelihood, an estimator of the posterior mean of the expected log-likelihood, is given by

$$\eta(G) \longleftarrow \frac{1}{n} \int \log f(\boldsymbol{X}_n|\boldsymbol{\theta})\pi(\boldsymbol{\theta}|\boldsymbol{X}_n)d\boldsymbol{\theta} - \hat{b}(G) \ .$$

This estimator is usually expressed in the form

$$\mathrm{IC} = -2 \int \log f(\boldsymbol{X}_n|\boldsymbol{\theta})\pi(\boldsymbol{\theta}|\boldsymbol{X}_n)d\boldsymbol{\theta} + 2n\hat{b}(G),$$

where $\hat{b}(G)$ is an estimator of the true bias $b(G)$. Ando (2007) obtained the asymptotic bias under the model mis-specification, i.e., a specified parametric family of probability distributions does not necessarily encompass the true model. The first term on the right-hand side of the IC equation measures the model fitness, and the second term is a penalty measuring the complexity of the model.

7.1.3 Definition of the Bayesian predictive information criterion

Let us consider the following situation: (a) The specified parametric model $f(x|\boldsymbol{\theta})$ contains the true distribution $g(x)$, that is $g(x) = f(x;\boldsymbol{\theta}_0)$ for some

$\boldsymbol{\theta}_0 \in \Theta$, and the specified parametric model is not far from the true model. (b) The order of the log-prior is $\log \pi(\boldsymbol{\theta}) = O_p(1)$, so the prior is dominated by the likelihood as n increases.

Under these two assumptions, together with some regularity conditions, Ando (2007) evaluated the asymptotic bias term as $\hat{b}(G) = p/n$, and proposed the Bayesian predictive information criterion (BPIC) for evaluating the Bayesian models:

$$\text{BPIC} = -2\int \log\{f(\boldsymbol{X}_n|\boldsymbol{\theta})\}\pi(\boldsymbol{\theta}|\boldsymbol{X}_n)d\boldsymbol{\theta} + 2p, \tag{7.3}$$

where p is the number of parameters in the Bayesian model. A theoretical derivation of this criterion and some regularity conditions are given in Chapter 8. The best model can be selected by minimizing the BPIC score.

In a practical situation, the analytical from of the posterior mean of the log-likelihood is not available. We therefore usually approximate this quantity using a Monte Carlo integration:

$$\int \log\{f(\boldsymbol{X}_n|\boldsymbol{\theta})\}\pi(\boldsymbol{\theta}|\boldsymbol{X}_n)d\boldsymbol{\theta} \approx \frac{1}{L}\sum_{j=1}^{L} \log f\left(\boldsymbol{X}_n|\boldsymbol{\theta}^{(j)}\right),$$

where $\{\boldsymbol{\theta}^{(1)}, ..., \boldsymbol{\theta}^{(L)}\}$ is a set of posterior samples generated from $\pi(\boldsymbol{\theta}|\boldsymbol{X}_n)$, and L is the number of posterior samples.

As shown in the BPIC score (7.3), the bias correction term can be applied in an automatic way in various situations. Moreover, the bias approximated by the number of parameters in the model is constant and does not depend on the observations.

The BPIC is a suitable criterion for evaluating Bayesian models estimated with weak prior information. On the other hand, practitioners often have strong prior information. In such cases one may employ the informative prior, i.e, the order of the log-prior is $\log \pi(\boldsymbol{\theta}) = O_p(n)$ (e.g., Konishi et al. (2004), Ando (2007, 2009c), Ando and Konishi (2009)). The BPIC is not theoretically justified for this kind of problem, even if the specified parametric family of probability distributions encompasses the true distribution $g(x)$. Furthermore, the specified model $f(x|\boldsymbol{\theta})$ does not necessarily contain the true model generating the data.

For problems where assumptions (a) and (b) do not hold, Ando (2007) showed how to evaluate the asymptotic bias more accurately. Given some regularity conditions, the asymptotic bias of $\hat{\eta}(G)$ is approximately

$$\begin{aligned}\hat{b}(G) \approx\ & \frac{1}{n}\int \left[\int \log\{f(\boldsymbol{X}_n|\boldsymbol{\theta})\pi(\boldsymbol{\theta})\}\pi(\boldsymbol{\theta}|\boldsymbol{X}_n)d\boldsymbol{\theta}\right] dG(\boldsymbol{X}_n) \\ & -\frac{1}{n}\log\{f(\boldsymbol{X}_n|\boldsymbol{\theta}_0)\pi(\boldsymbol{\theta}_0)\} + \frac{1}{n}\text{tr}\left\{S^{-1}(\boldsymbol{\theta}_0)Q(\boldsymbol{\theta}_0)\right\} + \frac{p}{2n},\end{aligned}$$

where the notation $\approx$ indicates that the difference between the two sides of the equation tends to zero as $n \to \infty$, p is the dimension of $\boldsymbol{\theta}$, and $\boldsymbol{\theta}_0$ is the mode of the expected penalized log-likelihood function

$$\int \{\log f(x|\boldsymbol{\theta}) + \log \pi_0(\boldsymbol{\theta})\}\, g(x)dx,$$

with $\log \pi_0(\boldsymbol{\theta}) = \lim_{n\to\infty} n^{-1} \log \pi(\boldsymbol{\theta})$. The matrices $Q_n(\hat{\boldsymbol{\theta}}_n)$ and $S_n(\hat{\boldsymbol{\theta}}_n)$ are given by

$$\begin{aligned} Q(\boldsymbol{\theta}) &= \int \left[\frac{\partial \log\{f(x|\boldsymbol{\theta})\pi_0(\boldsymbol{\theta})\}}{\partial \boldsymbol{\theta}} \cdot \frac{\partial \log\{f(x|\boldsymbol{\theta})\pi_0(\boldsymbol{\theta})\}}{\partial \boldsymbol{\theta}^T}\right] dG(x), \\ S(\boldsymbol{\theta}) &= -\int \left[\frac{\partial^2 \log\{f(x|\boldsymbol{\theta})\pi_0(\boldsymbol{\theta})\}}{\partial \boldsymbol{\theta}\partial \boldsymbol{\theta}^T}\right] dG(x), \end{aligned}$$

respectively. The prior distribution $\pi(\boldsymbol{\theta})$ may depend on n as long as $\lim_{n\to\infty} n^{-1} \log \pi(\boldsymbol{\theta})$ is limited. The regularity conditions, including the consistency of the parameter vector $\boldsymbol{\theta}$, and a derivation of this result are given in Chapter 8.

In practical situations, we have to replace the true distribution G in the bias estimator $\hat{b}(G)$ with the empirical distribution $\hat{G}$. After estimating the value of parameter $\boldsymbol{\theta}_0$ by the posterior mode $\hat{\boldsymbol{\theta}}_n$, and then replacing the matrices $S(\boldsymbol{\theta}_0)$ and $Q(\boldsymbol{\theta}_0)$ with $S_n(\hat{\boldsymbol{\theta}}_n)$ and $Q_n(\hat{\boldsymbol{\theta}}_n)$, we obtain an estimator of the bias. To correct the asymptotic bias of $\hat{\eta}$, Ando (2007) proposed the following Bayesian predictive information criterion:

$$\text{BPIC} = -2\int \log\{f(\boldsymbol{X}_n|\boldsymbol{\theta})\}\pi(\boldsymbol{\theta}|\boldsymbol{X}_n)d\boldsymbol{\theta} + 2n\hat{b}(\hat{G}), \tag{7.4}$$

where $\hat{b}(\hat{G})$ is given by

$$\begin{aligned} \hat{b}(\hat{G}) &= \frac{1}{n}\int \log\{f(\boldsymbol{X}_n|\boldsymbol{\theta})\pi(\boldsymbol{\theta})\}\pi(\boldsymbol{\theta}|\boldsymbol{X}_n)d\boldsymbol{\theta} \\ &\quad -\frac{1}{n}\log\left\{f\left(\boldsymbol{X}_n|\hat{\boldsymbol{\theta}}_n\right)\pi\left(\hat{\boldsymbol{\theta}}_n\right)\right\} + \frac{1}{n}\text{tr}\left\{S_n^{-1}\left(\hat{\boldsymbol{\theta}}_n\right)Q_n\left(\hat{\boldsymbol{\theta}}_n\right)\right\} + \frac{p}{2n}, \end{aligned}$$

and

$$\begin{aligned} Q_n(\hat{\boldsymbol{\theta}}_n) &= \frac{1}{n}\sum_{\alpha=1}^{n}\left[\frac{\partial\{\log f(x_\alpha|\boldsymbol{\theta}) + \log \pi(\boldsymbol{\theta})/n\}}{\partial \boldsymbol{\theta}}\right. \\ &\qquad \left. \cdot \left.\frac{\partial\{\log f(x_\alpha|\boldsymbol{\theta}) + \log \pi(\boldsymbol{\theta})/n\}}{\partial \boldsymbol{\theta}^T}\right|_{\boldsymbol{\theta}=\hat{\boldsymbol{\theta}}_n}\right], \\ S_n(\hat{\boldsymbol{\theta}}_n) &= -\frac{1}{n}\sum_{\alpha=1}^{n}\left[\left.\frac{\partial^2\{\log f(x_\alpha|\boldsymbol{\theta}) + \log \pi(\boldsymbol{\theta})/n\}}{\partial \boldsymbol{\theta}\partial \boldsymbol{\theta}^T}\right|_{\boldsymbol{\theta}=\hat{\boldsymbol{\theta}}_n}\right]. \end{aligned}$$

Again, we choose the predictive distribution that minimizes BPIC. If we impose assumptions (a) and (b) on the Bayesian model just described, the BPIC score in Equation (7.4) reduces to the score in (7.3).

BPIC has been applied to various Bayesian statistical model selection problems. To predict the hazard term structure or, equivalently, the term structure of the default probability, Ando (2009c) used BPIC to evaluate Bayesian inferences with functional predictors. With respect to the functional data analysis, we refer to Ramsay and Silverman (1997). The Bayesian inferences for several nonlinear, non-Gaussian stochastic volatility models with leverage effects were also studied by Ando (2006). More recently, Ando (2009b) used the BPIC to study Bayesian portfolio selection under a multifactor model.

7.1.4 Example: Bayesian generalized state space modeling

In practical marketing, one interesting problem is quantifying the impacts of marketing strategies and related factors on sales. Under a Bayesian generalized state space modeling approach, Ando (2008a) used BPIC to measure the sales promotion effect and baseline sales for incense products.

It is useful to begin with a brief review of the general state space representation (Kitagawa (1987), Kitagawa and Gersch (1996)). Its main advantage is that it accounts for not only linear Gaussian state space models but also nonlinear and non-Gaussian models.

Generalized state space models

The general state space model consists of two stochastic components: an observation equation and a system equation. We have

$$\begin{cases} \text{Observation equation :} & \boldsymbol{y}_t \sim f(\boldsymbol{y}_t|F_t,\boldsymbol{\theta}), \\ \text{System equation :} & \boldsymbol{h}_t \sim f(\boldsymbol{h}_t|F_{t-1},\boldsymbol{\theta}), \end{cases} \qquad t=1,...,n,$$

where n is the number of observations. The p-dimensional vector $\boldsymbol{y}_t = (y_{1t},...,y_{pt})^T$ is the observable time series, while the q-dimensional vector $\boldsymbol{h}_t = (h_{1t},...,h_{qt})^T$ is unobserved. The latter, $\boldsymbol{h}_t$, is called the state vector. Also, F_t denotes the history of the information sequence up to time t. It includes exogenous variables $X_t = \{\boldsymbol{x}_1,...,\boldsymbol{x}_t\}$, the observable time series $\boldsymbol{Y}_t = \{\boldsymbol{y}_1,...,\boldsymbol{y}_t\}$ up to time t, the unobservable time series of state vectors $H_t = \{\boldsymbol{h}_1,...,\boldsymbol{h}_t\}$, and so on. As an exception, however, F_t in the observation equation does not contain the information $\boldsymbol{y}_t$.

The densities $f(\boldsymbol{y}_t|F_t,\boldsymbol{\theta})$ and $f(\boldsymbol{h}_t|F_{t-1},\boldsymbol{\theta})$ are the conditional distributions of $\boldsymbol{y}_t$ given F_t and of $\boldsymbol{h}_t$ given F_{t-1} respectively. The main focus of a generalized state space model is constructing these densities so that the model captures the true structure governing the time series $\boldsymbol{y}_t$.

Maximum likelihood estimates of $\boldsymbol{\theta}$ can be obtained by maximizing the following likelihood function:

$$f(\boldsymbol{Y}_n|\boldsymbol{\theta}) = \prod_{t=1}^{n} f(\boldsymbol{y}_t|F_{t-1},\boldsymbol{\theta}) = \prod_{t=1}^{n}\left[\int f(\boldsymbol{y}_t|F_t,\boldsymbol{\theta})f(\boldsymbol{h}_t|F_{t-1},\boldsymbol{\theta})d\boldsymbol{h}_t\right], \quad (7.5)$$

which depends on high-dimensional integrals (e.g., Kitagawa, 1987; Kitagawa and Gersch, 1996). The crux of the problem is that we cannot express the density $f(\boldsymbol{h}_t|F_{t-1}, \boldsymbol{\theta})$ in closed form. Maximum likelihood estimation of general state space models is thus very difficult.

In contrast, a Bayesian treatment of this inference problem relies solely on probability theory. The model parameters can be estimated easily, without evaluating the likelihood function. In particular, the Markov Chain Monte Carlo (MCMC) algorithm is very useful for estimating model parameters. Under this approach, the unobserved states H_t are considered model parameters. An inference on the parameters is then conducted by producing a sample from the posterior distribution

$$\pi(\boldsymbol{\theta}, H_n|\boldsymbol{Y}_n) \propto \prod_{t=1}^{n} [f(\boldsymbol{y}_t|F_t, \boldsymbol{\theta})f(\boldsymbol{h}_t|F_{t-1}, \boldsymbol{\theta})]\, \pi(\boldsymbol{\theta}),$$

where $\pi(\boldsymbol{\theta})$ is a prior density for the model parameter vector $\boldsymbol{\theta}$. Using the posterior samples, we can develop the Bayesian models.

One of the most crucial issues is the choice of an optimal model that adequately expresses the dynamics of the time series. A tailor-made version of the Bayesian predictive information criterion (BPIC) given in (7.3) is as follows:

$$\text{BPIC} = -2\int \log f(\boldsymbol{Y}_n|\boldsymbol{\theta})\pi(\boldsymbol{\theta}|\boldsymbol{Y}_n)d\boldsymbol{\theta} + 2\text{dim}\{\boldsymbol{\theta}\}. \tag{7.6}$$

The likelihood function $f(\boldsymbol{Y}_n|\boldsymbol{\theta})$ is given in (7.5). Here, the best predictive distribution is selected by minimizing the Bayesian predictive information criterion (BPIC). Again, we would like to point out that other Bayesian model selection criteria can also be used.

Particle filtering

Here we review the particle filtering method (Kitagawa, 1996; Pitt and Shephard, 1999), which can be usesd to approximate the likelihood function numerically. For simplicity of explanation we assume that the noise vector $\boldsymbol{w}_t$ in the observation equation can be expressed as $\boldsymbol{w}_t = \boldsymbol{y}_t - \boldsymbol{m}(F_t)$, where $\boldsymbol{m}(F_t) = (m_1(F_t), ..., m_p(F_t))^T$ is a p-dimensional function of F_t. As shown in Equation (7.5), the likelihood function has no analytical form; it is marginalized over the latent variables H_t. The particle filtering procedure for likelihood estimation is generally given as follows:

Let us take a sample $\boldsymbol{h}_{t-1}^1, ..., \boldsymbol{h}_{t-1}^M \sim f(\boldsymbol{h}_{t-1}|F_{t-1}, \boldsymbol{\theta})$ from the filtered distribution. It follows that the one-step forward predictive distribution

$f(\boldsymbol{h}_t|F_{t-1},\boldsymbol{\theta})$ can be approximated as follows:

$$\begin{aligned} f(\boldsymbol{h}_t|F_{t-1},\boldsymbol{\theta}) &= \int f(\boldsymbol{h}_t|F_{t-1},\boldsymbol{\theta})f(\boldsymbol{h}_{t-1}|F_{t-1},\boldsymbol{\theta})d\boldsymbol{h}_{t-1} \\ &\approx \frac{1}{M}\sum_{j=1}^{M} f(\boldsymbol{h}_t|F_{t-1}^j,\boldsymbol{\theta}), \end{aligned}$$

where F_t^j denotes the history of the information sequence up to time t. Note that F_t^j also includes the historical time series of unobservable state vectors $H_t^j = \{\boldsymbol{h}_1^j,...,\boldsymbol{h}_t^j\}$, instead of $H_t = \{\boldsymbol{h}_1,...,\boldsymbol{h}_t\}$. The one-step forward density is then estimated by Monte Carlo averaging of $f(\boldsymbol{y}_t|F_{t-1},\boldsymbol{\theta})$ over the realizations of $\boldsymbol{h}_t^j \sim f(\boldsymbol{h}_t|F_{t-1}^j,\boldsymbol{\theta})$:

$$f(\boldsymbol{y}_t|F_{t-1},\boldsymbol{\theta}) = \int f(\boldsymbol{y}_t|F_t,\boldsymbol{\theta})f(\boldsymbol{h}_t|F_{t-1},\boldsymbol{\theta})d\boldsymbol{h}_t \approx \frac{1}{M}\sum_{j=1}^{M} f(\boldsymbol{y}_t|F_t^j,\boldsymbol{\theta}).$$

This procedure is recursive, each iteration requiring a new realization of the sequence $\boldsymbol{h}_t$ from the filtered distribution $f(\boldsymbol{h}_t|F_t,\boldsymbol{\theta})$. The problem is to obtain a filtered sample $\boldsymbol{h}_t^j \sim f(\boldsymbol{h}_t|F_t,\boldsymbol{\theta})$. From Bayes' theorem, we have

$$f(\boldsymbol{h}_t|F_t,\boldsymbol{\theta}) \propto f(\boldsymbol{y}_t|F_t,\boldsymbol{\theta})f(\boldsymbol{h}_t|F_{t-1},\boldsymbol{\theta}).$$

Thus, one simple way to obtain a sample $\boldsymbol{h}_t^1,...,\boldsymbol{h}_t^M$ from the filtered distribution $f(\boldsymbol{h}_t|F_t,\boldsymbol{\theta})$ is to resample the distribution $\boldsymbol{h}_t^j \sim f(\boldsymbol{h}_t|F_{t-1}^j,\boldsymbol{\theta})$ with probabilities proportional to $f(\boldsymbol{y}_t|F_t^j,\boldsymbol{\theta})$. We now summarize the process of likelihood estimation based on the particle filtering method (Kitagawa, 1996):

Particle filtering algorithm

Step 1. Initialize $\boldsymbol{\theta}$ and generate M samples $\boldsymbol{h}_0^1,...,\boldsymbol{h}_0^M \sim f(\boldsymbol{h}_0|\boldsymbol{\theta})$.
Step 2. Repeat the following steps for $t = 1 \sim n$.
Step 2-1. Generate M samples $\boldsymbol{h}_t^1,...,\boldsymbol{h}_t^M$ from $f(\boldsymbol{h}_t|F_{t-1}^j,\boldsymbol{\theta})$.
Step 2-2. Compute the density estimate

$$\hat{f}(\boldsymbol{y}_t|F_{t-1},\boldsymbol{\theta}) = M^{-1}\sum_{j=1}^{M} f(\boldsymbol{y}_t|F_t^j,\boldsymbol{\theta}).$$

Step 2-3. Resample $\{\boldsymbol{h}_t^1,..,\boldsymbol{h}_t^M\}$ with probabilities proportional to $f(\boldsymbol{y}_t|F_t^j,\boldsymbol{\theta})$ to produce the filtered samples $\boldsymbol{h}_t^1,..,\boldsymbol{h}_t^M \sim f(\boldsymbol{h}_t|F_t,\boldsymbol{\theta})$.
Step 2-4. Update the history of the filtered sample H_{t-1}^j to H_t^j, $(j = 1,...,M)$.
Step 3. Return the likelihood estimate $\hat{f}(\boldsymbol{Y}_n|\boldsymbol{\theta}) = \prod_{t=1}^{n} \hat{f}(\boldsymbol{y}_t|F_{t-1},\boldsymbol{\theta})$.

More efficient versions of the basic particle filter algorithm given above also exist, such as the auxiliary particle filter (Pitt and Shephard, 1999). The following section describes a practical example of Bayesian generalized state space modeling with the BPIC.

Forecasting daily sales of incense products

In this section, we focus on 2-dimensional time series data for daily sales of incense products in two department stores. More detailed information on this example can be found in Ando (2008a). Figures 7.1 (a) and (b) plot the time series of daily sales at Store 1 and Store 2, respectively. The units of y_{jt} are *thousands of yen*. Summary statistics of the two series are shown in Table 7.1. Since the kurtosis of the returns is greater than three, the true distribution of the data must be fat-tailed. Using the Shapiro-Wilk normality test (Patrick, 1982), the null hypothesis that sales were normally distributed was firmly rejected. The p-values for Store 1 and Store 2 were 2.58×10^{-16} and 2.62×10^{-14}, respectively.

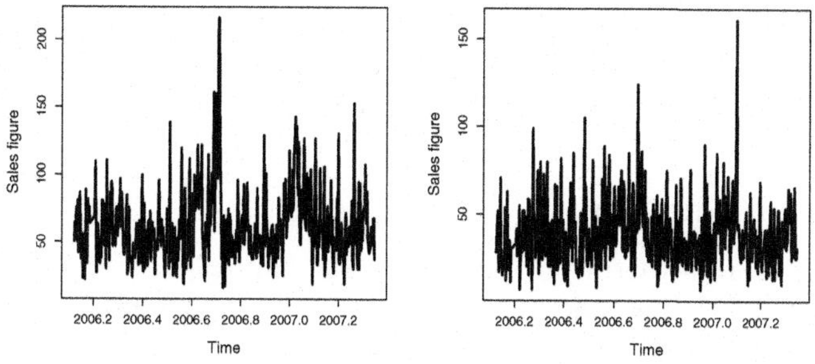

(a) (b)

FIGURE 7.1: (From Ando, T., *Ann. I. Stat. Math.*, 60, 2008a. With permission.) Daily sales figures for incense products from January, 2006 to March, 2007 at (a) Store 1 and (b) Store 2.

TABLE 7.1: *Source:* Ando, T., *Ann. I. Stat. Math.*, 60, 2008a. With permission. Summary statistics of the sales data for Store 1 and Store 2; μ: the mean, σ: standard deviation, s: skewness, and k: kurtosis.

	Store 1	Store 2
μ	61.990	38.605
σ	27.507	19.482
s	1.559	1.333
k	4.561	3.859

In addition to daily sales data, we tabulated the weather effect x_{j1t}, the weekly and holiday effect x_{j2t}, the sales promotion effect x_{j3t}, and the event effect x_{j4t}. The definitions of these variables are as follows:

$$x_{j1t} = \begin{cases} 1 & \text{(Fine)} \\ 0 & \text{(Cloudy)} \\ \text{-1} & \text{(Rain)} \end{cases}, \quad j = 1, 2,$$

$$x_{j2t} = \begin{cases} 1 & \text{(Sunday, Saturday, National holiday)} \\ 0 & \text{(Otherwise)} \end{cases}, \quad j = 1, 2,$$

$$x_{j3t} = \begin{cases} 1 & \text{(Promotional event)} \\ 0 & \text{(No promotional event)} \end{cases}, \quad j = 1, 2,$$

$$x_{j4t} = \begin{cases} 1 & \text{(Entertainment event)} \\ 0 & \text{(No entertainment event)} \end{cases}, \quad j = 1, 2.$$

Note that several other variables such as price levels, price discount percentages, features, displays, and post-promotional dips can have an important impact on sales. Unfortunately, due to the limitations of the dataset, we considered only the four variables defined above.

Observation equation

Given a mean structure of total sales y_{jt}, say $\mu_{jt} = E[y_{jt}|F_t]$, we shall decompose it into a baseline sales figure h_{jt} and other components by incorporating the covariate effects:

$$\mu_{jt}(h_{jt}, \boldsymbol{\beta}_j, \boldsymbol{x}_{jt}) = h_{jt} + \sum_{a=1}^{4} \beta_{ja} x_{jat} = h_{jt} + \boldsymbol{\beta}_j^T \boldsymbol{x}_{jt}, \quad j = 1, 2 . \tag{7.7}$$

Here $\boldsymbol{\beta}_j = (\beta_{j1}, ..., \beta_{j4})^T$ and $\boldsymbol{x}_{jt} = (x_{j1t}, ..., x_{j4t})^T$ are the 4-dimensional vector of unknown parameters to be estimated and the 4-dimensional covariate vector, respectively.

We are usually not sure about the underlying distribution of daily sales y_{jt}. We therefore consider several possible density functions. Because the sales data must be positive, we use truncated distributions:

Truncated normal :

$$f_N(y_{jt}|\mu_{jt}, \sigma_j^2) = I(y_{jt} > 0) \cdot \frac{1}{2\sqrt{2\pi\sigma_j^2}} \exp\left\{-\frac{(y_{jt} - \mu_{jt})^2}{2\sigma_j^2}\right\},$$

Truncated Student's t :

$$f_{St}(y_{jt}|\mu_{jt}, \sigma_j^2, \nu_j)$$

$$= I(y_{jt} > 0) \cdot \frac{\Gamma(\frac{\nu_j+1}{2})}{2\Gamma(\frac{1}{2})\Gamma(\frac{\nu_j}{2})\sqrt{\nu_j \sigma_j^2}} \left\{1 + \frac{(y_{jt} - \mu_{jt})^2}{\sigma_j^2 \nu_j}\right\}^{-\frac{\nu_j+1}{2}},$$

Truncated Cauchy :

$$f_C(y_{jt}|\mu_{jt}, \sigma_j^2) = I(y_{jt} > 0) \cdot \frac{1}{2\pi\sigma_j} \left\{1 + \frac{(y_{jt} - \mu_{jt})^2}{\sigma_j^2}\right\}^{-1}, \tag{7.8}$$

Poisson :

$$f_P(y_{jt}|\mu_{jt}) = \frac{\exp\{-\mu_{jt}\}\mu_{jt}^{y_{jt}}}{y_{jt}!}.$$

The indicator function $I(y_{jt} > 0)$ takes the value 1 if $y_{jt} > 0$, and zero otherwise. μ_{jt} is the mean parameter given in (7.7), s_j^2 is the variance parameter and ν_j is the number of degrees of freedom for Student's t-distribution. Hereafter, for simplicity of presentation, we denote these densities generically by $f(y_{jt}|\boldsymbol{x}_j, h_{jt}, \boldsymbol{\gamma}_j)$, where $\boldsymbol{\gamma}_j$ is the unknown parameter vector associated with a density function of the observation equation.

System equation

It is assumed that the state variable h_{jt}, the baseline sales effect for the j-th store, follows an r-th order trend model:

$$\Delta^r h_{jt} = \varepsilon_{jt}, \quad j = 1, 2,$$

where Δ ($\Delta h_{jt} = h_{jt} - h_{j,t-1}$) is the difference operator (e.g., Kitagawa and Gersch, 1996) and $\varepsilon_{jt} \sim N(0, \sigma_{jj})$ is a Gaussian white noise sequence. For $r = 1$, the baseline sales data follow the well-known random walk model,

$$h_{jt} = h_{j,t-1} + \varepsilon_{jt}.$$

For $k = 2$, the model becomes

$$h_{jt} = 2h_{j,t-1} - h_{j,t-1} + \varepsilon_{jt}.$$

Another expression of the r-th order trend model is

$$h_{jt} = \sum_{s=1}^{r} c_s \times B^s h_{jt} + \varepsilon_{jt},$$

where B ($B^1 h_{jt} = h_{j,t-1}$) is the backshift operator and $c_s = (-1)^{s-1} \times {}_rC_i$ are the binomial coefficients (Kitagawa and Gersch, 1996). It is natural to assume that the daily sales at each store are correlated. Ando (2008a) therefore looked for correlation between the noise terms ε_{jt} and ε_{kt}:

$$\text{Cov}(\varepsilon_{jt}, \varepsilon_{kt}) = \sigma_{jk}.$$

Summarizing the above specifications, we formulate the following system of equations:

$$\boldsymbol{h}_t \sim f(\boldsymbol{h}_t|\boldsymbol{h}_{t-1}, ..., \boldsymbol{h}_{t-r}; \Sigma), \ \Sigma = (\sigma_{ij}), \tag{7.9}$$

where $f(\boldsymbol{h}_t|\boldsymbol{h}_{t-1}, ..., \boldsymbol{h}_{t-r}; \Sigma)$ is the $p = 2$-dimensional normal density with mean $\boldsymbol{h}_t = \sum_{s=1}^{r} c_s \times B^s \boldsymbol{h}_t$ and covariance matrix Σ.

Bayesian inference via MCMC

Our next task is to estimate the unknown parameter vector $\boldsymbol{\theta} = (\boldsymbol{\gamma}^T, \text{vech}(\Sigma)^T)^T$ with $\boldsymbol{\gamma} = (\boldsymbol{\gamma}_1^T, \boldsymbol{\gamma}_2^T)^T$. To complete the Bayesian model, we formulate a prior distribution on the parameters. Independence of the parameters is assumed:

$$\pi(\boldsymbol{\theta}) = \pi(\Sigma)\pi(\boldsymbol{\gamma}), \quad \pi(\boldsymbol{\gamma}) = \prod_{j=1}^{2} \pi(\boldsymbol{\gamma}_j).$$

We begin by decomposing the covariance matrix Σ into a product of the variance matrix R and correlation matrix C, $\Sigma = RCR$, where $R = (r_{ij})$ is diagonal (Barnard et al., 2000). We can then formulate a prior distribution on the elements r_{ii} $(i = 1, 2)$ and $\{c_{ij}, i < j\}$. Ando (2008a) assumed that each of the elements $\{r_{ii}; i = 1, 2\}$ is independently and identically distributed. We then place a gamma prior with parameters a and b on the diagonal entries of Σ:

$$\pi(\sigma_{ii}) = \frac{b^a}{\Gamma(a)}(\sigma_{ii})^{a-1}\exp\{-b\sigma_{ii}\}, \qquad i = 1, 2,$$

which implies that

$$\pi(r_{ii}) = \pi(\sigma_{ii})\frac{d\sigma_{ii}}{dr_{ii}} = \frac{2b^a}{\Gamma(a)}(r_{ii})^{2a-1}\exp\{-br_{ii}^2\}.$$

To make the prior uninformative, we set $a = 10^{-10}$ and $b = 10^{-10}$. The elements $\{c_{ij}, i < j\}$ are drawn randomly from a uniform prior distribution $U[-1, 1]$.

When we specify Student's t-distribution for y_{jt}, the unknown parameter vector $\boldsymbol{\gamma}_j$ includes the degrees of freedom ν_j as well as the coefficients $\boldsymbol{\beta}_j$ and s_j^2. For $\boldsymbol{\beta}_j$, we use the 4-dimensional uninformative normal prior $N(\mathbf{0}, 10^{10} \times I)$. In addition to σ_{ii}, an inverse gamma prior with parameters $a = b = 10^{-10}$ is used for s_j^2. Finally, the uniform prior distribution $U[2, 100]$ is used for $\pi(\nu_j)$. The same prior distributions are employed for other density functions.

The MCMC sampling algorithm is then summarized as follows.

MCMC sampling algorithm:

Step 1. Initialize $\boldsymbol{\theta}$ and H_n.

Step 2. Sample $\boldsymbol{h}_t$ from $\boldsymbol{h}_t|\boldsymbol{\theta}, H_{-h_t}, \boldsymbol{Y}_n, X_n$, for $t = 1, ..., n$.
Step 3. Sample $\boldsymbol{\beta}_j$ from $\boldsymbol{\beta}_j|\boldsymbol{\theta}_{-\beta_j}, H_n, \boldsymbol{Y}_n, X_n$, for $j = 1, 2$.
Step 4. Sample r_{ii} from $r_{ii}|\boldsymbol{\theta}_{-r_{ii}}, H_n, \boldsymbol{Y}_n, X_n$, for $i = 1, 2$.
Step 5. Sample c_{ij} from $c_{ij}|\boldsymbol{\theta}_{-c_{ij}}, H_n, \boldsymbol{Y}_n, X_n$, for $i, j = 1, 2$ $(i < j)$.
Step 6. Sample s_j^2 from $s_j^2|\boldsymbol{\theta}_{-s_j^2}, H_n, \boldsymbol{Y}_n, X_n$, for $j = 1, 2$.
Step 7. Sample ν_j from $\nu_j|\boldsymbol{\theta}_{-\nu_j}, H_n, \boldsymbol{Y}_n, X_n$, for $j = 1, 2$.
Step 8. Repeat Step 2–Step 7 for sufficient iterations.

Here H_{-h_t} denotes the rest of a set of state vectors H_n; that is, all state vectors other than $\boldsymbol{h}_t$. Note that $\boldsymbol{Y}_n$ and X_n are vectors: $\boldsymbol{Y}_n = \{\boldsymbol{y}_1, ..., \boldsymbol{y}_n\}$ and $X_n = \{\boldsymbol{x}_1, ..., \boldsymbol{x}_n\}$. By sampling from a random walk proposal density, the Metropolis-Hastings (MH) algorithm implements Steps 2–7.

For instance, let us assume a first-order random walk model for baseline sales. In Step 2, the conditional posterior density function of $\boldsymbol{h}_t$ is

$$\pi(\boldsymbol{h}_t|\boldsymbol{\theta}, H_{-h_t}, \boldsymbol{Y}_n, X_n)$$
$$\propto \begin{cases} f(\boldsymbol{h}_2|\boldsymbol{h}_1, \Sigma) \times \prod_{j=1}^2 f(y_{j1}|h_{j1}; \boldsymbol{x}_{j1}, \boldsymbol{\gamma}_j), \ (t = 1), \\ f(\boldsymbol{h}_{t+1}|\boldsymbol{h}_t, \Sigma) \times f(\boldsymbol{h}_t|\boldsymbol{h}_{t-1}, \Sigma) \times \prod_{j=1}^2 f(y_{jt}|h_{jt}; \boldsymbol{x}_{jt}, \boldsymbol{\gamma}_j), \ (t \neq 1, n), \\ f(\boldsymbol{h}_n|\boldsymbol{h}_{n-1}, \Sigma) \times \prod_{j=1}^2 f(y_{jn}|h_{jn}; \boldsymbol{x}_{jn}, \boldsymbol{\gamma}_j), \ (t = n). \end{cases}$$

At the k-th iteration, we draw a candidate value of $\boldsymbol{h}_t^{(k+1)}$ using the Gaussian prior density function centered at the current value $\boldsymbol{h}_t^{(k)}$ with the variance matrix $0.01 \times I$. We then accept the candidate value with probability

$$\alpha = \min\left\{1, \frac{\pi(\boldsymbol{h}_t^{(k+1)}|\boldsymbol{\theta}, H_{-h_t}, \boldsymbol{Y}_n, X_n)}{\pi(\boldsymbol{h}_t^{(k)}|\boldsymbol{\theta}, H_{-h_t}, \boldsymbol{Y}_n, X_n)}\right\}.$$

The conditional posterior density functions are then

$$\pi(\boldsymbol{\beta}_j|\boldsymbol{\theta}_{-\beta_j}, H_n, \boldsymbol{Y}_n, X_n) \propto \prod_{t=1}^n \prod_{j=1}^2 f(y_{jt}|h_{jt}; \boldsymbol{x}_{jt}, \boldsymbol{\gamma}_j) \times \pi(\boldsymbol{\beta}_j),$$

$$\pi(r_{ii}|\boldsymbol{\theta}_{-r_{ii}}, H_n, \boldsymbol{Y}_n, X_n) \propto \prod_{t=2}^n f(\boldsymbol{h}_t|\boldsymbol{h}_{t-1}, \Sigma) \times \pi(r_{ii}),$$

$$\pi(c_{ij}|\boldsymbol{\theta}_{-c_{ij}}, H_n, \boldsymbol{Y}_n, X_n) \propto \prod_{t=2}^n f(\boldsymbol{h}_t|\boldsymbol{h}_{t-1}, \Sigma) \times \pi(c_{ij}),$$

$$\pi(s_j^2|\boldsymbol{\theta}_{-s_j^2}, H_n, \boldsymbol{Y}_n, X_n) \propto \prod_{t=1}^n \prod_{j=1}^2 f(y_{jt}|h_{jt}; \boldsymbol{x}_{jt}, \boldsymbol{\gamma}_j) \times \pi(s_j^2),$$

$$\pi(\nu_j|\boldsymbol{\theta}_{-\nu_j}, H_n, \boldsymbol{Y}_n, X_n) \propto \prod_{t=1}^n \prod_{j=1}^2 f(y_{jt}|h_{jt}; \boldsymbol{x}_{jt}, \boldsymbol{\gamma}_j) \times \pi(\nu_j).$$

In addition to implementing Step 2, the MH algorithm implements Steps 3–7. The outcomes of the MH algorithm can be regarded as a sample drawn from the posterior density functions after a burn-in period.

Our final task is to evaluate whether the estimated model is a good approximation of the true structure. For example, we have to select the sampling density function f among a set of models in (7.8). Also, we have to select the lag r of the baseline sales $\boldsymbol{h}_t$. The BPIC score given in (7.6) is useful for this purpose. We have

$$\text{BPIC}(f, r) = -2 \int \log f(\boldsymbol{Y}_n | X_n, \boldsymbol{\theta}) \pi(\boldsymbol{\theta} | X_n, \boldsymbol{Y}_n) d\boldsymbol{\theta} + 2\text{dim}\{\boldsymbol{\theta}\},$$

where f is the sampling density function for the observation model, $\boldsymbol{\theta} = (\gamma_1^T, \gamma_2^T, \text{vech}(\Sigma)^T)^T$, and the likelihood function $f(\boldsymbol{Y}_n | X_n, \boldsymbol{\theta})$ is given by

$$f(\boldsymbol{Y}_n | X_n, \boldsymbol{\theta}) = \prod_{t=1}^{n} \left[\int \prod_{j=1}^{2} f(y_{jt} | h_{jt}; \boldsymbol{x}_{jt}, \gamma_j) f(\boldsymbol{h}_t | \boldsymbol{h}_{t-1}, ..., \boldsymbol{h}_{t-r}, \Sigma) d\boldsymbol{h}_t \right].$$

The best predictive distribution is selected by minimizing the Bayesian predictive information criterion (BPIC).

Estimation results

Ando (2008a) fit the various statistical models given in (7.8). In principle one can evaluate any number of distributional assumptions on $\boldsymbol{y}_t$, any possible lag of the baseline sales r, and any combination of the covariates $\boldsymbol{x}_{jt}$. Because one of the aims was to quantify the impacts of each covariate, Ando (2008a) considered all four of the distributional assumptions mentioned above and three values for the lag of the baseline sales: $r = \{1, 2, 3\}$.

The total number of MCMC iterations was 6,000, of which the first 1,000 are discarded as a burn-in period. To test the convergence of the MCMC sampling algorithm, every fifth iterations was stored after the burn-in period. All inferences are therefore based on 1,000 generated samples. It is also necessary to check whether the generated posterior sample is taken from a stationary distribution. This was accomplished by calculating Geweke's (1992) convergence diagnostic (CD) test. Based on this statistic, we evaluated the equality of the means in the first and last parts of the Markov chains. If the samples are drawn from a stationary distribution, the two means should be statistically consistent. It is known that the CD test statistic has an asymptotic standard normal distribution. All of the results reported in this paper are based on samples that have passed Geweke's convergence test at a significance level of 5% for all parameters.

The most adequate model for baseline sales is Student's t-distribution with a two-day lag ($r = 2$), as this model yielded the smallest value of the BPIC, BPIC = 9948.833. Table 7.2 reports the posterior means, standard deviations,

TABLE 7.2: *Source:* Ando, T., *Ann. I. Stat. Math.*, 60, 2008a. With permission. Summary of the estimation results. posterior means, standard deviations (SDs), 95% confidence intervals, inefficiency factors (INEFs) and Geweke's CD test statistic (1992) for each parameter.

	Mean	SDs	95% Conf. Interval		INEFs	CD
β_{11}	-1.883	0.944	[-3.740	0.053]	2.385	-0.584
β_{21}	10.028	0.845	[8.398	11.810]	2.692	-0.839
β_{12}	2.223	0.893	[0.596	3.742]	2.452	-0.335
β_{22}	3.127	0.763	[1.739	4.624]	2.547	-1.332
β_{13}	-0.596	0.831	[-2.243	1.126]	2.193	0.550
β_{23}	10.099	0.742	[8.573	11.604]	2.849	1.032
β_{14}	24.396	0.966	[22.592	26.105]	2.309	-0.697
β_{24}	11.670	0.864	[9.841	13.421]	2.325	-1.725
s_1^2	25.472	0.080	[25.216	25.762]	2.604	-1.814
s_2^2	17.061	0.049	[16.964	17.155]	2.270	-1.745
σ_{11}	25.472	0.063	[25.243	25.653]	5.857	-0.995
σ_{22}	17.006	0.046	[16.960	17.155]	5.935	0.056
σ_{12}	0.185	0.010	[0.169	0.201]	2.783	0.967
ν_1	26.106	0.602	[24.998	27.042]	25.092	0.653
ν_2	5.001	0.483	[4.049	6.012]	24.330	0.976

95% confidence intervals, the inefficiency factor and the values of Geweke's CD test statistics. Based on 1,000 draws for each of the parameters, their posterior means, standard errors, and 95% confidence intervals are calculated. The 95% confidence intervals reported here are estimated using the 2.5th and 97.5th percentiles of the posterior samples. The inefficiency factor is a useful measure for evaluating the efficiency of the MCMC sampling algorithm. 1,000 lags was used to estimate the inefficiency factors. As shown in Table 7.2, our sampling procedure achieved a good efficiency.

As shown in Table 7.2, weather appears to affect the demand for lifestyle incense. In Store 2, sales rose during the rainy season. (The posterior mean of β_{21} is greater than 0.) The estimated coefficients of the weekly effects β_{12} and β_{22} indicate that working days have a negative impact on sales. This is to be expected, since employed persons rarely visit department stores during a workday. There is a significant difference between the promotion effects observed in Store 1 and Store 2. The posterior mean of β_{13} is close to zero, while that of β_{23} is far from zero. Moreover, the 95% confidence interval around β_{13} includes zero. This suggests that sales will not increase at Store 1 even during a promotion. On the other hand, daily sales do increase during promotional events at Store 2.

Figure 7.2 plots the posterior means of the baseline sales for each store. Both stores clearly exhibit time-varying sales figures over the observed period. We can also see that the baseline sales function is different for each store. In Ando (2008a), more detailed information is provided.

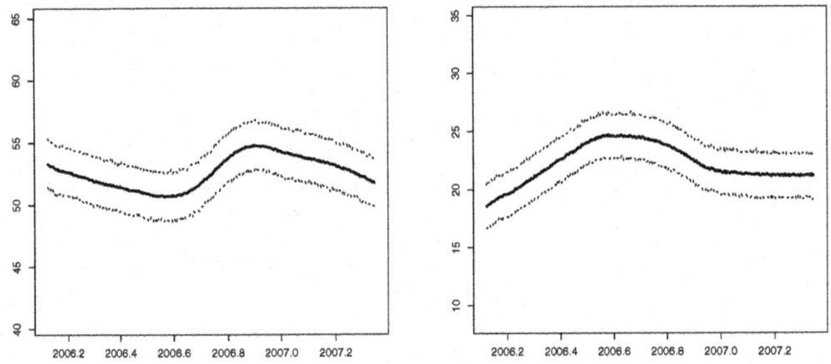

(a) Store 1. (b) Store 2.

FIGURE 7.2: (From Ando, T., *Ann. I. Stat. Math.*, 60, 2008a. With permission.) Fluctuations in the posterior means of baseline sales for each store. The dashed lines are 95% confidence intervals.

7.2 Deviance information criterion

Spiegelhalter et al. (2002) pointed out that the posterior mean of the expected log-likelihood can be taken as a Bayesian measure of fit. Defining a Bayesian measure of model complexity p_D, the posterior mean of the deviance minus the deviance evaluated at the posterior mean $\bar{\boldsymbol{\theta}}_n$ (or mode, median),

$$P_D = 2\log\{f(\boldsymbol{X}_n|\bar{\boldsymbol{\theta}}_n)\} - 2\int \log\{f(\boldsymbol{X}_n|\boldsymbol{\theta})\}\pi(\boldsymbol{\theta}|\boldsymbol{X}_n)d\boldsymbol{\theta},$$

a deviance information criterion DIC is proposed.

$$\mathrm{DIC} = -2\int \log\{f(\boldsymbol{X}_n|\boldsymbol{\theta})\}\pi(\boldsymbol{\theta}|\boldsymbol{X}_n)d\boldsymbol{\theta} + P_D. \tag{7.10}$$

An advantage of DIC is that the DIC is easily calculated from the generated MCMC samples.

Let us expand the log-likelihood around the posterior mean $\bar{\boldsymbol{\theta}}_n$

$$\log f(\boldsymbol{X}_n|\boldsymbol{\theta}) \approx \log f\left(\boldsymbol{X}_n|\bar{\boldsymbol{\theta}}_n\right) + \left(\boldsymbol{\theta} - \bar{\boldsymbol{\theta}}_n\right)^T \left.\frac{\partial \log f(\boldsymbol{X}_n|\boldsymbol{\theta})}{\partial \boldsymbol{\theta}}\right|_{\boldsymbol{\theta}=\bar{\boldsymbol{\theta}}_n}$$
$$-\frac{n}{2}\left(\boldsymbol{\theta} - \bar{\boldsymbol{\theta}}_n\right)^T J_n(\bar{\boldsymbol{\theta}}_n)\left(\boldsymbol{\theta} - \bar{\boldsymbol{\theta}}_n\right),$$

with

$$J_n(\bar{\boldsymbol{\theta}}_n) = -\frac{1}{n}\left.\frac{\partial^2 \log f(\boldsymbol{X}_n|\boldsymbol{\theta})}{\partial \boldsymbol{\theta}\partial \boldsymbol{\theta}^T}\right|_{\boldsymbol{\theta}=\bar{\boldsymbol{\theta}}_n}$$

the observed information matrix evaluated at the posterior mean $\bar{\boldsymbol{\theta}}_n$. Taking the expectation with respect to the posterior distribution of parameter gives

$$\int \log\{f(\boldsymbol{X}_n|\boldsymbol{\theta})\}\pi(\boldsymbol{\theta}|\boldsymbol{X}_n)d\boldsymbol{\theta} \approx \log\{f(\boldsymbol{X}_n|\bar{\boldsymbol{\theta}}_n)\} - \frac{n}{2}\mathrm{tr}\left\{J_n(\bar{\boldsymbol{\theta}}_n)V(\boldsymbol{\theta})\right\},$$

where $V(\boldsymbol{\theta})$ is the posterior covariance matrix of $\boldsymbol{\theta}$. Note that, under a large sample situation with $\log\pi(\boldsymbol{\theta}) = O_p(1)$, we can approximate the Bayesian posterior distribution of the model parameters $\pi(\boldsymbol{\theta}|\boldsymbol{X}_n)$ by multivariate normal distribution with mean the posterior mode $\hat{\boldsymbol{\theta}}_n$ and covariance matrix $n^{-1}J_n^{-1}(\hat{\boldsymbol{\theta}}_n)$.

Noting also that the posterior mode $\hat{\boldsymbol{\theta}}_n$ and the posterior mean $\bar{\boldsymbol{\theta}}_n$ converges to the same value as $n \to \infty$, a Bayesian measure of model complexity p_D, can be approximated as

$$P_D \approx n\mathrm{tr}\left\{J_n(\bar{\boldsymbol{\theta}}_n)V(\boldsymbol{\theta})\right\} \approx \mathrm{tr}\left\{I\right\} = p,$$

where p is the dimension of $\boldsymbol{\theta}$. Thus, P_D reduces to the number of parameters in the Bayesian models.

DIC is applied to various Bayesian model selection problems. Berg et al. (2004) illustrated the performance of DIC in discriminating between various different stochastic volatility models using daily returns data on the Standard & Poors (S&P) 100 index. Ando (2006) applied DIC for selecting stochastic volatility models and regime switching stochastic volatility models using daily returns data on the Nikkei 225 index. Celeux et al. (2006) extended DIC so that we compare models for incomplete data and proposed a number of adaptations. van der Linde (2005) applied DIC for variable selection problems.

Spiegelhalter et al. (2002) gave an asymptotic justification for deriving the effective number of parameters and showed that it is a natural extension of AIC (Akaike (1974)). However, as pointed out by Robert and Titterington (2002), the same data were used twice in the construction of P_D. As a result, the predictive distribution chosen by DIC overfits the observed data.

7.2.1 Example: Hierarchical Bayesian modeling for logistic regression

To illustrate DIC, we consider a hierarchical Bayesian modeling for logistic regression that retains the ability to take the characteristics of specific individuals into account. For an individual α ($\alpha = 1, ..., n$), the model is expressed as

$$\Pr(y_{\alpha i} = 1|\boldsymbol{x}_\alpha) = \frac{\exp(\boldsymbol{x}_{\alpha i}^T\boldsymbol{\beta}_\alpha)}{1 + \exp(\boldsymbol{x}_{\alpha i}^T\boldsymbol{\beta}_\alpha)},$$

where $\boldsymbol{\beta}_h$ is a p-dimensional vector of regression coefficients for an individual α, $\boldsymbol{x}_\alpha$ is a p-dimensional vector of attributes, and an index $i = 1, ..., n_i$ is the number of questions to each individual.

An advantage of hierarchical approach is heterogeneity that is incorporated into the model with a random-effect whose mean is a function of observable covariates ($\boldsymbol{z}$):

$$\boldsymbol{\beta}_\alpha = \Gamma \boldsymbol{z}_\alpha + \boldsymbol{e}_\alpha, \quad \boldsymbol{e}_\alpha \sim N(\mathbf{0}, \Sigma_\beta),$$

where Γ is $p \times q$ matrix of regression coefficients. and Σ_β is the $q \times q$ covariance matrix that characterizes the extent of unobserved heterogeneity. For prior densities of $\boldsymbol{\gamma} = \text{vec}\Gamma$ and Σ_β, we use

$$\pi(\boldsymbol{\gamma}|\Sigma_\beta) = N(\boldsymbol{\gamma}_0, A^{-1} \otimes \Sigma_\beta) \quad \text{and} \quad \pi(\Sigma_\beta) = IW(\nu_0, \Sigma_0).$$

An implementation of the MCMC algorithm for this model can be done using the **R** function **rhierBinLogit** in the package **bayesm**. Setting $p = 2$, $q = 3$, $n_i = 5$ and

$$\Gamma = \begin{pmatrix} -2.0 & -1.0 & 0.0 \\ -1.0 & 1.1 & -0.5 \end{pmatrix},$$

we generated a set of $n = 50$ observations. The first element of $\boldsymbol{z}_\alpha$ is 1, the second element is uniform between 20 and 50 (age), and the third element is 0 or 1 (gender), respectively.

Using the **R** function **rhierBinLogit**, we generated 1,000 posterior samples. First 1,000 samples are discarded as a burn in period, and then we stored every 10th iterations. For hyperparameters, we set $\boldsymbol{\gamma}_0 = \mathbf{0}$, $\nu_0 = 10$ and $\Sigma_0 = 10I$. A Bayesian measure of model complexity p_D, is calculated by using the posterior mean of the deviance (=64.02) minus the deviance evaluated at the posterior mode (=42.85), and thus we obtained $P_D = 21.17$. Then the DIC score is 85.19 ($= 64.02 + 21.17$).

In marketing research, we often want to quantify customer preferences. If customers take one from alternative, the above model provides an useful tool. In this case, $\boldsymbol{x}_{\alpha i}$ may contain characteristics of a product $\boldsymbol{x}$, including brand, service quality, price, etc. Also, an observable covariates $\boldsymbol{z}$ can include customer's age, gender, living area, etc. Once we apply the above method to such dataset, we may quantify customer's preferences. Various applications of Bayesian methods for marketing topics are covered by Rossi et al. (2005).

7.3 A minimum posterior predictive loss approach

As well as BPIC (Ando (2007)), Gelfand and Ghosh (1998) also considered the Bayesian model selection from predictive perspectives. Based on the idea of evaluating models by comparing observed data to predictions (Kadane and Dickey (1980)), Gelfand and Ghosh (1998) proposed to minimize the expected posterior loss over all possible predictions of future observations. Assuming that the future observations have the same distribution as the observed data, this approach aims to obtain good predictions for future observations.

In this framework, a loss function has the general form

$$L(\boldsymbol{X}_{\text{rep}}, a; \boldsymbol{X}_{\text{obs}}) = L(\boldsymbol{X}_{\text{rep}}, a) + kL(\boldsymbol{X}_{\text{obs}}, a),$$

where $k \geq 0$ is the weight parameter, $\boldsymbol{X}_{\text{obs}}$ denotes the observed data, $\boldsymbol{X}_{\text{rep}}$ denotes the future observations assumed to come from the same distribution as the observed data, and a is the "action", usually an estimate. The parameter k makes the tradeoff between the two loss functions $L(\boldsymbol{X}_{\text{rep}}, a)$, and $L(\boldsymbol{X}_{\text{obs}}, a)$. In constrast to Ando (2007), in Gelfand and Ghosh (1998), the true model $g(x)$ that generates future observations is assumed to be the predictive density under the specified Bayesian model.

Using an example from Gelfand and Ghosh (1998), we illustrate their approach. Let us define

$$D_k(M) = \sum_{\alpha=1}^{n} \min_{a_\alpha} E_{x_{\alpha,\text{rep}}|\boldsymbol{X}_{\text{obs}}} \left[L(x_{\alpha,\text{rep}}, a_\alpha; \boldsymbol{X}_{\text{obs}})\right],$$

where M represents one of the competing models. Then the general form of the described loss becomes

$$D_k(M) = \sum_{\alpha=1}^{n} \min_{a_\alpha} E_{x_{\alpha,\text{rep}}|\boldsymbol{X}_{\text{obs}}} \left[L(x_{\alpha,\text{rep}}, a) + kL(\boldsymbol{X}_{\text{obs}}, a)\right].$$

When we take the squared loss function $L(x, a) = (x - a)^2$, for a fixed a_α, the αth term in this sum is

$$\sigma_\alpha^2 + (a_\alpha - \mu_\alpha)^2 + k(a_\alpha - x_{\alpha,\text{obs}}),$$

where σ_α^2 is the variance of $x_{\alpha,\text{rep}}$ given $\boldsymbol{X}_{\text{obs}}$ and the Bayesian model M, and μ_α is the expected value of $x_{\alpha,\text{rep}}$ given $\boldsymbol{X}_{\text{obs}}$ and the Bayesian model M. Note that these quantities are depending on the specified Bayesian model M. We suppressed the dependence on the specified Bayesian model M in the notation for simplicity.

The minimizing estimate a_α is then

$$\frac{\mu_\alpha + kx_{\alpha,\text{obs}}}{1 + k}.$$

Putting this estimate into the expression for $D_k(M)$, we have

$$D_k(M) = \frac{k}{1+k} \sum_{\alpha=1}^{n} (\mu_\alpha - x_{\alpha,\text{obs}})^2 + \sum_{\alpha=1}^{n} \sigma_\alpha^2.$$

We can see that the first term as a goodness-of-fit measure which quantifies a closeness between the predicted value and the observation. Gelfand and Ghosh (1998) regarded the second term as a type of penalty. Gelfand and Ghosh (1998) further pointed out that, under the assumption that x_α comes from a normal distribution, the first term is equivalent to the likelihood ratio statistic with μ_α replacing the maximum likelihood estimate of the mean of x_α.

Suppose that $\boldsymbol{y}_n$ comes from a normal linear model

$$\boldsymbol{y}_n = X_n\boldsymbol{\beta} + \boldsymbol{\varepsilon}, \quad \boldsymbol{\varepsilon} \sim N(\boldsymbol{0}, \sigma^2 I),$$

with σ^2 known. We use a normal prior $\pi(\boldsymbol{\beta}) = N(\boldsymbol{\mu}_\beta, \Sigma_\beta)$. Gelfand and Ghosh (1998) investigated that if the prior variance is very large, then the predictive density of $\boldsymbol{y}_{\text{rep}}$ given $\boldsymbol{y}_n$ and X_n has an approximate

$$\boldsymbol{y}_{\text{rep}} = N\left(X_n\hat{\boldsymbol{\beta}}, \sigma^2\left(X_n(X_n^TX_n)^{-1}X_n^T + I\right)\right),$$

where $\hat{\boldsymbol{\beta}} = (X_n^TX_n)^{-1}X_n^T\boldsymbol{y}_n$. Then we approximately have

$$D_k(M) = \left(\boldsymbol{y}_n - X_n\hat{\boldsymbol{\beta}}\right)^T\left(\boldsymbol{y}_n - X_n\hat{\boldsymbol{\beta}}\right) + \sigma^2(n+p).$$

Although we appriximatly made the calculation of $D_k(M)$ explicitly, as pointed out by Gelfand and Ghosh (1998), in general, however, we have to employ a combination of asymptotic expansions and Monte Carlo simulation for the evaluation of integrals.

7.4 Modified Bayesian information criterion

Schwarz (1978) proposed the Bayesian information criterion, BIC, from Bayesian aspects. The number of parameters is a measure of the complexity of the model. However, in nonlinear models, especially nonlinear models estimated by the penalized maximum likelihood method, the number of parameters is not a suitable measure of model complexity, since the complexity may depend on the penalty term. The concept of number of parameters was extended to the effective number of parameters by Hastie and Tibshirani (1990), Moody (1992), Eilers and Marx (1996), Hurvich et al. (1998), Spiegelhalter et al. (2002) and others.

Recall the Gaussian nonlinear regression model $f(y_\alpha|\boldsymbol{x}_\alpha;\boldsymbol{\theta})$ based on the linear combination of basis functions in (5.23) estimated by the maximum penalized likelihood method

$$\hat{\boldsymbol{\theta}}_n = \text{argmax}_\theta\left[\sum_{\alpha=1}^n \log f(y_\alpha|\boldsymbol{x}_\alpha;\boldsymbol{\theta}) - \frac{n\lambda}{2}\boldsymbol{w}^TD\boldsymbol{w}\right].$$

Then the fitted value $\hat{\boldsymbol{y}}_n$ can be expressed as $\hat{\boldsymbol{y}}_n = H\boldsymbol{y}_n$ for the prescribed value of $\lambda_0 = \sigma^2\lambda$, where H is the smoother matrix given by

$$H = B(B^TB + n\lambda_0 K)^{-1}B^T,$$

where $B = (\boldsymbol{b}(\boldsymbol{x}_1), ..., \boldsymbol{b}(\boldsymbol{x}_n))^T$ is the $n \times m$ dimensional design matrix. Hastie and Tibshirani (1990) used the trace of the smoother matrix as an approximation to the effective number of parameters:

$$\text{Effective number of parameters} \equiv \text{tr}\{H\},$$

the sum of the diagonal elements of H.

The basic idea of the effective number of parameters is described as follows. Consider the parametric model $\boldsymbol{y}_n = X_n\boldsymbol{\beta} + \boldsymbol{\varepsilon}, \quad \boldsymbol{\varepsilon} \sim N(\boldsymbol{0}, \sigma^2 I)$, $\boldsymbol{\beta} \in R^p$. When $\boldsymbol{\beta}$ is estimated by maximum likelihood method, the smoother matrix reduces to the hat matrix $R = X_n(X_n'X_n)^{-1}X_n'$. In this case, the trace of the hat matrix is equivalent to the number of free parameters.

$$\text{tr}\{R\} = \text{tr}\left\{X_n(X_n^TX_n)^{-1}X_n^T\right\} = \text{tr}\left\{(X_n^TX_n)^{-1}(X_n^TX_n)\right\} = p. \tag{7.11}$$

Hence the trace of the smoother matrix can be used as an approximation to the effective number of parameters.

By replacing the number of free parameters in BIC by $\text{tr}\{H\}$, we formally obtain Modified BIC (Eilers and Marx (1998)) for evaluating the nonlinear regression model estimated by the penalized maximum likelihood method in the form

$$\text{MBIC} = -2\log f\left(\boldsymbol{Y}_n|X_n, \hat{\boldsymbol{\theta}}_n\right) + \log n \times \text{tr}\{H\}, \tag{7.12}$$

where $\log f\left(\boldsymbol{Y}_n|X_n, \hat{\boldsymbol{\theta}}_n\right)$ is the log-likelihood value evaluated at the maximum penalized likelihood estimate $\hat{\boldsymbol{\theta}}_n$ or, equivalently, the posterior mode that maximizes $\pi(\boldsymbol{\theta}|\boldsymbol{Y}_n, X_n)$.

An advantage of the criterion MBIC is that they can be applied in an automatic way to each practical situation where there is a smoother matrix. There is however no theoretical justification, since BIC is a criterion for evaluating models estimated by the maximum likelihood method (Konishi et al. (2004)).

Modified BIC can also be used for the model evaluation of the logistic and Poisson regression models respectively given as

$$f(y_\alpha|\boldsymbol{x}_\alpha; \boldsymbol{w}) = \pi(\boldsymbol{x}_\alpha; \boldsymbol{w})^{y_\alpha}\{1 - \pi(\boldsymbol{x}_\alpha; \boldsymbol{w})\}^{1-y_\alpha},$$

and

$$f(y_\alpha|\boldsymbol{x}_\alpha; \boldsymbol{w}) = \frac{\exp\{-\gamma(\boldsymbol{x}_\alpha; \boldsymbol{w})\}\gamma(\boldsymbol{x}_\alpha; \boldsymbol{w})^{y_\alpha}}{y_\alpha!},$$

where the conditional expectations of y_α are

$$\pi(\boldsymbol{x}_\alpha) = \frac{\exp\left\{\boldsymbol{w}^T\boldsymbol{b}(\boldsymbol{x}_\alpha)\right\}}{1 + \exp\left\{\boldsymbol{w}^T\boldsymbol{b}(\boldsymbol{x}_\alpha)\right\}},$$

and

$$\gamma(\boldsymbol{x}_\alpha) = \exp\left\{\boldsymbol{w}^T \boldsymbol{b}(\boldsymbol{x}_\alpha)\right\}.$$

In this case, the smoother matrix can be expressed as

$$H = B(B^T W B + n\lambda K)^{-1} B^T W,$$

where the $n \times n$ diagonal matrix W with α-th element is given as

$$\text{Logistic regression}: \quad W_{\alpha\alpha} = \hat{\pi}(\boldsymbol{x}_\alpha)\{1 - \hat{\pi}(\boldsymbol{x}_\alpha)\},$$

with

$$\hat{\pi}(\boldsymbol{x}_\alpha) = \frac{\exp\left\{\hat{\boldsymbol{w}}_n^T \boldsymbol{b}(\boldsymbol{x}_\alpha)\right\}}{1 + \exp\left\{\hat{\boldsymbol{w}}_n^T \boldsymbol{b}(\boldsymbol{x}_\alpha)\right\}},$$

and

$$\text{Poisson regression}: \quad W_{\alpha\alpha} = \hat{\gamma}(\boldsymbol{x}_\alpha),$$

with

$$\hat{\gamma}(\boldsymbol{x}_\alpha) = \exp\left\{\hat{\boldsymbol{w}}_n^T \boldsymbol{b}(\boldsymbol{x}_\alpha)\right\}.$$

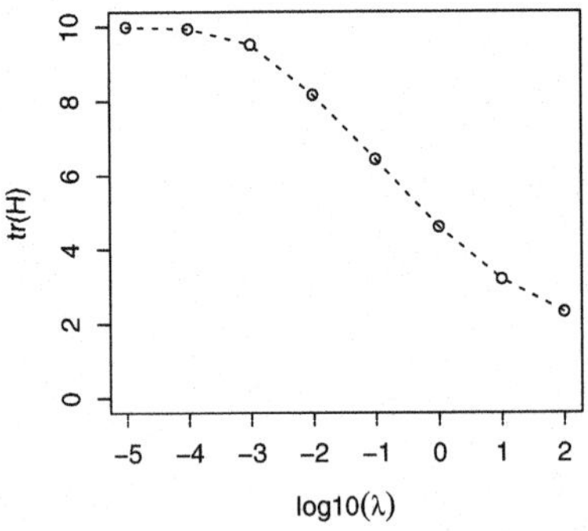

FIGURE 7.3: A behavior of MBIC under various values of λ.

7.4.1 Example: *P*-spline regression model with Gaussian noise

We generated a set of $n = 50$ observations from

$$y_\alpha = \exp\{-x_\alpha \sin(2\pi x_\alpha)\} + 1 + \varepsilon_\alpha, \quad \alpha = 1, \ldots, 50,$$

where $x_\alpha = (50 - \alpha)/49$ $(\alpha = 1, ..., 50)$ and $\varepsilon_\alpha \sim N(0, 0.3^2)$. To estimate the true curve, B-spline regression model

$$y_\alpha = \sum_{j=1}^{p} \beta_j \phi_j(x_\alpha) + \varepsilon_\alpha, \quad \varepsilon_\alpha \sim N(0, \sigma^2)$$

is considered. We set the number of basis to be $p = 10$ for a simplicity of explanation.

Figure 7.3 shows the trace of the smoother matrix, an approximation to the effective number of parameters. As shown in Figure 7.3, the smaller value of λ gives larger value of effective number of parameters. Figures 7.4 and 7.5 show the fitted curve and behavior of MBIC under various values of λD MBIC attained its minimum around $\lambda = 0.1$ and the corresponding fitted curve captures the true curve well.

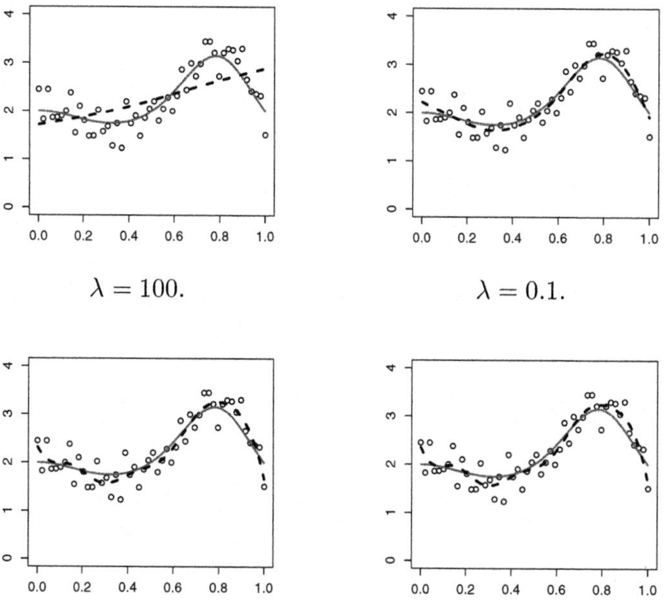

$\lambda = 100.$ $\lambda = 0.1.$

$\lambda = 0.001.$ $\lambda = 0.00001.$

FIGURE 7.4: Fitted curves under the various settings of λ.

7.4.2 Example: P-spline logistic regression

In Section 2.4.7, we described a nonlinear logistic regression to investigate the relationship between the kyphosis (y) and age (x). Here we consider P-

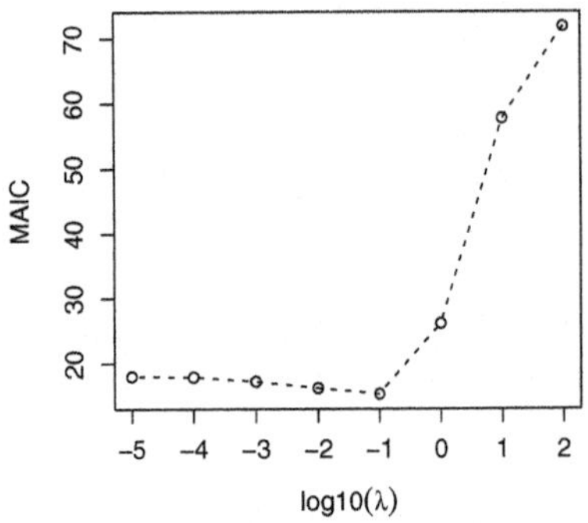

FIGURE 7.5: MBIC score.

spline logistic regression model

$$\frac{\Pr(y_\alpha = 1)}{\Pr(y_\alpha = 1)} = \sum_{j=1}^{p} \beta_j \phi_j(x_\alpha).$$

with the number of basis functions to be $p = 13$. Preparing various values of the smoothing parameter λ, we estimated the parameter vector $\boldsymbol{\beta}$. Figure 7.6 shows the estimated conditional probabilities $\hat{\pi}(\text{kyphosis} = 1|\text{age})$. The corresponding MBIC scores are also shown in the figure. The MBIC scores are minimized at $\lambda = 0.01$ among the candidate values $\lambda = 10, 0.1, 0.001, 0.00001$. Thus, we conclude that the model with $\lambda = 0.01$ is the best model among the candidates. Figure 7.6 (b) indicates that the operation risk has a peak around the 100th month after birth.

7.5 Generalized information criterion

This section explains the generalized information criteria through the generalized linear regression modeling. Suppose that we have n independent observations y_α with respect to the p-dimensional design points $\boldsymbol{x}_\alpha$ for $\alpha = 1, ..., n$. It is assumed that the responses y_α are generated from an unknown true distribution $G(y|\boldsymbol{x})$ with density $g(y|\boldsymbol{x})$. In practice we select an approximating model to the true model $g(y|\boldsymbol{x})$ generating the data. In generalized linear models y_α are assumed to be drawn from the exponential family of distributions

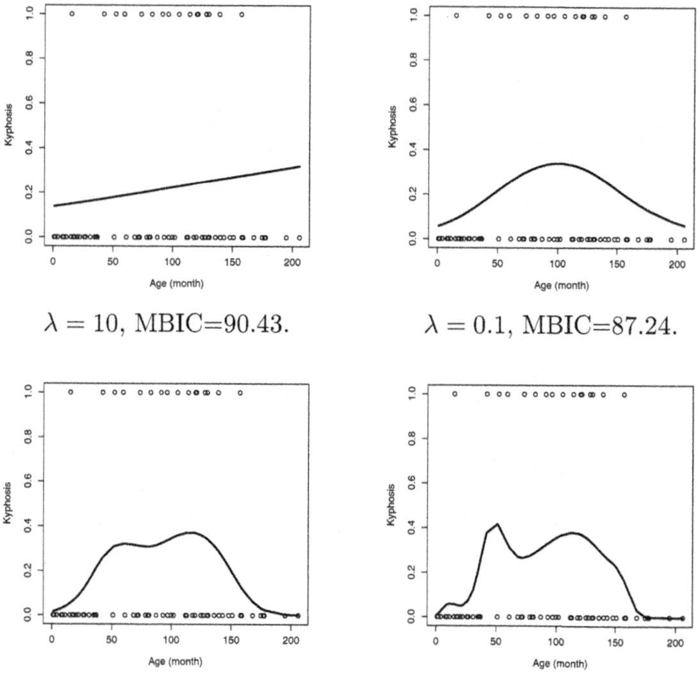

$\lambda = 10$, MBIC=90.43.

$\lambda = 0.1$, MBIC=87.24.

$\lambda = 0.001$, MBIC=99.11.

$\lambda = 10^{-5}$, MBIC=112.38.

FIGURE 7.6: The estimated conditional probabilities $\hat{\pi}(\text{kyphosis} = 1|\text{age})$ under the various settings of λ.

with densities

$$f(y_\alpha|\boldsymbol{x}_\alpha; \xi_\alpha, \phi) = \exp\left\{\frac{y_\alpha \xi_\alpha - u(\xi_\alpha)}{\phi} + v(y_\alpha, \phi)\right\},$$

where $u(\cdot)$ and $v(\cdot, \cdot)$ are functions specific to each distribution and ϕ is an unknown scale parameter.

For the predictor η_α, we assume a linear combination of basis functions

$$\eta_\alpha = \sum_{k=1}^{m} w_k b_k(\boldsymbol{x}_\alpha), \qquad \alpha = 1, 2, ..., n.$$

Combining the random component and the systematic component, we have a generalized linear model with basis expansion predictors

$$f(y_\alpha|\boldsymbol{x}_\alpha; \boldsymbol{\theta}) = \exp\left[\frac{y_\alpha r\left\{\boldsymbol{w}^T \boldsymbol{b}(\boldsymbol{x}_\alpha)\right\} - s\left\{\boldsymbol{w}^T \boldsymbol{b}(\boldsymbol{x}_\alpha)\right\}}{\phi} + v(y_\alpha, \phi)\right],$$

where $\boldsymbol{\theta} = (\boldsymbol{w}^T, \phi)^T$. Using a singular multivariate normal prior density (Konishi et al. (2004), Ando (2007))

$$\pi(\boldsymbol{\theta}) = \left(\frac{n\lambda}{2\pi}\right)^{(m-d)/2} |R|_+^{1/2} \exp\left\{-n\lambda\frac{\boldsymbol{\theta}^T R\boldsymbol{\theta}}{2}\right\},$$

the model is constructed by maximizing the penalized log-likelihood function

$$\log f(\boldsymbol{Y}_n|X_n, \boldsymbol{\theta}) + \log \pi(\boldsymbol{\theta}) = \sum_{\alpha=1}^{n} \log f(y_\alpha|\boldsymbol{x}_\alpha; \boldsymbol{\theta}) - \frac{n\lambda}{2}\boldsymbol{w}^T D\boldsymbol{w},$$

where $f(\boldsymbol{Y}_n|\boldsymbol{\theta}, X_n)$ is the likelihood function, $\boldsymbol{Y}_n = \{y_1, ..., y_n\}$ and $X_n = \{\boldsymbol{x}_1, ..., X_n\}$. With respect to the matrix R, see Section 5.6.1. Thus the posterior mode $\hat{\boldsymbol{\theta}}_n$ is corresponding to the the penalized maximum likelihood estimate.

To evaluate the estimated model, we outline the derivation of an information criterion for the statistical model $f(y|\boldsymbol{x}; \hat{\boldsymbol{\theta}}_n)$ with estimated weight parameter vector $\hat{\boldsymbol{w}}_n$ and scale parameter $\hat{\phi}_n$ estimated by the penalized maximum likelihood method.

Generalized information criteria

In order to assess the closeness of $f(y|\boldsymbol{x}; \hat{\boldsymbol{\theta}}_n)$ to $g(y|\boldsymbol{x})$, we use the Kullback-Leibler information (Kullback and Leibler (1951))

$$\begin{aligned} I\{g, f\} &= E_{G(z|\boldsymbol{x})}\left[\log \frac{g(z|\boldsymbol{x})}{f(z|\boldsymbol{x}; \hat{\boldsymbol{\theta}}_n)}\right] \\ &= E_{G(z|\boldsymbol{x})}\left[\log g(z|\boldsymbol{x})\right] - E_{G(z|\boldsymbol{x})}\left[\log f(z|\boldsymbol{x}; \hat{\boldsymbol{\theta}}_n)\right]. \end{aligned} \quad (7.13)$$

An information criterion is obtained as an estimator of the Kullback-Leibler information or, equivalently, the expected log-likelihood

$$E_{G(z|\boldsymbol{x})}\left[\log f(z|\boldsymbol{x}; \hat{\boldsymbol{\theta}}_n)\right]$$

and is, in general, given by

$$\log f(\boldsymbol{y}_n|X_n; \hat{\boldsymbol{\theta}}_n) - \hat{\text{bias}}(G). \quad (7.14)$$

Here $\hat{\text{bias}}(G)$ is an estimator of the bias of the log-likelihood, $\log f(\boldsymbol{y}_n|X_n; \hat{\boldsymbol{\theta}}_n)$, in estimating the expected log-likelihhod defined by

$$\text{bias}(G) = E_{G(\boldsymbol{y}_n|X_n)}\left[\log f(\boldsymbol{y}_n|X_n; \hat{\boldsymbol{\theta}}_n) - E_{G(\boldsymbol{z}_n|X_n)}[\log f(\boldsymbol{z}_n|X_n; \hat{\boldsymbol{\theta}}_n)]\right] (7.15)$$

where $g(\boldsymbol{z}_n|X_n) = \Pi_{\alpha=1}^{n} g(z_\alpha|\boldsymbol{x}_\alpha)$ for the future observations $z_1, ..., z_n$.

Konishi and Kitagawa (1996) considered an asymptotic bias for a statistical model with functional estimator and approximated the bias by a function of the empirical influence function of estimators and the score function of a specified parametric model. It may be seen that the regularized estimator $\hat{\boldsymbol{\theta}}_n$ can be expressed as $\hat{\boldsymbol{\theta}}_n = \boldsymbol{T}(\hat{G})$ for the functional $\boldsymbol{T}(\cdot)$ defined by

$$\int \frac{\partial}{\partial \boldsymbol{\theta}} \left(\log f(y|\boldsymbol{x};\boldsymbol{\theta}) - \frac{\lambda}{2}\boldsymbol{w}^T D\boldsymbol{w} \right)\Bigg|_{\boldsymbol{\theta}=\boldsymbol{T}(G)} dG(y,\boldsymbol{x}) = \boldsymbol{0}, \tag{7.16}$$

where G and $\hat{G}$ are respectively the joint distribution of $(\boldsymbol{x}, y)$ and the empirical distribution function based on the data. Replacing G in (7.16) by $G_\varepsilon = (1-\varepsilon)G + \varepsilon\delta_{(y,\boldsymbol{x})}$ with $\delta(y,\boldsymbol{x})$ being a point of mass at $(y,\boldsymbol{x})$ and differentiating with respect to ε yield the influence function of the regularized estimator $\hat{\boldsymbol{\theta}}_n = \boldsymbol{T}(\hat{G})$ in the form

$$\boldsymbol{T}^{(1)}(y|\boldsymbol{x};G) = R(G)^{-1}\frac{\partial}{\partial \boldsymbol{\theta}}\left\{\log f(z|\boldsymbol{x};\boldsymbol{\theta}) - \frac{\lambda}{2}\boldsymbol{w}^T D\boldsymbol{w}\right\}\Bigg|_{\boldsymbol{\theta}=\boldsymbol{T}(G)}, \tag{7.17}$$

where

$$R(G) = -\int \frac{\partial^2\left\{\log f(y|\boldsymbol{x};\boldsymbol{\theta}) - \frac{\lambda}{2}\boldsymbol{w}^T D\boldsymbol{w}\right\}}{\partial\boldsymbol{\theta}\partial\boldsymbol{\theta}^T} dG(y,\boldsymbol{x}).$$

It follows from Theorem 2.1 in Konishi and Kitagawa (1996) that the bias in (7.15) is asymptotically given by

$$b(G) = \mathrm{tr}\left\{Q(G)R(G)^{-1}\right\} + o\left(\frac{1}{n}\right), \tag{7.18}$$

where

$$Q(G) = \int\left\{\frac{\partial\{\log f(y|\boldsymbol{x};\boldsymbol{\theta}) - \frac{\lambda}{2}\boldsymbol{w}^T D\boldsymbol{w}\}}{\partial\boldsymbol{\theta}}\frac{\partial \log f(y|\boldsymbol{x};\boldsymbol{\theta})}{\partial\boldsymbol{\theta}^T}\right\}\Bigg|_{\boldsymbol{\theta}=\boldsymbol{T}(G)} dG(y,\boldsymbol{x}).$$

By replacing the unknown distribution G by the empirical distribution $\hat{G}$, we have a generalized information criterion

$$\mathrm{GIC} = -2\sum_{\alpha=1}^{n}\log f(y_\alpha|\boldsymbol{x}_\alpha;\hat{\boldsymbol{\theta}}_n) + 2\mathrm{tr}\left\{Q(\hat{G})R^{-1}(\hat{G})\right\}. \tag{7.19}$$

Substituting the sampling density into the Equation (7.19) and differentiating the results with respect to $\boldsymbol{\theta}$, we obtain an information criterion for evaluating the statistical model $f(y_\alpha|\boldsymbol{x}_\alpha;\hat{\boldsymbol{\theta}}_n)$. Here $Q(\hat{G})$ and $R(\hat{G})$ are $(m+1)\times(m+1)$

matrices given by

$$Q(\hat{G}) = \frac{1}{n\hat{\phi}_n}\begin{pmatrix} B^T\Lambda/\hat{\phi}_n - \lambda D\hat{\boldsymbol{w}}_n\mathbf{1}_n^T \\ \boldsymbol{p}^T \end{pmatrix}\left(\Lambda B,\ \hat{\phi}_n\boldsymbol{p}\right),$$

$$R(\hat{G}) = \frac{1}{n\hat{\phi}_n}\begin{pmatrix} B^T\Gamma B + n\hat{\phi}_n\lambda D & B^T\Lambda\mathbf{1}_n/\hat{\phi}_n \\ \mathbf{1}_n^T\Lambda & \\ B/\hat{\phi}_n & -\hat{\phi}_n\boldsymbol{q}^T\mathbf{1}_n \end{pmatrix}.$$

Here Λ and Γ are $n \times n$ diagonal matrices with i-th diagonal elements

$$\Lambda_{ii} = \frac{y_i - \hat{\mu}_i}{u''(\hat{\xi}_i)h'(\hat{\mu}_i)},$$

$$\Gamma_{ii} = \frac{(y_i - \hat{\mu}_i)\{u'''(\hat{\xi}_i)h'(\hat{\mu}_i) + u''(\hat{\xi}_i)^2 h''(\hat{\mu}_i)\}}{\{u''(\hat{\xi}_i)h'(\hat{\mu}_i)\}^3} + \frac{1}{u''(\hat{\xi}_i)h'(\hat{\mu}_i)^2},$$

respectively, and $\boldsymbol{p}$ and $\boldsymbol{q}$ are n-dimentinal vectors with i-th elements

$$p_i = -\frac{y_i r(\hat{\boldsymbol{w}}_n^T\boldsymbol{b}(\boldsymbol{x}_i)) - s(\boldsymbol{w}^T\boldsymbol{b}(\boldsymbol{x}_i))}{\hat{\phi}^2} + \frac{\partial}{\partial\phi}v(y_i,\phi)\bigg|_{\phi=\hat{\phi}_n} \quad \text{and} \quad q_i = \frac{\partial p_i}{\partial\phi}\bigg|_{\phi=\hat{\phi}_n}.$$

Also, $r(\cdot)$ and $s(\cdot)$ are defined by $r(\cdot) = u'^{-1} \circ h^{-1}(\cdot)$ and $s(\cdot) = u \circ u'^{-1} \circ h^{-1}(\cdot)$, respectively. See Exercise 3.3.5, where the basics of a generalized linear model is described.

For theoretical and practical work on information-theoretic criteria, we refer to Akaike (1973, 1974), Linhart and Zucchini (1986), Rao and Wu (2001), Konishi and Kitagawa (2003, 2008), Burnham and Anderson (2002) etc. Section 8.4 provides a derivation of generalized information criterion.

7.5.1 Example: Heterogeneous error model for the analysis motorcycle impact data

The motorcycle impact data were simulated to investigate the efficacy of crash helmets and comprise a series of measurements of head acceleration in units of gravity and times in milliseconds after impact (Silverman (1985), Härdle (1990)). This data set has been extensively used to examine smoothing techniques. The common weakness of the previous approach appears to lie in the assumption for error variance, since the motorcycle impact data show clear heteroscedasticity. Ando et al. (2008) proposed a nonlinear regression model with heterogeneous Gaussian noise.

For design points $\{x_\alpha; \alpha = 1, ..., n\}$ we divide the interval $R = (\min_\alpha\{x_\alpha\}, \max_\alpha\{x_\alpha\})$ into, e.g., three parts $R_1 = \{x | x \leq a\}$, $R_2 = \{x | a < x \leq b\}$ and $R_3 = \{x | b < x\}$ so that $R = R_1 \cup R_2 \cup R_3$, where a and b are unknown location parameters. Let $\boldsymbol{\delta}(x_\alpha) = (\delta(x_\alpha, R_1), \delta(x_\alpha, R_2), \delta(x_\alpha, R_3))^T$, where $\delta(x_\alpha, R_j) = 1$ if $x_\alpha \in R_j$, $= 0$ otherwise, and put $\boldsymbol{\sigma} = (\sigma_1^2, \sigma_2^2, \sigma_3^2)^T$. By replacing the variance σ^2 in the nonlinear Gaussian regression model (5.23) with $\boldsymbol{\delta}^T\boldsymbol{\sigma}$, Ando et al. (2008) considered a nonlinear regression model with heterogeneous error variance

$$f(y_\alpha|x_\alpha;\boldsymbol{\theta}) = \frac{1}{\sqrt{2\pi\boldsymbol{\delta}^T\boldsymbol{\sigma}}}\exp\left\{-\frac{(y_\alpha - \boldsymbol{w}^T\boldsymbol{b}(x_\alpha))^2}{2\boldsymbol{\delta}^T\boldsymbol{\sigma}}\right\},$$

where $\boldsymbol{\theta} = (\boldsymbol{w}^T, \boldsymbol{\sigma})^T$ and $\boldsymbol{b}(x_\alpha)$ is a vector of basis functions given by Equation (5.21). We can estimate the parameters $\boldsymbol{w}$ and $\boldsymbol{\sigma}$ by finding the posterior mode $\hat{\boldsymbol{w}}_n$ and $\hat{\boldsymbol{\sigma}}_n$. The estimated model depends on λ, ν, m and also the positions a and b that change when error variance occurs. Ando et al. (2008) proposed an information criterion in the form:

$$\text{GIC}(m,\lambda,\nu,a,b) = -2\sum_{\alpha=1}^{n}\log f(y_\alpha|x_\alpha;\hat{\boldsymbol{\theta}}_n) + 2\text{tr}\left\{Q(\hat{G})R(\hat{G})^{-1}\right\},$$

where $Q(\hat{G})$ and $R(\hat{G})$ are $(m+4)\times(m+4)$ matrices given by

$$\begin{aligned} Q(\hat{G}) &= \frac{1}{n}\begin{pmatrix} B^T\Lambda - \lambda Q\hat{\boldsymbol{w}}_n\boldsymbol{1}_n^T \\ D^T(\Lambda^2-\Gamma)/2 \end{pmatrix}\left(\Lambda B,\ (\Lambda^2-\Gamma)D/2\right), \\ R(\hat{G}) &= \frac{1}{n}\begin{pmatrix} B^T\Gamma B + n\lambda Q & \Phi_\nu^T\Lambda\Gamma D \\ D^T\Gamma\Lambda B & D^T(2\Lambda^2-\Gamma)\Gamma D/2 \end{pmatrix}, \end{aligned}$$

with $\Lambda = \text{diag}[(y_1 - \hat{\boldsymbol{w}}_n^T\boldsymbol{b}(x_1))/\hat{\boldsymbol{\sigma}}^T\boldsymbol{\delta}(x_1), ..., (y_n - \hat{\boldsymbol{w}}_n^T\boldsymbol{b}(x_n))/\hat{\boldsymbol{\sigma}}^T\boldsymbol{\delta}(x_n)]$, $D = (\boldsymbol{\delta}(x_1), ..., \boldsymbol{\delta}(x_n))^T$ and $\Gamma = \text{diag}[1/\hat{\boldsymbol{\sigma}}^T\boldsymbol{\delta}(x_1), ..., 1/\hat{\boldsymbol{\sigma}}^T\boldsymbol{\delta}(x_n)]$. The values of m, λ, ν and the change points a, b were chosen as the minimizers of the criterion GIC.

Ando et al. (2008) found that the heteroscedasticity has occurred at $\hat{a} = 14.6$ ms and at $\hat{b} = 25.6$ ms. Figure 7.7 shows the motorcycle impact data with smoothed estimate. The value of GIC for the heteroscedastic model is 1124.42 which is smaller than 1221 for the homoscedastic model and also for the heteroscedastic model with one change point at $\hat{a} = 14.6$ ms. We therefore conclude that the heteroscedastic model with two change points at $\hat{a} = 14.6$ ms and $\hat{b} = 25.6$ ms would be more likely than homoscedastic model.

7.5.2 Example: Microarray data analysis

The DNA microarray measures the activities of several thousand genes simultaneously and the gene expression profiles are increasingly being performed in biological and medical researchers. Since transcriptional changes accurately reflect the status of cancers, the expression level of genes contains the keys to address fundamental problems relating to the prevention and cure of tumors, biological evolution mechanisms and drug discovery.

The gene expression data has very unique characteristics. First, it has very high-dimensionality and usually contains up to tens of thousands of genes. Second, the publicly available data size is very small; some have sizes below 100.

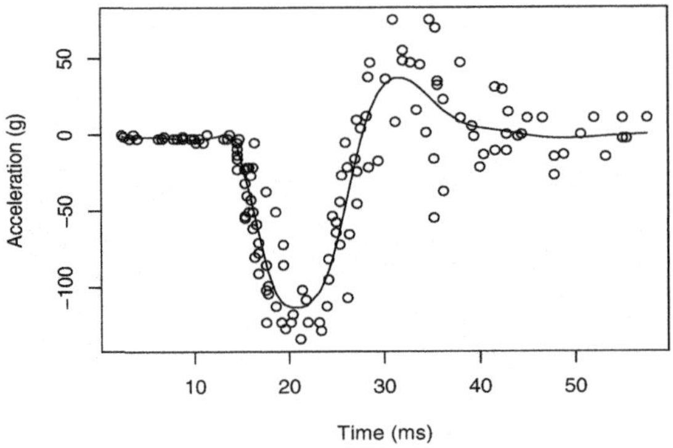

FIGURE 7.7: (From Ando, T. et al., *J. Stat. Plan. Infer.*, 138, 2008. With permission.) The motorcycle impact data with nonlinear smoothed estimate based on heteroscedastic model.

Third, most genes are not related to cancer classification. It is obvious that those existing statistical classification methods were not designed to handle this kind of data efficiently.

To deal with high-dimensional gene expression data, we here use radial basis function network multi-class classification models given in Section 5.6.2. Attention is not only focused on the classification of tumors but on the identification of "marker" genes that characterize the different tumor classes. We describe the results of applying the proposed multi-class classification method for four gene expression data sets. The first data set to be analyzed is small round blue cell tumors (Khan et al. (2001)).

Khan et al. (2001) successfully diagnosed the small round blue cell tumors (SRBCTs) of childhood into four classes; neuroblastoma (NB), rhabdomyosarcoma (RMS), non-Hodgkin lymphoma (NHL) and the Ewing family of tumors (EWS), using neural networks. The data set contains 2308 genes, out of 6567, after filtering for a minimal level of expression. The training set consists of 63 samples (NB: 12, RMS: 20, BL: 8, EWS: 23), and the test set has 20 SRBCT samples (NB: 6, RMS: 5, BL: 3, EWS: 6) and five non-SRBCTs. A logarithm base 10 of the expression levels was taken and standardized arrays are used before applying our classification method.

The multinomial logistic regression model based on radial basis functions was applied by Ando (2004). Estimation was done by finding the mode $\hat{\boldsymbol{w}}_n$ as illustrated in example 5.6.2. In order to identify the important genes for cancer classification, each of the genes were ranked by using the information criterion GIC (Ando and Konishi (2008)):

$$\mathrm{GIC} = -2\sum_{\alpha=1}^{n}\sum_{k=1}^{G} y_k^{(\alpha)} \log \pi_k(\boldsymbol{x}_\alpha;\hat{\boldsymbol{w}}_n) + 2\mathrm{tr}\left\{Q(\hat{\boldsymbol{w}}_n)R(\hat{\boldsymbol{w}}_n)^{-1}\right\} \tag{7.20}$$

where $Q(\hat{G})$ and $R(\hat{G})$ are given respectively by

$$\begin{aligned} Q(\hat{G}) &= -\frac{1}{n}(C\oplus A)^T(C\oplus A) + \frac{1}{n}D + \lambda I_{(G-1)(m+1)}; \\ R(\hat{G}) &= \frac{1}{n}((B-C)\oplus A)^T((B-C)\oplus A) - \frac{\lambda}{n}\hat{\boldsymbol{w}}_n\boldsymbol{1}_n^T((B-C)\oplus A); \end{aligned}$$

$A = (B,\ldots,B)$ is the $n\times(m+1)(G-1)$ dimensional matrix, $B = (\boldsymbol{y}_{(1)}\boldsymbol{1}_{m+1}^T,\ldots,\boldsymbol{y}_{(G-1)}\boldsymbol{1}_{m+1}^T)$, $C = (\boldsymbol{\pi}_{(1)}\boldsymbol{1}_{m+1}^T,\ldots,\boldsymbol{\pi}_{(G-1)}\boldsymbol{1}_{m+1}^T)$, $D = \mathrm{diag}\{B^T\mathrm{diag}\{\boldsymbol{\pi}_{(1)}\}B,\ldots,B^T\mathrm{diag}\{\boldsymbol{\pi}_{(G-1)}\}B\}$, $B = (\boldsymbol{b}(\boldsymbol{x}_1),\ldots,\boldsymbol{b}(\boldsymbol{x}_n))^T$, $\boldsymbol{y}_{(k)} = (y_k^{(1)},\ldots,y_k^{(n)})^T$, and $\boldsymbol{\pi}_{(k)} = (\pi_k(\boldsymbol{x}_1;\hat{\boldsymbol{w}}_n),\ldots,\pi_k(\boldsymbol{x}_n;\hat{\boldsymbol{w}}_n))^T$. Here the operator $\oplus$ means the elementwise product (suppose that the arbitrary matrices $A_{ij} = (a_{ij})$, $B_{ij} = (b_{ij})$ are given; then $A_{ij}\oplus B_{ij} = (a_{ij}\times b_{ij})$). We choose the optimum values of the smoothing parameter λ, the hyperparameter ν, and the number of basis functions m which minimize the value of the information criterion GIC in (7.20).

Based on these ranks, the step-wise variable selection procedure was used. Then the multi-class classification model that utilizes 35 genes was selected. Figure 7.8 shows the predicted posterior probabilities $\pi_1 = P(\mathrm{EWS}|\boldsymbol{x})$, $\pi_2 = P(\mathrm{BL}|\boldsymbol{x})$, $\pi_3 = P(\mathrm{NB}|\boldsymbol{x})$, $\pi_4 = P(\mathrm{RMS}|\boldsymbol{x})$ for the 20 test samples. The sample numbers 1–6, 7–9, 10–15 and 16–20 correspond to EWS, BL, NB and RMS samples. A sample is classified to a diagnostic category if it provides the highest posterior probability. The plot confirms that all the 20 test examples from four classes are classified correctly.

Hierarchical clustering results of the 83 samples with the selected 35 genes are shown in Figure 7.9. All 63 training and the 20 test SRBCTs correctly clustered within their diagnostic categories. Lee and Lee (2003) introduced the multi-category support vector machines, which is a recently proposed extension of the binary SVM, and applied it to these data. In their study, a perfect classification was achieved in testing the blind 20 samples. We also applied a full leave-one-out-cross-validation procedure and again achieved perfect classification. Reader can obtain the R program code from the book website.

Exercises

1. *Let us assume that we have n independent observations* $\boldsymbol{X}_n = \{x_1,...,x_n\}$*, each drawn from a normal distribution with true mean* μ_t

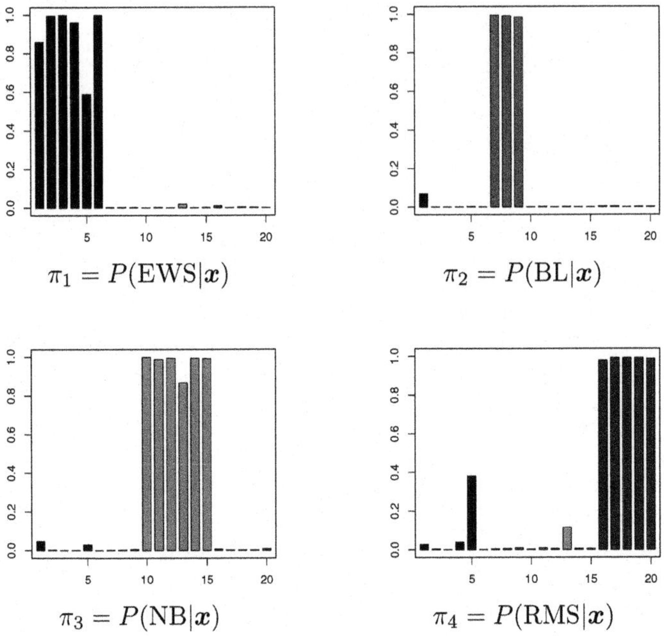

FIGURE 7.8: Classification of the test samples. The sample numbers 1–6, 7–9, 10–15 and 16–20 correspond to EWS, BL, NB and RMS samples.

and known variance σ^2, i.e., $g(z|\mu_t) = N(\mu_t, \sigma^2)$. We also assume that the data are generated from a normal distribution $f(z|\mu) = N(\mu, \sigma^2)$. Thus, the model is correctly specified.

Show that the use of a normal prior $\mu \sim N(\mu_0, \tau_0^2)$ leads to the posterior distribution of μ being normal with mean

$$\hat{\mu}_n = \frac{\mu_0/\tau_0^2 + \sum_{\alpha=1}^n x_\alpha/\sigma^2}{1/\tau_0^2 + n/\sigma^2}$$

and variance

$$\sigma_n^2 = \frac{1}{1/\tau_0^2 + n/\sigma^2}.$$

2. *(Continued) Show that the asymptotic bias estimate of BPIC is*

$$n\hat{b}(\hat{G}) = -\left(\frac{n\sigma_n^2}{2\sigma^2} + \frac{\sigma_n^2}{2\tau_0^2}\right) + S_n^{-1}(\hat{\mu}_n)Q_n(\hat{\mu}_n) + \frac{1}{2}$$

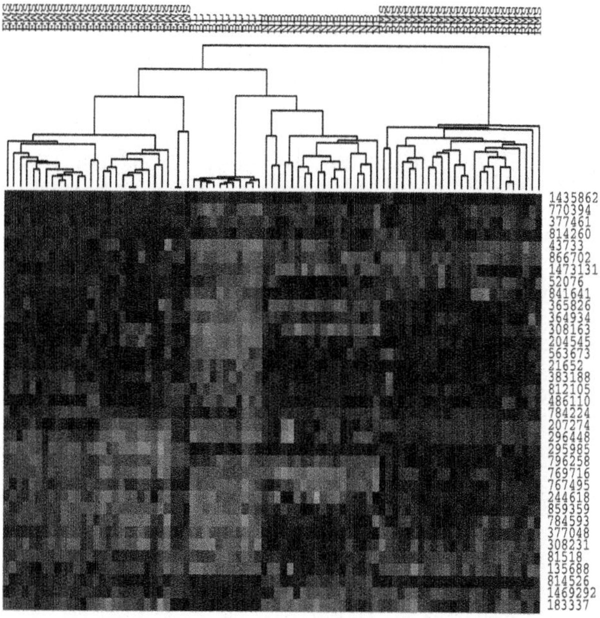

FIGURE 7.9: Comparison of 35 gene expressions for distinguishing the small round blue cell tumors. The top 35 genes as ranked by the proposed criterion GIC were used for the hierarchical clustering analysis.

with

$$\begin{aligned} Q_n(\hat{\mu}_n) &= \frac{1}{n}\sum_{\alpha=1}^{n}\left[\frac{\partial\{\log f(x_\alpha|\mu)+\log\pi(\mu)/n\}}{\partial\mu}\right. \\ &\qquad \left.\cdot\frac{\partial\{\log f(x_\alpha|\mu)+\log\pi(\mu)/n\}}{\partial\mu}\bigg|_{\mu=\hat{\mu}_n}\right] \\ &= \frac{1}{n}\sum_{\alpha=1}^{n}\left(\frac{(x_\alpha-\mu)}{\sigma^2}+\frac{(\mu_0-\mu)}{n\tau_0^2}\right)^2\bigg|_{\mu=\hat{\mu}_n}, \\ S_n(\hat{\mu}_n) &= -\frac{1}{n}\sum_{\alpha=1}^{n}\left[\frac{\partial^2\{\log f(x_\alpha|\mu)+\log\pi(\mu)/n\}}{\partial\mu^2}\bigg|_{\mu=\hat{\mu}_n}\right] \\ &= \frac{1}{\sigma^2}+\frac{1}{n\tau_0^2}=\frac{1}{n\sigma_n^2}. \end{aligned}$$

3. *(Continued) Since the specified parametric family of probability distributions encompasses the true model and the prior is dominated by the likelihood as n increases, we expect the bias term to reduce to the number of parameters:* $\hat{b}(\hat{G}) \to 1$ *as* $n \to \infty$*.*

This intuition can be confirmed as follows. Noting that $\sigma_n^2 = (1/\tau_0^2 + n/\sigma^2)^{-1}$*, the asymptotic behavior of the first term can be evaluated as*

$$\begin{aligned}\frac{n\sigma_n^2}{2\sigma^2} + \frac{\sigma_n^2}{2\tau_0^2} &= \frac{n}{2\sigma^2\left[\frac{1}{\tau_0^2} + \frac{n}{\sigma^2}\right]} + \frac{1}{2\tau_0^2\left[\frac{1}{\tau_0^2} + \frac{n}{\sigma^2}\right]} \\ &= \frac{n}{2n\left[\frac{\sigma^2}{n\tau_0^2} + 1\right]} + \frac{1}{2n\left[\frac{1}{n} + \frac{\tau_0^2}{\sigma^2}\right]} \\ &\to \frac{1}{2} + 0 = \frac{1}{2} \quad \text{as} \quad n \to \infty\end{aligned}$$

under the fixed, very large value of τ_0^2*. Also, we have*

$$S_n^{-1}(\hat{\mu}_n)Q_n(\hat{\mu}_n) \simeq S_n^{-1}(\hat{\mu}_n)S_n(\hat{\mu}_n) = 1.$$

In this case, therefore, the asymptotic bias estimate of BPIC is approximated by the dimension of parameter.

4. *(Continued) Show that the asymptotic behavior of the penalty term of DIC (Spiegelhalter et al. (2002)) is*

$$\frac{P_D}{2} = \frac{n\sigma_n^2}{2\sigma^2} \to \frac{1}{2} \quad \text{as} \quad n \to \infty.$$

In this case, the BPIC asymptotic bias estimate is double the penalty term used in the DIC.

5. *In Section 5.8.3, we considered Bayesian spatial data analysis. Reanalyze the Bartlett experimental forest inventory data using DIC approach. Especially, compare the following three models:*

$$\begin{aligned}\mu(\boldsymbol{s}) &= \beta_1\text{ELEV}(\boldsymbol{s}) + \beta_2\text{SLOPE}(\boldsymbol{s}), \\ \mu(\boldsymbol{s}) &= \beta_0 + \beta_1\text{ELEV}(\boldsymbol{s}) + \beta_2\text{SLOPE}(\boldsymbol{s}), \\ \mu(\boldsymbol{s}) &= \beta_0 + \beta_1\text{ELEV}(\boldsymbol{s}) + \beta_2\text{SLOPE}(\boldsymbol{s}) + \beta_3\text{TC1}(\boldsymbol{s}) \\ &\quad + \beta_4\text{TC2}(\boldsymbol{s}) + \beta_5\text{TC3}(\boldsymbol{s}).\end{aligned}$$

You can implement MCMC sampling through the **R** *package* **spBayes**.

6. *Generate a set of* $n = 100$ *observations from* $y_\alpha = \exp\{-x_\alpha \sin(2\pi x_\alpha)\} - 1 + \varepsilon_\alpha$, $\alpha = 1, ..., 100$*, where* x_α *are uniformly distributed in* $[0, 1]$*, and* $\varepsilon_\alpha \sim N(0, 0.3^2)$*. To estimate the true curve, we consider the 15th polynomial regression model given in Section 2.8.2. We estimate the parameter vector* $\boldsymbol{\theta} = (\boldsymbol{\beta}^T, \sigma^2)^T$ *by using the maximum penalized likelihood method*

$$\hat{\boldsymbol{\theta}}_n = \text{argmax}_\theta \left[\sum_{\alpha=1}^{n} \log f\left(\boldsymbol{y}_n | X_n, \boldsymbol{\beta}, \sigma^2\right) - \frac{\lambda}{2}\boldsymbol{\beta}^T\boldsymbol{\beta}\right],$$

where $f\left(\boldsymbol{y}_n|X_n, \boldsymbol{\beta}, \sigma^2\right)$ *is the likelihood function. Fix the value of* λ_0*; the maximum penalized likelihood estimate of* $\boldsymbol{\beta}$ *is*

$$\hat{\boldsymbol{\beta}} = (B_n^T B_n + \lambda_0 I)^{-1} B_n^T,$$

with $\lambda_0 = \hat{\sigma}_n^2 \lambda$*,* I *is the unit diagonal matrix and* $\hat{\sigma}_n^2 = \sum_{\alpha=1}^n (y_\alpha - \hat{\boldsymbol{\beta}}^T \boldsymbol{x}_\alpha)/n$ *is the estimated variance parameter.*

Using the smoother matrix $H = B_n(B_n^T B_n + \lambda_0 I)^{-1} B_n^T$*, calculate MBIC score for the diffecent value of* $\lambda_0 = \{1, 10^{-3}, 10^{-6}, 10^{-9}\}$*. Then select the best value of* λ_0 *among them.*

7. *Suppose that we have a set of n independent observations* y_α*,* $\alpha = q, ..., n$*, a sequence of independent binary random variables taking the values 0 and 1 with conditional probabilities*

$$\mathrm{P}(Y_\alpha = 1|x_\alpha) = \pi(x_\alpha) \quad \text{and} \quad \mathrm{P}(Y_\alpha = 0|x_\alpha) = 1 - \pi(x_\alpha).$$

Assume that the conditional probability $\pi(\boldsymbol{x}_\alpha)$ *can be rewritten as*

$$\pi(x_\alpha) = \frac{\exp\left\{\boldsymbol{w}^T \boldsymbol{b}(x_\alpha)\right\}}{1 + \exp\left\{\boldsymbol{w}^T \boldsymbol{b}(x_\alpha)\right\}}, \tag{7.21}$$

where $\boldsymbol{w} = (w_0, ..., w_m)^T$ *and* $\boldsymbol{b}(x_\alpha) = (1, b_1(\boldsymbol{x}_\alpha), ..., b_m(\boldsymbol{x}_\alpha))^T$*. Show that the log-likelihood function is expressed as*

$$\log f(\boldsymbol{y}_n|X_n, \boldsymbol{w}) = \sum_{\alpha=1}^{n} \left(y_\alpha \boldsymbol{w}^T \boldsymbol{b}(x_\alpha) - \log\left[1 + \exp\{\boldsymbol{w}^T \boldsymbol{b}(x_\alpha)\}\right]\right).$$

8. *(Continued) Generate a set of n independent binay observations* $\boldsymbol{y}_n = \{y_1, ..., y_n\}$ *with conditional probability* $\pi(\boldsymbol{x}_\alpha)$

$$\pi(\boldsymbol{x}_\alpha) = 1/\left[1 + \exp(-\sin(2\pi x_\alpha^2))\right]$$

where the design points x_α *are uniformly distributed in* $[-1, 1]$*.* **R** *function* **rbinom** *generates random samples. Then using a singular multivariate normal prior density,*

$$\pi(\boldsymbol{w}) = \left(\frac{n\lambda}{2\pi}\right)^{(m-2)/2} |R|_+^{1/2} \exp\left\{-\frac{n\lambda}{2} \boldsymbol{w}^T D_2^T D_2 \boldsymbol{w}\right\},$$

find the posterior mode $\hat{\boldsymbol{w}}_n$*, which can be estimated by Fisher scoring iterations:*

$$\boldsymbol{w}^{new} = \left(B^T W B + n\lambda D_2^T D_2\right)^{-1} B^T W \boldsymbol{\zeta},$$

where W *is an* $n \times n$ *diagonal matrix, and* $\boldsymbol{\zeta}$ *is an* n*-dimensional vector:*

$$\begin{aligned} W_{\alpha\alpha} &= \pi(x_\alpha)(1 - \pi(x_\alpha)), \\ \zeta_\alpha &= \{y_\alpha - \pi(x_\alpha)\}/[\pi(x_\alpha)(1 - \pi(x_\alpha))] + \boldsymbol{w}^T \boldsymbol{b}(x_\alpha). \end{aligned}$$

B-spline basis function can be constructed by using the **R** *package* **splines**. *Also, you update the parameter vector* $\boldsymbol{w}$ *until a suitable convergence criterion is satisfied.*

9. *(Continued) Optimize the number of B-spline basis functions and the value of smoothing aparameter* λ *using the MBIC score*

$$\text{MBIC} = 2\sum_{\alpha=1}^{n}\sum_{\alpha=1}^{n}\left(\log\left[1+\exp\{\hat{\boldsymbol{w}}_n^T\boldsymbol{b}(x_\alpha)\}\right] - y_\alpha\hat{\boldsymbol{w}}_n^T\boldsymbol{b}(x_\alpha)\right) + \log(n)\text{tr}\{H\},$$

where the smoother matrix H *is given as*

$$H = B(B^TWB + n\lambda D_2^TD_2)^{-1}B^TW.$$

Chapter 8

Theoretical development and comparisons

In this section, we provide the theoretical derivations of the Bayesian information criterion (BIC), the generalized Bayesian information criterion (GBIC), the Bayesian predictive information criterion (BPIC), and the generalized information criterion.

8.1 Derivation of Bayesian information criteria

In this section, we consider the case $\log \pi(\boldsymbol{\theta}) = O_p(1)$, i.e., the prior information can be ignored for a sufficiently large n. As shown by Konishi et al. (2004), the order of the prior distribution has a large influence on the calculation of the Bayes factor. In this case, the posterior mode $\hat{\boldsymbol{\theta}}$ converges to the pseudo parameter value $\boldsymbol{\theta}_0$, which minimizes the Kullback-Leibler distance between the specified model $f(y|\boldsymbol{\theta})$ and the true model $g(y)$ that generates data. Hereafter, we restrict our attention to a proper situation in which the Fisher information matrix of the specified model $f(y|\boldsymbol{\theta})$ is nonsingular at $\boldsymbol{\theta}_{k0}$, which is uniquely determined and interior to Θ.

Noting that the first derivative of the log-likelihood function $f(\boldsymbol{X}_n|\boldsymbol{\theta})$ evaluated at the maximum likelihood estimator $\hat{\boldsymbol{\theta}}_{\text{MLE}}$ equals to zero, we have the following Taylor expansion of the log-likelihood function:

$$\begin{aligned}\log f(\boldsymbol{X}_n|\boldsymbol{\theta}) &= \log f(\boldsymbol{X}_n|\hat{\boldsymbol{\theta}}_{\text{MLE}}) \\ &\quad -\frac{n}{2}(\boldsymbol{\theta}-\hat{\boldsymbol{\theta}}_{\text{MLE}})^T J_n(\hat{\boldsymbol{\theta}}_{\text{MLE}})(\boldsymbol{\theta}-\hat{\boldsymbol{\theta}}_{\text{MLE}}) + o_p(1) \quad (8.1)\end{aligned}$$

with

$$J_n\left(\hat{\boldsymbol{\theta}}_{\text{MLE}}\right) = -\frac{1}{n}\left.\frac{\partial^2 \log f(\boldsymbol{X}_n|\boldsymbol{\theta})}{\partial\boldsymbol{\theta}\partial\boldsymbol{\theta}^T}\right|_{\boldsymbol{\theta}=\hat{\boldsymbol{\theta}}_{\text{MLE}}}.$$

Similarly, we have the Taylor expansion of the prior density

$$\pi(\boldsymbol{\theta}) = \pi(\hat{\boldsymbol{\theta}}_{\text{MLE}}) + (\boldsymbol{\theta}-\hat{\boldsymbol{\theta}}_{\text{MLE}})^T \left.\frac{\partial\pi(\boldsymbol{\theta})}{\partial\boldsymbol{\theta}}\right|_{\boldsymbol{\theta}=\hat{\boldsymbol{\theta}}_{\text{MLE}}} + o_p(1). \quad (8.2)$$

Substituting the Equations (8.1) and (8.2) into (5.3), the marginal likelihood $P(\boldsymbol{X}_n|M)$ can be re-expressed as

$$\begin{aligned}
&P(\boldsymbol{X}_n|M)\\
&\approx \int \exp\left\{\log f(\boldsymbol{X}_n|\hat{\boldsymbol{\theta}}_{\mathrm{MLE}}) - \frac{n}{2}(\boldsymbol{\theta}-\hat{\boldsymbol{\theta}}_{\mathrm{MLE}})^T J_n\left(\hat{\boldsymbol{\theta}}_{\mathrm{MLE}}\right)(\boldsymbol{\theta}-\hat{\boldsymbol{\theta}}_{\mathrm{MLE}})\right\}\\
&\qquad \times \left\{\pi(\hat{\boldsymbol{\theta}}_{\mathrm{MLE}}) + (\boldsymbol{\theta}-\hat{\boldsymbol{\theta}}_{\mathrm{MLE}})^T \left.\frac{\partial \pi(\boldsymbol{\theta})}{\partial \boldsymbol{\theta}}\right|_{\boldsymbol{\theta}=\hat{\boldsymbol{\theta}}_{\mathrm{MLE}}}\right\} d\boldsymbol{\theta}\\
&\approx f(\boldsymbol{X}_n|\hat{\boldsymbol{\theta}}_{\mathrm{MLE}})\pi(\hat{\boldsymbol{\theta}}_{\mathrm{MLE}})\\
&\qquad \times \int \exp\left\{-\frac{n}{2}(\boldsymbol{\theta}-\hat{\boldsymbol{\theta}}_{\mathrm{MLE}})^T J_n\left(\hat{\boldsymbol{\theta}}_{\mathrm{MLE}}\right)(\boldsymbol{\theta}-\hat{\boldsymbol{\theta}}_{\mathrm{MLE}})\right\} d\boldsymbol{\theta}.
\end{aligned}$$

Here the term of order $o_p(1)$ and higher-order terms are ignored. Also, we have used the fact that

$$\int (\boldsymbol{\theta}-\hat{\boldsymbol{\theta}}_{\mathrm{MLE}}) \exp\left\{-\frac{n}{2}(\boldsymbol{\theta}-\hat{\boldsymbol{\theta}}_{\mathrm{MLE}})^T J_n\left(\hat{\boldsymbol{\theta}}_{\mathrm{MLE}}\right)(\boldsymbol{\theta}-\hat{\boldsymbol{\theta}}_{\mathrm{MLE}})\right\} d\boldsymbol{\theta} = \boldsymbol{0},$$

which follows from the property of the multivariate normal distribution.

Noting that the integrand

$$\int \exp\left\{-\frac{n}{2}(\boldsymbol{\theta}-\hat{\boldsymbol{\theta}}_{\mathrm{MLE}})^T J_n\left(\hat{\boldsymbol{\theta}}_{\mathrm{MLE}}\right)(\boldsymbol{\theta}-\hat{\boldsymbol{\theta}}_{\mathrm{MLE}})\right\} d\boldsymbol{\theta}$$

is the density function of the multivariate normal density with mean $\hat{\boldsymbol{\theta}}_{\mathrm{MLE}}$ and covariance matrix $n^{-1}J_n^{-1}(\hat{\boldsymbol{\theta}}_{\mathrm{MLE}})$, we have

$$P(\boldsymbol{X}_n|M) \approx f(\boldsymbol{X}_n|\hat{\boldsymbol{\theta}}_{\mathrm{MLE}})\pi(\hat{\boldsymbol{\theta}}_{\mathrm{MLE}}) \times \frac{(2\pi)^{p/2}}{n^{p/2}|J_n(\hat{\boldsymbol{\theta}}_{\mathrm{MLE}})|^{1/2}}.$$

Substituting this approximation in Equation (5.2) and taking the logarithm of the resulting formula, we obtaion

$$\begin{aligned}
&-2\log\left\{P(M)\int f(\boldsymbol{X}_n|\boldsymbol{\theta})\pi(\boldsymbol{\theta})d\boldsymbol{\theta}\right\}\\
&= -2\log\{P(M)P(\boldsymbol{X}_n|M)\}\\
&\approx -2\log f(\boldsymbol{X}_n|\hat{\boldsymbol{\theta}}_{\mathrm{MLE}}) - 2\log\pi(\hat{\boldsymbol{\theta}}_{\mathrm{MLE}}) + p\log n\\
&\quad + \log|J_n(\hat{\boldsymbol{\theta}}_{\mathrm{MLE}})| - 2\log P(M) - p\log 2\pi.
\end{aligned}$$

Ignoring the term of order $O(1)$ and higher-order terms in this equation, and assuming that the prior probabilities, $P(M_k)$, are all equal, we have Schwarz's (1978) Bayesian information criterion,

$$\mathrm{BIC} = -2\log f(\boldsymbol{X}_n|\hat{\boldsymbol{\theta}}_{\mathrm{MLE}}) + p\log n.$$

The BIC is a criterion for evaluating models estimated by the maximum likelihood method.

8.2 Derivation of generalized Bayesian information criteria

This considers the case $\log \pi(\boldsymbol{\theta}) = O_p(n)$, i.e., the prior information grows with the sample size. Thus the prior information cannot be ignored even when the sample size n is large. Therefore, the posterior mode $\hat{\boldsymbol{\theta}}$ does not converge to the pseudo parameter value $\boldsymbol{\theta}_0$.

The marginal likelihood of the data $\boldsymbol{X}_n$ under the model M can be rewritten as

$$P(\boldsymbol{X}_n|M) = \int \exp\{s(\boldsymbol{\theta}|\boldsymbol{X}_n)\} \, d\boldsymbol{\theta}, \tag{8.3}$$

where

$$s(\boldsymbol{\theta}|\boldsymbol{X}_n) = \log f(\boldsymbol{X}_n|\boldsymbol{\theta}) + \log \pi(\boldsymbol{\theta}).$$

Consider the case $\log \pi(\boldsymbol{\theta}) = O(n)$. Let $\hat{\boldsymbol{\theta}}_n$ be the mode of $s(\boldsymbol{\theta}|\boldsymbol{X}_n)$. Then, using the Laplace method for integrals in the Bayesian framework developed by Tierney and Kadane (1986), Tierney et al. (1989), Kass et al. (1990), and Konishi et al. (2004) under some regularity conditions, we obtain the Laplace approximation to the marginal distribution (8.3) in the form

$$\begin{aligned} P(\boldsymbol{X}_n|M) &= \int \exp\{s(\boldsymbol{\theta}|\boldsymbol{X}_n)\} \, d\boldsymbol{\theta} \\ &\approx \frac{(2\pi)^{p/2}}{n^{p/2}|S_n(\hat{\boldsymbol{\theta}}_n)|^{1/2}} \exp\left\{ns(\hat{\boldsymbol{\theta}}_n|\boldsymbol{X}_n)\right\}, \end{aligned} \tag{8.4}$$

where

$$S_n(\hat{\boldsymbol{\theta}}_n) = -\frac{1}{n}\frac{\partial^2 s(\boldsymbol{\theta}|\boldsymbol{X}_n)}{\partial\boldsymbol{\theta}\partial\boldsymbol{\theta}^T}\bigg|_{\boldsymbol{\theta}=\hat{\boldsymbol{\theta}}_n} = J_n(\hat{\boldsymbol{\theta}}_n) - \frac{1}{n}\frac{\partial^2 \log \pi(\boldsymbol{\theta})}{\partial\boldsymbol{\theta}\partial\boldsymbol{\theta}^T}\bigg|_{\boldsymbol{\theta}=\hat{\boldsymbol{\theta}}_n}.$$

Substituting the Laplace approximation in Equation (5.2) and taking the logarithm of the resulting formula, we obtain an approximation to the posterior probability, the generalized Bayesian information criterion:

$$\begin{aligned} \text{GBIC} &= -2\log\{P(M)P(\boldsymbol{X}_n|M)\} \\ &= -2\log f(\boldsymbol{X}_n|\hat{\boldsymbol{\theta}}_n) - 2\log \pi(\hat{\boldsymbol{\theta}}_n) + p\log n + \log|S_n(\hat{\boldsymbol{\theta}}_n)| \\ &\quad -p\log 2\pi - 2\log P(M) \end{aligned}$$

Choosing the model with the largest posterior probability among a set of candidate models is equivalent to choosing the model that minimizes the criterion GBIC.

Konishi et al. (2004) also showed that GBIC reduces to the original

BIC and also derived an improved version of BIC. Consider the case where $\log \pi(\boldsymbol{\theta}) = O_p(1)$. Then the mode $\hat{\boldsymbol{\theta}}_n$ of $s(\boldsymbol{\theta}|\boldsymbol{X}_n)$ can be expanded as

$$\hat{\boldsymbol{\theta}}_n = \hat{\boldsymbol{\theta}}_{\text{MLE}} + \frac{1}{n} J_n^{-1}\left(\hat{\boldsymbol{\theta}}_{\text{MLE}}\right) \frac{\partial}{\partial \boldsymbol{\theta}} \log \pi(\boldsymbol{\theta})\bigg|_{\boldsymbol{\theta}=\hat{\boldsymbol{\theta}}_{\text{MLE}}} + O_p(n^{-2}), \tag{8.5}$$

where $\hat{\boldsymbol{\theta}}_{\text{MLE}}$ is the maximum likelihood estimate of $\boldsymbol{\theta}$. Substituting the stochastic expansion (8.5) in the equation of GBIC yields

$$\begin{aligned} & -2\log\left\{P(M)P(\boldsymbol{X}_n|M)\right\} \\ = & -2\log f(\boldsymbol{X}_n|\hat{\boldsymbol{\theta}}_{\text{MLE}}) - 2\log \pi(\hat{\boldsymbol{\theta}}_{\text{MLE}}) + p\log n \\ & + \log|J_n(\hat{\boldsymbol{\theta}}_{\text{MLE}})| - 2\log \text{pr}(M) - p\log 2\pi + O_p(n^{-1}). \end{aligned}$$

Ignoring the term of order $O(1)$ and higher-order terms in this equation, we have Schwarz's (1978) Bayesian information criterion, BIC.

Suppose that the prior probabilities, $P(M)$, are all equal, and that the prior density $\pi(\boldsymbol{\theta})$ is sufficiently flat in the neighborhood of $\hat{\boldsymbol{\theta}}_n$. These conditions lead to the modification of Equation (8.6) to the following:

$$\text{IBIC} = -2\log f(\boldsymbol{X}_n|\hat{\boldsymbol{\theta}}_{\text{MLE}}) + p\log n + \log|J_n(\hat{\boldsymbol{\theta}}_{\text{MLE}})| - p\log 2\pi.$$

This variant, based on the inclusion of the term $\log|J_n(\hat{\boldsymbol{\theta}}_{\text{MLE}})|$, is regarded as an improved version of the Bayesian information criterion.

8.3 Derivation of Bayesian predictive information criterion

The critical issue with Bayesian modeling is how to evaluate the goodness of the predictive distributions. Bayesian predictive information criterion (Ando (2007)) considered the maximization of the posterior mean of the expected log-likelihood

$$\eta(G) = \int\left\{\int \log f(z|\boldsymbol{\theta})\pi(\boldsymbol{\theta}|\boldsymbol{X}_n)d\boldsymbol{\theta}\right\}dG(z) \tag{8.6}$$

to measure the deviation of the predictive distribution from the true model $g(z)$. The posterior mean of the expected log-likelihood η in (8.6) depends on the model fitted, and on the unknown true model $g(z)$. A natural estimator of η is the posterior mean of the log-likelihood

$$\eta(\hat{G}) = \frac{1}{n}\int \log f(\boldsymbol{X}_n|\boldsymbol{\theta})\pi(\boldsymbol{\theta}|\boldsymbol{X}_n)d\boldsymbol{\theta}. \tag{8.7}$$

which is formally obtained by replacing the unknown distribution $G(z)$ in (8.6) by the empirical distribution, $\hat{G}$, putting math $1/n$ on each observation.

The quantity, $\hat{\eta}$, is generally positively biased as an estimator of η, because the same data $\boldsymbol{X}_n$ are used both to construct the posterior distributions $\pi(\boldsymbol{\theta}|y)$ and to evaluate η. Therefore, bias correction should be considered, where the bias is

$$\begin{aligned} b(G) &= \int \left\{ \eta(\hat{G}) - \eta(G) \right\} dG(\boldsymbol{X}_n) \\ &= \int \left(\frac{1}{n} \int \log f(\boldsymbol{X}_n|\boldsymbol{\theta})\pi(\boldsymbol{\theta}|\boldsymbol{X}_n) d\boldsymbol{\theta} \right. \\ &\quad \left. - \int \left\{ \int \log f(z|\boldsymbol{\theta})\pi(\boldsymbol{\theta}|\boldsymbol{X}_n) d\boldsymbol{\theta} \right\} dG(z) \right) dG(\boldsymbol{X}_n), \end{aligned}$$

where $G(\boldsymbol{X}_n)$ is the joint density of $\boldsymbol{X}_n$.

If the bias can be estimated, by $\hat{b}(G)$, the bias-corrected posterior mean of the log-likelihood is given by $n^{-1} \int \log f(\boldsymbol{X}_n|\boldsymbol{\theta})\pi(\boldsymbol{\theta}|\boldsymbol{X}_n) d\boldsymbol{\theta} - \hat{b}(G)$, which is usually used in the form

$$\mathrm{IC} = -2 \int \log f(\boldsymbol{X}_n|\boldsymbol{\theta})\pi(\boldsymbol{\theta}|\boldsymbol{X}_n) d\boldsymbol{\theta} + 2n\hat{b}(G).$$

Ando (2007) obtained the asymptotic bias under the model misspecification. The following sections first describe some asymptotic aspects of the parameter and then the derivation of the bias term of the Bayesian predictive information criterion.

8.3.1 Derivation of BPIC

Hereafter, we restrict our attention to a proper situation in which the Hessian of the expected penalized log-likelihood function $\int \log\{f(z|\boldsymbol{\theta})\pi_0(\boldsymbol{\theta})\} dG(z)$ is nonsingular at $\boldsymbol{\theta}_0$, which is uniquely determined and interior to Θ.

We decompose the bias b(G) defined in the previous section as

$$b(G) = \int \left\{ \eta(\hat{G}) - \eta(G) \right\} dG(\boldsymbol{X}_n) = D_1 + D_2 + D_3,$$

where

$$D_1 = \int \left[\frac{1}{n} \int \log f(\boldsymbol{X}_n|\boldsymbol{\theta})\pi(\boldsymbol{\theta}|\boldsymbol{X}_n) d\boldsymbol{\theta} - \frac{1}{n} \log\{f(\boldsymbol{X}_n|\boldsymbol{\theta}_0)\pi(\boldsymbol{\theta}_0)\} \right] dG(\boldsymbol{X}_n),$$

$$D_2 = \int \left[\frac{1}{n} \log\{f(\boldsymbol{X}_n|\boldsymbol{\theta}_0)\pi(\boldsymbol{\theta}_0)\} - \int \log\{f(z|\boldsymbol{\theta}_0)\pi_0(\boldsymbol{\theta}_0)\} dG(z) \right] dG(\boldsymbol{X}_n),$$

$$D_3 = \int \left[\int \log\{f(z|\boldsymbol{\theta}_0)\pi_0(\boldsymbol{\theta}_0)\} dG(z) \right.$$

$$\left. - \int \left\{ \int \log f(z|\boldsymbol{\theta})\pi(\boldsymbol{\theta}|\boldsymbol{X}_n) d\boldsymbol{\theta} \right\} dG(z) \right] dG(\boldsymbol{X}_n).$$

In the next step, we calculate these three terms: D_1, D_2 and D_3.

Calculation of D_1

Noting that the first derivative of the penalized log-likelihood function evaluated at the posterior mode is zero, $\partial \log\{f(\boldsymbol{X}_n|\hat{\boldsymbol{\theta}}_n)\pi(\hat{\boldsymbol{\theta}}_n)\}/\partial\boldsymbol{\theta} = \boldsymbol{0}$, the Taylor expansion of the penalized log-likelihood function $\log\{f(\boldsymbol{X}_n|\boldsymbol{\theta}_0)\pi(\boldsymbol{\theta}_0)\}$ around the posterior mode $\hat{\boldsymbol{\theta}}_n$ gives

$$\begin{aligned} &\log\{f(\boldsymbol{X}_n|\boldsymbol{\theta}_0)\pi(\boldsymbol{\theta}_0)\} \\ &= \log\{f(\boldsymbol{X}_n|\hat{\boldsymbol{\theta}}_n)\pi(\hat{\boldsymbol{\theta}}_n)\} + \frac{n}{2}(\boldsymbol{\theta}_0 - \hat{\boldsymbol{\theta}}_n)^T S_n(\hat{\boldsymbol{\theta}}_n)(\boldsymbol{\theta}_0 - \hat{\boldsymbol{\theta}}_n) + O_p(n^{-1/2}), \end{aligned}$$

where

$$S_n(\hat{\boldsymbol{\theta}}_n) = -\frac{1}{n} \frac{\partial^2 \log\{f(\boldsymbol{X}_n|\boldsymbol{\theta})\pi(\boldsymbol{\theta})\}}{\partial\boldsymbol{\theta}\partial\boldsymbol{\theta}^T}\bigg|_{\boldsymbol{\theta}=\hat{\boldsymbol{\theta}}_n}.$$

Thus, we have

$$\begin{aligned} &D_1 \\ &= \frac{1}{n}\int \left[\int \log f(\boldsymbol{X}_n|\boldsymbol{\theta})\pi(\boldsymbol{\theta}|\boldsymbol{X}_n) d\boldsymbol{\theta} - \frac{1}{n} \log\{f(\boldsymbol{X}_n|\hat{\boldsymbol{\theta}}_n)\pi(\hat{\boldsymbol{\theta}}_n)\} \right] dG(\boldsymbol{X}_n) \\ &\quad + \frac{1}{2n}\mathrm{tr}\left[\int \left\{ S_n(\hat{\boldsymbol{\theta}}_n)\sqrt{n}(\boldsymbol{\theta}_0 - \hat{\boldsymbol{\theta}}_n)\sqrt{n}(\boldsymbol{\theta}_0 - \hat{\boldsymbol{\theta}}_n)^T \right\} \right] dG(\boldsymbol{X}_n) + O_p(n^{-3/2}). \end{aligned}$$

From the Bayesian central limit theorem, the covariance matrix of $\sqrt{n}(\hat{\boldsymbol{\theta}}_n - \boldsymbol{\theta}_0)$ is asymptotically given by $S^{-1}(\boldsymbol{\theta}_0)Q(\boldsymbol{\theta}_0)S^{-1}(\boldsymbol{\theta}_0)$. With this result and since $S_n(\hat{\boldsymbol{\theta}}_n) \to S(\boldsymbol{\theta}_0)$ and $\hat{\boldsymbol{\theta}}_n \to \boldsymbol{\theta}_0$ in probability as $n \to \infty$, D_1 can be approximated by

$$\begin{aligned} D_1 &\simeq \frac{1}{n}\int \left[\int \log f(\boldsymbol{X}_n|\boldsymbol{\theta})\pi(\boldsymbol{\theta}|\boldsymbol{X}_n) d\boldsymbol{\theta} - \frac{1}{n} \log\{f(\boldsymbol{X}_n|\boldsymbol{\theta}_0)\pi(\boldsymbol{\theta}_0)\} \right] dG(\boldsymbol{X}_n) \\ &\quad + \frac{1}{2n}\mathrm{tr}\left\{ S^{-1}(\boldsymbol{\theta}_0)Q(\boldsymbol{\theta}_0) \right\}. \end{aligned}$$

Calculation of D_2

The term D_2 can be regarded as zero. Noting that $\log \pi_0(\boldsymbol{\theta})$ can be approximated by $n^{-1} \log \pi(\boldsymbol{\theta})$ for a moderate sample size, we obtain

$$\begin{aligned} D_2 &= \int \left[\frac{1}{n} \log\{f(\boldsymbol{X}_n|\boldsymbol{\theta}_0)\} - \int \log\{f(z|\boldsymbol{\theta}_0)\} dG(z) \right] dG(\boldsymbol{X}_n) \\ &\quad - \log \pi_0(\boldsymbol{\theta}_0) + \frac{1}{n} \log \pi(\boldsymbol{\theta}_0) \\ &= \frac{1}{n} \log \pi(\boldsymbol{\theta}_0) - \log \pi_0(\boldsymbol{\theta}_0) \simeq 0. \end{aligned}$$

Calculation of D_3

The term D_3 can be modified as follows:

$$\begin{aligned} D_3 &= \int \left[\int \log\{f(z|\boldsymbol{\theta}_0)\pi_0(\boldsymbol{\theta}_0)\} dG(z) \right] dG(\boldsymbol{X}_n) \\ &\quad + \int \left[\int \log \pi_0(\boldsymbol{\theta})\pi(\boldsymbol{\theta}|\boldsymbol{X}_n) d\boldsymbol{\theta} \right] dG(\boldsymbol{X}_n) \\ &\quad - \int \left[\int \left\{ \int \log\{f(z|\boldsymbol{\theta})\pi_0(\boldsymbol{\theta})\}\pi(\boldsymbol{\theta}|\boldsymbol{X}_n) d\boldsymbol{\theta} \right\} dG(z) \right] dG(\boldsymbol{X}_n). \end{aligned}$$

Expanding $\log\{f(z|\boldsymbol{\theta})\pi_0(\boldsymbol{\theta})\}$ around $\boldsymbol{\theta}_0$ and taking an expectation, we obtain

$$\begin{aligned} \int \left[\log\{f(z|\boldsymbol{\theta})\pi_0(\boldsymbol{\theta})\} \right] dG(z) &= \int \left[\log\{f(z|\boldsymbol{\theta}_0)\pi_0(\boldsymbol{\theta}_0)\} \right] dG(z) \\ &\quad - \frac{1}{2} \mathrm{tr} \left\{ S(\boldsymbol{\theta}_0)(\boldsymbol{\theta} - \boldsymbol{\theta}_0)(\boldsymbol{\theta} - \boldsymbol{\theta}_0)^T \right\}. \end{aligned}$$

Together with the approximation $\log \pi_0(\boldsymbol{\theta}) \simeq n^{-1} \log \pi(\boldsymbol{\theta})$, the term D_3 can be evaluated as

$$\begin{aligned} D_3 &\simeq \frac{1}{2n} \mathrm{tr} \left(S(\boldsymbol{\theta}_0) \int \left\{ \int n(\boldsymbol{\theta} - \boldsymbol{\theta}_0)(\boldsymbol{\theta} - \boldsymbol{\theta}_0)^T \pi(\boldsymbol{\theta}|\boldsymbol{X}_n) d\boldsymbol{\theta} \right\} dG(\boldsymbol{X}_n) \right) \\ &\quad + \frac{1}{n} \int \left[\int \log \pi(\boldsymbol{\theta})\pi(\boldsymbol{\theta}|\boldsymbol{X}_n) d\boldsymbol{\theta} \right] dG(\boldsymbol{X}_n) \end{aligned}$$

Considering the Remark given later, the posterior covariance matrix of $(\boldsymbol{\theta} - \boldsymbol{\theta}_0)$ is approximated as

$$\frac{1}{n} S^{-1}(\boldsymbol{\theta}_0) + \frac{1}{n} S^{-1}(\boldsymbol{\theta}_0) Q(\boldsymbol{\theta}_0) S^{-1}(\boldsymbol{\theta}_0).$$

We finally have

$$D_3 \simeq \frac{1}{2n} \mathrm{tr} \left\{ S^{-1}(\boldsymbol{\theta}_0) Q(\boldsymbol{\theta}_0) \right\} + \frac{p}{2n} + \frac{1}{n} \int \left[\int \log \pi(\boldsymbol{\theta})\pi(\boldsymbol{\theta}|\boldsymbol{X}_n) d\boldsymbol{\theta} \right] dG(\boldsymbol{X}_n),$$

where p is the dimension of $\boldsymbol{\theta}$.

Evaluation of the bias

When the above results are combined, the asymptotic bias is given by

$$\begin{aligned} b(G) &= \int \left\{ \eta(\hat{G}) - \eta(G) \right\} dG(\boldsymbol{X}_n) \\ &\simeq \frac{1}{n} \int \left[\int \log\{f(\boldsymbol{X}_n|\boldsymbol{\theta})\pi(\boldsymbol{\theta})\}\pi(\boldsymbol{\theta}|\boldsymbol{X}_n)d\boldsymbol{\theta} \right] dG(\boldsymbol{X}_n) \\ &\quad -\frac{1}{n}\log\{f(\boldsymbol{X}_n|\boldsymbol{\theta}_0)\pi(\boldsymbol{\theta}_0)\} + \frac{1}{n}\mathrm{tr}\left\{S^{-1}(\boldsymbol{\theta}_0)Q(\boldsymbol{\theta}_0)\right\} + \frac{p}{2n}. \end{aligned}$$

Replacing the expectation of $G(\boldsymbol{X}_n)$ by the empirical distribution and estimating the matrices $S(\boldsymbol{\theta}_0)$ and $Q(\boldsymbol{\theta}_0)$ by $S_n(\hat{\boldsymbol{\theta}}_n)$ and $Q_n(\hat{\boldsymbol{\theta}}_n)$, we obtain an estimator of the bias.

Remark

Assuming the regularity conditions of the Bayesian central limit theorem, i.e., that the posterior distribution $\pi(\boldsymbol{\theta}|\boldsymbol{X}_n)$ can be approximated by the normal distribution with mean $\hat{\boldsymbol{\theta}}_n$ and covariance matrix $n^{-1}S_n^{-1}(\hat{\boldsymbol{\theta}}_n)$, we obtain the following approximation

$$\begin{aligned} &\int \left[\int \left\{(\boldsymbol{\theta} - \boldsymbol{\theta}_0)(\boldsymbol{\theta} - \boldsymbol{\theta}_0)^T\right\} \pi(\boldsymbol{\theta}|\boldsymbol{X}_n)d\boldsymbol{\theta} \right] dG(\boldsymbol{X}_n) \\ &\simeq \frac{1}{n}S^{-1}(\boldsymbol{\theta}_0) + \frac{1}{n}S^{-1}(\boldsymbol{\theta}_0)Q(\boldsymbol{\theta}_0)S^{-1}(\boldsymbol{\theta}_0). \end{aligned}$$

Outline of the Proof

A simple modification leads to

$$\begin{aligned} &\int \left\{ \int \sqrt{n}(\boldsymbol{\theta} - \boldsymbol{\theta}_0)\sqrt{n}(\boldsymbol{\theta} - \boldsymbol{\theta}_0)^T \pi(\boldsymbol{\theta}|\boldsymbol{X}_n)d\boldsymbol{\theta} \right\} dG(\boldsymbol{X}_n) \\ &= \int \left\{ \int \sqrt{n}(\boldsymbol{\theta} - \hat{\boldsymbol{\theta}}_n + \hat{\boldsymbol{\theta}}_n - \boldsymbol{\theta}_0) \right. \\ &\qquad \left. \times \sqrt{n}(\boldsymbol{\theta} - \hat{\boldsymbol{\theta}}_n + \hat{\boldsymbol{\theta}}_n - \boldsymbol{\theta}_0)^T \pi(\boldsymbol{\theta}|\boldsymbol{X}_n)d\boldsymbol{\theta} \right\} dG(\boldsymbol{X}_n) \\ &= \int \left\{ \int (\boldsymbol{\theta} - \hat{\boldsymbol{\theta}}_n)(\boldsymbol{\theta} - \hat{\boldsymbol{\theta}}_n)^T \pi(\boldsymbol{\theta}|\boldsymbol{X}_n)d\boldsymbol{\theta} \right\} dG(\boldsymbol{X}_n) \\ &\quad + \frac{1}{n}\int \left\{ \sqrt{n}(\hat{\boldsymbol{\theta}}_n - \boldsymbol{\theta}_0)\sqrt{n}(\hat{\boldsymbol{\theta}}_n - \boldsymbol{\theta}_0)^T \right\} dG(\boldsymbol{X}_n) \\ &\quad + \frac{1}{n\sqrt{n}}\int \left\{ n(\bar{\boldsymbol{\theta}}_n - \hat{\boldsymbol{\theta}}_n)\sqrt{n}(\hat{\boldsymbol{\theta}}_n - \boldsymbol{\theta}_0)^T \right\} dG(\boldsymbol{X}_n) \\ &\quad + \frac{1}{n\sqrt{n}}\int \left\{ \sqrt{n}(\hat{\boldsymbol{\theta}}_n - \boldsymbol{\theta}_0)n(\bar{\boldsymbol{\theta}}_n - \hat{\boldsymbol{\theta}}_n)^T \right\} dG(\boldsymbol{X}_n) \\ &= E_1 + E_2 + E_3 + E_4 \end{aligned}$$

where $\bar{\boldsymbol{\theta}}_n$ and $\hat{\boldsymbol{\theta}}_n$ are the posterior mean and the posterior mode, respectively.

From the Bayesian central limit theorem, the posterior distribution $\pi(\boldsymbol{\theta}|\boldsymbol{X}_n)$ can be approximated by the normal distribution with mean $\hat{\boldsymbol{\theta}}_n$ and covariance matrix $n^{-1}S_n^{-1}(\hat{\boldsymbol{\theta}}_n)$. Thus, the first term E_1 can be approximated by $n^{-1}S^{-1}(\boldsymbol{\theta}_0)$. From the asymptotic normality of the posterior mode, the second term E_2 is asymptotically evaluated as $n^{-1}S^{-1}(\boldsymbol{\theta}_0)Q(\boldsymbol{\theta}_0)S^{-1}(\boldsymbol{\theta}_0)$.

From $\partial \log\{f(\boldsymbol{X}_n|\hat{\boldsymbol{\theta}}_n)\pi(\hat{\boldsymbol{\theta}}_n)\}/\partial\boldsymbol{\theta} = 0$, the posterior mode $\hat{\boldsymbol{\theta}}_n$ can be expanded as

$$\hat{\boldsymbol{\theta}}_n = \bar{\boldsymbol{\theta}}_n + \frac{1}{n}S_n^{-1}(\bar{\boldsymbol{\theta}}_n)\left.\frac{\partial \log\{f(\boldsymbol{X}_n|\boldsymbol{\theta})\pi(\boldsymbol{\theta})\}}{\partial\boldsymbol{\theta}}\right|_{\boldsymbol{\theta}=\bar{\boldsymbol{\theta}}_n} + O_p(n^{-2}).$$

Thus, we obtain

$$\hat{\boldsymbol{\theta}}_n - \bar{\boldsymbol{\theta}}_n = O_p(n^{-1}).$$

Noting that $\hat{\boldsymbol{\theta}}_{nj} - \boldsymbol{\theta}_{0j} = O_p(n^{-1/2})$ and $\bar{\boldsymbol{\theta}}_{nj} - \hat{\boldsymbol{\theta}}_{nj} = O_p(n^{-1})$ for $j = 1, \cdots, p$, the third term can be ignored given a moderate sample size. Combination of these results verifies the proof. When the posterior mean $\bar{\boldsymbol{\theta}}_n$ and the posterior mode $\hat{\boldsymbol{\theta}}_n$ are identical, the third term drops out completely.

8.3.2 Further simplification of BPIC

Consider a model $f(y|\boldsymbol{\theta})$ with $\log \pi(\boldsymbol{\theta}) = O_p(1)$, where the prior is assumed to be dominated by the likelihood as n increases. We assume further that the specified parametric models contain the true model, or are similar to the true model. In this case, it can be shown that the bias term of BPIC reduces to the number of model parameters p.

Outline of the Proof

The estimated bias term of BPIC can be modified as follows:

$$\begin{aligned} n \times \hat{b}(G) &= \int \log f(\boldsymbol{X}_n|\boldsymbol{\theta})\pi(\boldsymbol{\theta}|\boldsymbol{X}_n)d\boldsymbol{\theta} - \log f(\boldsymbol{X}_n|\hat{\boldsymbol{\theta}}_n) \\ &\quad + \left[\int \log \pi(\boldsymbol{\theta})\pi(\boldsymbol{\theta}|\boldsymbol{X}_n)d\boldsymbol{\theta} - \log \pi(\hat{\boldsymbol{\theta}}_n) + \text{tr}\left\{S_n(\hat{\boldsymbol{\theta}}_n)Q_n(\hat{\boldsymbol{\theta}}_n)\right\} + \frac{p}{2}\right] \\ &= E_1 + E_2. \end{aligned}$$

In the next step, we calculate these two terms E_1 and E_2.

Calculation of E_1

We may expand the log-likelihood function $\log f(\boldsymbol{X}_n|\boldsymbol{\theta})$ around the posterior mode $\hat{\boldsymbol{\theta}}_n$ to give, to second order,

$$\begin{aligned}\log f(\boldsymbol{X}_n|\boldsymbol{\theta}) &= \log f(\boldsymbol{X}_n|\hat{\boldsymbol{\theta}}_n) + (\boldsymbol{\theta}-\hat{\boldsymbol{\theta}}_n)^T \left.\frac{\partial \log f(\boldsymbol{X}_n|\boldsymbol{\theta})}{\partial \boldsymbol{\theta}}\right|_{\boldsymbol{\theta}=\hat{\boldsymbol{\theta}}_n} \\ &\quad + \frac{n}{2}\mathrm{tr}\left\{S_n(\hat{\boldsymbol{\theta}}_n)(\boldsymbol{\theta}-\hat{\boldsymbol{\theta}}_n)(\boldsymbol{\theta}-\hat{\boldsymbol{\theta}}_n)^T\right\}.\end{aligned}$$

Taking expectations of this equation with respect to the posterior distribution gives

$$\int \log f(\boldsymbol{X}_n|\boldsymbol{\theta})\pi(\boldsymbol{\theta}|\boldsymbol{X}_n)d\boldsymbol{\theta} = \log f(\boldsymbol{X}_n|\hat{\boldsymbol{\theta}}_n) + \frac{n}{2}\mathrm{tr}\left\{S_n(\hat{\boldsymbol{\theta}}_n)V_n(\hat{\boldsymbol{\theta}}_n)\right\},$$

where

$$V_n(\bar{\boldsymbol{\theta}}_n) = \int (\boldsymbol{\theta}-\hat{\boldsymbol{\theta}}_n)(\boldsymbol{\theta}-\hat{\boldsymbol{\theta}}_n)^T\pi(\boldsymbol{\theta}|\boldsymbol{X}_n)d\boldsymbol{\theta},$$

is the posterior covariance matrix. Under approximate posterior normality, the posterior covariance matrix can be approximated as $V_n(\hat{\boldsymbol{\theta}}_n) \simeq n^{-1}S_n^{-1}(\hat{\boldsymbol{\theta}}_n)$. Thus, E_1 can be approximated by $E_1 \simeq -p/2$.

Calculation of E_2

Since the specified parametric models contain the true model, or are similar to the true model, we have $S_n(\hat{\boldsymbol{\theta}}_n) \simeq Q_n(\hat{\boldsymbol{\theta}}_n)$ from the Remark below. Thus

$$\mathrm{tr}\left\{S_n(\hat{\boldsymbol{\theta}}_n)Q_n(\hat{\boldsymbol{\theta}}_n)\right\} \simeq p.$$

Ignoring the term of order $o(1)$, $\int \log\pi(\boldsymbol{\theta})\pi(\boldsymbol{\theta}|\boldsymbol{X}_n)d\boldsymbol{\theta} - \log\pi(\hat{\boldsymbol{\theta}}_n)$, E_1 can be approximated by $E_1 \simeq p + p/2 = 3p/2$.

Combining the above results, the asymptotic bias is asymptotically given as the number of parameters.

Remark

Since the prior is assumed to be dominated by the likelihood as n increases, $\log\pi(\boldsymbol{\theta}) = O_p(1)$, under a large sample situation, we can ignore the effect of the prior distribution. Thus, the matrices $Q(\boldsymbol{\theta})$ and $S(\boldsymbol{\theta})$ in the bias term of BPIC reduce to

$$\begin{aligned}Q(\boldsymbol{\theta}) &= \int \frac{\partial \log f(\boldsymbol{x}|\boldsymbol{\theta})}{\partial \boldsymbol{\theta}}\frac{\partial \log f(\boldsymbol{x}|\boldsymbol{\theta})}{\partial \boldsymbol{\theta}^T}dG(\boldsymbol{x}), \\ S(\boldsymbol{\theta}) &= -\int \frac{\partial^2 \log f(\boldsymbol{x}|\boldsymbol{\theta})}{\partial\boldsymbol{\theta}\partial\boldsymbol{\theta}^T}dG(\boldsymbol{x}).\end{aligned}$$

Also, the following equality generally holds with respect to the second derivative of the log-likelihood function:

$$\begin{aligned}\frac{\partial^2 \log f(\boldsymbol{x}|\boldsymbol{\theta})}{\partial\boldsymbol{\theta}\partial\boldsymbol{\theta}^T} &= \frac{\partial}{\partial\boldsymbol{\theta}}\left[\frac{\partial \log f(\boldsymbol{x}|\boldsymbol{\theta})}{\partial\boldsymbol{\theta}^T}\right] \\ &= \frac{\partial}{\partial\boldsymbol{\theta}}\left[\frac{1}{f(\boldsymbol{x}|\boldsymbol{\theta})}\frac{\partial f(\boldsymbol{x}|\boldsymbol{\theta})}{\partial\boldsymbol{\theta}^T}\right] \\ &= \frac{1}{f(\boldsymbol{x}|\boldsymbol{\theta})}\frac{\partial^2 f(\boldsymbol{x}|\boldsymbol{\theta})}{\partial\boldsymbol{\theta}\partial\boldsymbol{\theta}^T} - \frac{1}{f(\boldsymbol{x}|\boldsymbol{\theta})^2}\frac{\partial f(\boldsymbol{x}|\boldsymbol{\theta})}{\partial\boldsymbol{\theta}}\frac{\partial f(\boldsymbol{x}|\boldsymbol{\theta})}{\partial\boldsymbol{\theta}^T} \\ &= \frac{1}{f(\boldsymbol{x}|\boldsymbol{\theta})}\frac{\partial^2 f(\boldsymbol{x}|\boldsymbol{\theta})}{\partial\boldsymbol{\theta}\partial\boldsymbol{\theta}^T} - \frac{\partial \log f(\boldsymbol{x}|\boldsymbol{\theta})}{\partial\boldsymbol{\theta}}\frac{\partial \log f(\boldsymbol{x}|\boldsymbol{\theta})}{\partial\boldsymbol{\theta}^T}.\end{aligned}$$

When the specified parametric models contain the true model, there exists a parameter value $\boldsymbol{\theta}_0$ such that $g(\boldsymbol{x}) = f(\boldsymbol{x}|\boldsymbol{\theta}_0)$. Therefore, we obtain

$$\begin{aligned}\int \frac{1}{f(\boldsymbol{x}|\boldsymbol{\theta}_0)}\frac{\partial^2 f(\boldsymbol{x}|\boldsymbol{\theta}_0)}{\partial\boldsymbol{\theta}\partial\boldsymbol{\theta}^T} g(\boldsymbol{x})d\boldsymbol{x} &= \int \frac{1}{f(\boldsymbol{x}|\boldsymbol{\theta}_0)}\frac{\partial^2 f(\boldsymbol{x}|\boldsymbol{\theta}_0)}{\partial\boldsymbol{\theta}\partial\boldsymbol{\theta}^T} f(\boldsymbol{x}|\boldsymbol{\theta}_0)d\boldsymbol{x} \\ &= \frac{\partial^2}{\partial\boldsymbol{\theta}\partial\boldsymbol{\theta}^T}\int f(\boldsymbol{x}|\boldsymbol{\theta}_0)d\boldsymbol{x} \\ &= \frac{\partial^2}{\partial\boldsymbol{\theta}\partial\boldsymbol{\theta}^T}\int f(\boldsymbol{x}|\boldsymbol{\theta}_0)d\boldsymbol{x} = O.\end{aligned}$$

Therefore, the equality $Q(\boldsymbol{\theta}_0) = S(\boldsymbol{\theta}_0)$ holds. Even when the specified parametric model doesn't contain the true model, we can approximately have the equality $Q(\boldsymbol{\theta}_0) \simeq S(\boldsymbol{\theta}_0)$ if the specified parametric model is similar to the true model.

8.4 Derivation of generalized information criterion

This section reviews a general framework of the information theoretic approach (Akaike (1973, 1974), Konishi and Kitagawa (1996)) through the evaluation of regression models.

8.4.1 Information theoretic approach

Suppose the independent responses $x_1, ..., x_n$ are generated from unknown true distribution $G(\boldsymbol{x})$ having probability density $g(x)$. We regard $g(x)$ to be a target probability mechanism generating the data. In practical situations, it is difficult to obtain precise information on the structure of a system or a process from a finite number of observed data. Therefore, one uses a parametric family of distributions with densities $\{f(x|\boldsymbol{\theta}); \boldsymbol{\theta} \in \Theta \subset R^p\}$ as an approximating model to the true model $g(x|\boldsymbol{x})$. Estimation of unknown parameter vector $\boldsymbol{\theta}$ in the approximating model is done by a suitable procedure such as maximum likelihood method, maximizing penalized likelihood method, Bayes approach and so on. The predictive density function $f(z|\hat{\boldsymbol{\theta}})$ for a future observation z can be constructed by replacing the unknown parameter vector $\boldsymbol{\theta}$ by its sample estimate $\hat{\boldsymbol{\theta}}$. A future observation z from the true model $g(z)$ is predicted by using the constructed probability density function $f(z|\hat{\boldsymbol{\theta}})$. After constructing a statistical model, one would like to assess the closeness of $f(z|\hat{\boldsymbol{\theta}})$ to the true model $g(z|\boldsymbol{x})$ from a predictive point of view.

Suppose that $z_1, ..., z_n$ are future observations for the response variable Y drawn from $g(y|\boldsymbol{x})$. Let $f(\boldsymbol{z}|\hat{\boldsymbol{\theta}}) = \Pi_{\alpha=1}^n f(z_\alpha|\hat{\boldsymbol{\theta}})$ and $g(\boldsymbol{z}) = \Pi_{\alpha=1}^n g(z_\alpha)$. In order to assess the closeness of $f(y|\hat{\boldsymbol{\theta}})$ to $g(y)$, the deviation of $f(\boldsymbol{z}|\hat{\boldsymbol{\theta}})$ from the true model $g(\boldsymbol{z})$ is measured by Kullback–Leibler information (Kullback and Leibler (1951))

$$\begin{aligned} I\{g(\boldsymbol{z}), f(\boldsymbol{z}|\hat{\boldsymbol{\theta}})\} &= E_{G(\boldsymbol{z}|X)}\left[\log \frac{g(\boldsymbol{z})}{f(\boldsymbol{z}|\hat{\boldsymbol{\theta}})}\right] \\ &= \int \log g(\boldsymbol{z}) dG(\boldsymbol{z}) - \int \log f(\boldsymbol{z};\hat{\boldsymbol{\theta}}) dG(\boldsymbol{z}), \end{aligned} \tag{8.8}$$

where $dG(\boldsymbol{z})$ is the Lebesgue measure with respect to a probability density $g(\boldsymbol{z})$. The Kullback–Leibler information (8.8) takes positive values, unless $f(\boldsymbol{z}|\hat{\boldsymbol{\theta}}) = g(\boldsymbol{z})$ holds almost everywhere. The best model is chosen by minimizing Kullback-Leibler information among different statistical models.

Since the first term $E_{G(\boldsymbol{z})}[\log g(\boldsymbol{z})]$ in the right-hand side of Equation (8.8) does not depend on the model, one can measure the relative deviation of $f(\boldsymbol{z}|\hat{\boldsymbol{\theta}})$ from $g(\boldsymbol{z})$ by the second term, the expected log-likelihood:

$$\eta(G;\hat{\boldsymbol{\theta}}) := E_{G(\boldsymbol{z})}\left[\log f(\boldsymbol{z}|\hat{\boldsymbol{\theta}})\right] = \int \log f(\boldsymbol{z}|\hat{\boldsymbol{\theta}}) dG(\boldsymbol{z}), \tag{8.9}$$

where the expectation is taken over the true distribution. Hence, instead of minimizing the Kullback–Leibler information (8.8), one maximizes the expected log-likelihood (8.9). Note here that the expected log-likelihood depends on the unknown true distribution $G(\boldsymbol{z})$ and on the observed data $y_1, ..., y_n$ taken from $G(y)$.

A natural estimator of the expected log-likelihood is the sample based log-likelihood

$$\eta(\hat{G};\hat{\boldsymbol{\theta}}) = \int \log f(\boldsymbol{z}|\hat{\boldsymbol{\theta}}) d\hat{G}(\boldsymbol{z}) = \frac{1}{n}\sum_{\alpha=1}^{n} \log f(x_\alpha|\hat{\boldsymbol{\theta}}), \tag{8.10}$$

which is formally obtained by replacing the unknown distribution $G(\boldsymbol{z})$ in (8.9) by the empirical distribution, $\hat{G}(\boldsymbol{z})$, putting math $1/n$ on each observation x_α. The log-likelihood $\eta(\hat{G};\hat{\boldsymbol{\theta}})$ generally provides a positive bias as an estimator of the expected log-likelihood $\eta(G;\hat{\boldsymbol{\theta}})$, because the same data are used both to estimate the parameters of the model and to evaluate the expected log-likelihood. Therefore, the bias correction of the log-likelihood should be considered.

The bias $b(G)$ of the log-likelihood in estimating the expected log-likelihood $\eta(G;\hat{\boldsymbol{\theta}})$ is given by

$$\begin{aligned} b(G) &:= E_{G(\boldsymbol{x})}[\eta(\hat{G};\hat{\boldsymbol{\theta}}) - \eta(G;\hat{\boldsymbol{\theta}})] \\ &= \int \left[\frac{1}{n}\sum_{\alpha=1}^{n} \log f(x_\alpha|\hat{\boldsymbol{\theta}}) - \int \log f(\boldsymbol{z}|\hat{\boldsymbol{\theta}})dG(\boldsymbol{z}) \right] \prod_{\alpha=1}^{n} dG(x_\alpha). \end{aligned} \tag{8.11}$$

If the bias $b(G)$ can be estimated by appropriate procedures, the bias corrected log-likelihood is given by

$$\frac{1}{n}\sum_{\alpha=1}^{n} \log f(x_\alpha|\hat{\boldsymbol{\theta}}) - \hat{b}(G),$$

which is usually used in the form

$$\begin{aligned} \text{IC} &= -2n\left\{ \frac{1}{n}\sum_{\alpha=1}^{n} \log f(x_\alpha|\hat{\boldsymbol{\theta}}) - \hat{b}(G) \right\} \\ &= -2\sum_{\alpha=1}^{n} \log f(x_\alpha|\hat{\boldsymbol{\theta}}) + 2n\hat{b}(G), \end{aligned} \tag{8.12}$$

where $\hat{b}(G)$ is an estimator of the bias $b(G)$. The first term on the right-hand side of (8.12) measures the model fitness and the second term is the penalty that measures the complexity of the statistical model.

Consider the situations that the specified parametric model $f(x|\boldsymbol{\theta})$ contains the true distribution $g(x)$, that is $g(x) = f(x|\boldsymbol{\theta}_0)$ for some $\boldsymbol{\theta}_0 \in \Theta$, and that the model is estimated by the maximum likelihood method. Under these assumptions, Akaike (1973, 1974) proposed Akaike's information criterion, known as AIC, for evaluating the constructed models:

$$\text{AIC} = -2\sum_{\alpha=1}^{n} \log f(x_\alpha|\hat{\boldsymbol{\theta}}_{\text{MLE}}) + 2p, \tag{8.13}$$

where $\hat{\boldsymbol{\theta}}_{\text{MLE}}$ is the maximum likelihood estimate and p is the number of free parameters in the model. The bias correction term can be applied in an automatic way in various situations. Moreover, the bias approximated by the number of parameters in the model is constant and has no variability, that is, it does not depend on the given observation (Konishi (1999)).

AIC is, however, a criterion for evaluating models estimated by the maximum likelihood method. AIC is, therefore, not theoretically justified for the evaluation of a model estimated by the Bayesian method, even if the specified parametric family of probability distributions encompasses the true distribution $g(x)$.

The problem is how to construct an information criterion that can be applied to a wider class of statistical models. In the face of this difficulty, Konishi and Kitagawa (1996), Konishi (1999) proposed the generalized information criterion, GIC, for evaluating the models constructed by various types of estimation procedures. They observe that the bias is approximated as a function of the empirical influence function of the estimator and the score function of the parametric model.

8.4.2 Derivation of GIC

Let $\hat{\boldsymbol{\theta}} = (\hat{\theta}_1, \hat{\theta}_2, \cdots, \hat{\theta}_p)^T$ be the p-dimensional statistical function such that $\hat{\boldsymbol{\theta}} = \boldsymbol{T}(\hat{G})$. Now, the stochastic expansion of each of the elements of $\boldsymbol{T}(\hat{G})$ around $\boldsymbol{T}(G)$ is

$$\hat{\theta}_i = T_i(G) + \frac{1}{n}\sum_{\alpha=1}^{n} T_i^{(1)}(x_\alpha; G) + \frac{1}{2n^2}\sum_{\alpha=1}^{n}\sum_{\beta=1}^{n} T_i^{(2)}(x_\alpha, x_\beta; G). \tag{8.14}$$

where $T_i^{(1)}(x_\alpha; G)$ and $T_i^{(2)}(x_\alpha, x_\beta; G)$ are the first and second order derivatives of the functional $\boldsymbol{T}(\cdot)$. It is known that

$$\int [\hat{\boldsymbol{\theta}} - \boldsymbol{T}(G)]dG(z) = \frac{1}{n}\boldsymbol{b} + o(n^{-1}), \quad b_i = \frac{1}{2}\int T_i^{(2)}(z, z; G)dG(z).$$

Also, the variance covariance matrix of the estimator of $n^{1/2}[\hat{\boldsymbol{\theta}} - \boldsymbol{T}(G)]$ is asymptotically given by $\Sigma = (\sigma_{ij})$ with

$$\sigma_{ij} = \int T_i^{(1)}(z; G)T_j^{(1)}(z; G)dG(z).$$

(See Konishi and Kitagawa (1996), Konishi and Kitagawa (2009)).

Putting the Equation (8.14) into a Talyor expansion of $\log f(z|\boldsymbol{\theta})$ around $\hat{\boldsymbol{\theta}} = T(G)$ gives

$$\begin{aligned}
&\eta(G; \hat{\boldsymbol{\theta}}) \\
&\approx \int \log f(z|T(G))dG(z) + \sum_{i=1}^{p}(\hat{\theta}_i - T_i(G)) \int \left.\frac{\partial \log f(z|\boldsymbol{\theta})}{\partial \theta_i}\right|_{\boldsymbol{\theta}=T(G)} dG(z) \\
&+\frac{1}{2}\sum_{i=1}^{p}\sum_{j=1}^{p}(\hat{\theta}_i - T_i(G))(\widehat{\theta}_j - T_j(G)) \int \left.\frac{\partial^2 \log f(z|\boldsymbol{\theta})}{\partial \theta_i \partial \theta_j}\right|_{\boldsymbol{\theta}=T(G)} dG(z)
\end{aligned}$$

$$= \int g(z) \log f(z|T(G)) dG(z)$$
$$+ \frac{1}{n} \sum_{i=1}^{p} \sum_{\alpha=1}^{n} T_i^{(1)}(x_\alpha; G) \int g(z) \left. \frac{\partial \log f(z|\boldsymbol{\theta})}{\partial \theta_i} \right|_{\boldsymbol{\theta}=T(G)} dG(z)$$
$$+ \frac{1}{2n^2} \sum_{\alpha=1}^{n} \sum_{\beta=1}^{n} \left[\sum_{i=1}^{p} T_i^{(2)}(x_\alpha, x_\beta; G) \int g(z) \left. \frac{\partial \log f(z|\boldsymbol{\theta})}{\partial \theta_i} \right|_{\boldsymbol{\theta}=T(G)} dG(z) \right.$$
$$\left. + \sum_{i=1}^{p} \sum_{j=1}^{p} T_i^{(1)}(x_\alpha; G) T_j^{(1)}(x_\beta; G) \int g(z) \left. \frac{\partial^2 \log f(z|\boldsymbol{\theta})}{\partial \theta_i \partial \theta_j} \right|_{\boldsymbol{\theta}=T(G)} dz \right].$$

Also, we have

$$\eta(\hat{G}; \hat{\boldsymbol{\theta}})$$
$$\approx \frac{1}{n} \sum_{\alpha=1}^{n} \log f(x_\alpha|T(G)) + \frac{1}{n} \sum_{i=1}^{p} \sum_{\alpha=1}^{n} (\hat{\theta}_i - T_i(G)) \left. \frac{\partial \log f(x_\alpha|\boldsymbol{\theta}\ )}{\partial \theta_i} \right|_{T(G)}$$
$$+ \frac{1}{2n} \sum_{i=1}^{p} \sum_{j=1}^{p} \sum_{\alpha=1}^{n} (\hat{\theta}_i - T_i(G))(\hat{\theta}_j - T_j(G)) \left. \frac{\partial^2 \log f(x_\alpha|\boldsymbol{\theta})}{\partial \theta_i \partial \theta_j} \right|_{\boldsymbol{\theta}=T(G)}$$
$$= \frac{1}{n} \sum_{\alpha=1}^{n} \log f(x_\alpha|T(G))$$
$$+ \frac{1}{n^2} \sum_{i=1}^{p} \sum_{\alpha=1}^{n} \sum_{\beta=1}^{n} T_i^{(1)}(x_\alpha; G) \left. \frac{\partial \log f(x_\beta|\boldsymbol{\theta}\ )}{\partial \theta_i} \right|_{\boldsymbol{\theta}=T(G)}$$
$$+ \frac{1}{2n^3} \sum_{\alpha=1}^{n} \sum_{\beta=1}^{n} \sum_{\gamma=1}^{n} \left[\sum_{i=1}^{p} T_i^{(2)}(x_\alpha, x_\beta; G) \left. \frac{\partial \log f(x_\gamma|\boldsymbol{\theta}\ )}{\partial \theta_i} \right|_{\boldsymbol{\theta}=T(G)} \right.$$
$$\left. + \sum_{i=1}^{p} \sum_{j=1}^{p} T_i^{(1)}(x_\alpha; G) T_j^{(1)}(x_\beta; G) \left. \frac{\partial^2 \log f(x_\gamma|\boldsymbol{\theta}\ )}{\partial \theta_i \partial \theta_j} \right|_{\boldsymbol{\theta}=T(G)} \right].$$

Taking the expectations yields

$$\int \eta(\hat{G}; \hat{\boldsymbol{\theta}}) dG(\boldsymbol{y}) = \int \log f(z|T(G)) dG(z) + \frac{1}{n} \left[\boldsymbol{b}^T \boldsymbol{a} - \frac{1}{2} \mathrm{tr}[\Sigma(G) J(G)] \right],$$
$$\int \eta(G; \hat{\boldsymbol{\theta}}) dG(\boldsymbol{y}) = \int \log f(z|T(G)) dG(z) + \frac{1}{n} \left[\boldsymbol{b}^T \boldsymbol{a} - \frac{1}{2} \mathrm{tr}[\Sigma(G) J(G)] \right.$$
$$\left. + \sum_{i=1}^{p} \int T_i^{(1)}(z; G) \left. \frac{\partial \log f(z|\boldsymbol{\theta})}{\partial \theta_i} \right|_{\boldsymbol{\theta}=T(G)} dG(z) \right]$$

with

$$\boldsymbol{a} = \int \left. \frac{\partial \log f(z|\boldsymbol{\theta})}{\partial \boldsymbol{\theta}} \right|_{\boldsymbol{\theta}=T(G)} dG(z)$$

and

$$J(G) = -\int \left. \frac{\partial^2 \log f(z|\boldsymbol{\theta})}{\partial\boldsymbol{\theta}\partial\boldsymbol{\theta}^T} \right|_{\boldsymbol{\theta}=T(G)} dG(z).$$

Finally, we have

$$\begin{aligned}
&\int \left[\eta(\hat{G};\hat{\boldsymbol{\theta}}) - \eta(G;\hat{\boldsymbol{\theta}})\right] dG(\boldsymbol{x}) \\
&= \frac{1}{n}\left[\sum_{i=1}^{p} \int T_i^{(1)}(z;G) \left.\frac{\partial \log f(z|\boldsymbol{\theta})}{\partial\theta_i}\right|_{\boldsymbol{\theta}=T(G)} dG(z)\right] \\
&= \frac{1}{n}\mathrm{tr}\left[\int T^{(1)}(z;G) \left.\frac{\partial \log f(z|\boldsymbol{\theta})}{\partial\boldsymbol{\theta}^T}\right|_{T(G)} dG(z)\right].
\end{aligned}$$

Replacing the expectation $G(\boldsymbol{x})$ with the empirical distribution $\hat{G}(\boldsymbol{x})$, the bias term of GIC is obtained.

8.5 Comparison of various Bayesian model selection criteria

In this section, we compare the properties of various Bayesian model selection criteria from several aspects, including the use of improper prior, computational amount, etc. Table 8.1 summarizes (1) whether the criteria are applicable under the improper prior, (2) the utility functions employed by the criteria, (3) their computational cost, and (4) the Bayesian estimation methods that can be combined with these criteria, (5) whether the criteria is applicable even when the model is mis-specified. For example, BIC is applicable even if we use the improper prior. Its utility function is the marginal likelihood and its computation is very easy. However, it can only be used for evaluating the models estimated by the maximum likelihood method. Also, it is based on the marginal likelihood and thus is applicable under the model mis-specification situation.

8.5.1 Utility function

To evaluate the estimated Bayesian models, we need a utility function, or loss function, that quantifies the goodness of the models. For example, the marginal likelihood

$$\int f(\boldsymbol{X}|\boldsymbol{\theta})\pi(\boldsymbol{\theta})d\boldsymbol{\theta},$$

covered in Chapter 5 is one of the measures. The marginal likelihood is employed as the original utility function of the Bayes factor, the Bayesian information criterion (BIC, Schwarz (1978)), the generalized Bayesian information criteria (GBIC; Konishi et al. (2004)) and other simulation based marginal likelihood approaches covered in Chapter 6. These criteria quantifies how well the prior density fits to the observations $\boldsymbol{X}_n$.

A disadvantage of the marginal likelihood is it is not well-defined under the improper prior. As alternative approaches, a variety of pseudo-Bayes factors have been proposed. A natural alternative to the marginal likelihood of the model is the following pseudo marginal likelihood:

$$\prod_{\ell=1}^{N} \int f(\boldsymbol{X}_{n(\ell)}|\boldsymbol{\theta})\pi(\boldsymbol{\theta}|\boldsymbol{X}_{-n(\ell)})d\boldsymbol{\theta}, \tag{8.15}$$

which measures the predictive ability of the Bayesian models. Therefore, Intrinsic Bayes factors (Berger and Pericchi (1996)), Fractional Bayes factors (O'Hagan (1995)), Cross validation predictive density approach (Gelfand et al. (1992)) might be useful from a forecasting point of view. An advantage of this method is that it can be applied in an automatic way to various practical situations. The computational time is, however, an enormous amount for a large sample size.

From a predictive point of view, Konishi and Kitagawa (1996) and Ando and Tsay (2009) proposed to evaluate the predictive ability of a given model based on the expected log-likelihood of the Bayesian predictive distributions

$$\int \log f(\boldsymbol{Z}_n|\boldsymbol{X}_n)g(\boldsymbol{Z}_n)d\boldsymbol{Z}_n,$$

with $f(\boldsymbol{Z}_n|\boldsymbol{X}_n) = \int f(\boldsymbol{Z}_n|\boldsymbol{\theta})\pi(\boldsymbol{\theta}|\boldsymbol{X}_n)d\boldsymbol{\theta}$. From an information theoretic point of view, the measure is a well-known statistic for model evaluation. Konishi and Kitagawa (1996) constructed the Bayesian models using the maximum likelihood estimate. On the other hand, Ando and Tsay's (2009) criteria evaluates the general Bayesian models estimated both by an empirical and full Bayesian approach.

As a Bayesian version of fitness criterion, Bayesian predictive information criterion (Ando (2007)) uses the posterior mean of the expected log-likelihood.

$$\int \left\{ \int \log f(z|\boldsymbol{\theta})\pi(\boldsymbol{\theta}|\boldsymbol{X}_n)d\boldsymbol{\theta} \right\} dG(z).$$

This quantity also measures the deviation of the predictive distribution from the true model $g(z)$.

In summary, these utility functions that quantify a predictive measure would be well-suited when we focus on the forecasting.

8.5.2 Robustness to the improper prior

We have seen that the marginal likelihood is not well defined when we use the improper priors. However, under the no (weak) prior knowledge, we often use a noninfonnative prior of some sort, including a Jeffreys (1961) prior, a Bernardo (1979) reference prior, or one of the many other possibilities. Thus, it is important to discuss whether the Bayesian model selection criteria are applicable to such situations.

In such a case, Bayesian information criterion (BIC, Schwarz (1978)) is available when one wants to employ the general framework of the Bayesian approach for model selection discussed in Chapter 5. Thus, the criteria that compute the marginal likelihood directly cannot be applied, including, generalized Bayesian information criteria (GBIC; Konishi et al. (2004)), the Laplace-Metropolis estimator (Lewis and Raftery, 1997), the so-called candidate formula (Chib, 1995), the harmonic mean estimator (Newton and Raftery, 1994), Gelfand and Dey's estimator (Gelfand and Dey, 1994) and so on.

As alternative criteria, we can use the Bayesian predictive information criterion (BPIC, Ando (2007)), the deviance information criteria (Spiegelhalter et al. (2002), the predictive likelihood approach (Ando and Tsay (2009)), and the generalized information criteria (Konishi and Kitagawa (1996)). Since these criteria are not based on the marginal likelihood, they are applicalbe for a wide range of Bayesian models with improper priors.

8.5.3 Computational cost

It is also important to argue about the amount of computation. Generally, we can estimate the Bayesian models based on 1. Analytical approach, 2. Asymptotic approach, 3, Simulation approach. If the Bayesian model can be estimated analytically, the required computational time for each criterion might be almost the same. However, the criteria that requires use of cross-validation and its variants, including the cross validation predictive density approach (Gelfand et al. (1992)) results in an enormous computational amount for a large sample size. Moreover, the difference of their computational time would become bigger when the Bayesian models are estimated by the asymptotic and simulation approaches.

Also, several criteria, like BIC, have a simple form, though that of GBIC is complex. This computational complexity might be one of the biggest factors that affect the users' behavior. Generally, we can obtain a simple form of model selection criterion if we impose strong assumptions on the model. For example, there are two versions of BPIC (7.3) and (7.4). In the Equation (7.3), the penalty term is the number of parameters. If we impose assumptions (i.e., a specified sampling density contains the true model and the prior information becomes weak as the sample size becomes large), the penalty term of (7.4) reduces to the number of model parameters. If such assumptions seem to be satisfied in practical situations, users may use the simple version of BPIC. On the other hand, if such assumptions are not acceptable, one would be forced to use a more complicated formula (7.4).

8.5.4 Estimation methods

It is also important to clarify the Bayesian estimation methods that can be combined with the criteria. For example, we cannot use the BIC if the model is estimated by the penalized likelihood method. Generally speaking, the criteria are preferable if they cover a wide range of Bayesian estimation methods. From this perspective, the simulation approaches that estimate the marginal likelihood, given in Chapter 6, are only available when the model is estimated by the posterior simulation approaches, including MCMC. Also, BPIC and DIC cover such estimation methods. On the other hand, BIC, GBIC, MBIC and GIC do not cover the Bayesian models estimated by the posterior simulation approaches. In summary, model selection criteria that can treat a wide range of estimation methods are favorable.

8.5.5 Misspecified models

If a specified parametric family of probability distributions $f(x|\boldsymbol{\theta})$ encompasses the true model $g(x)$, we say that the model is correctly specified. On the other hand, the model is mis-specified if the true model is not included in the specified model. It is also worth the time to investigate various criteria whether they are applicable — even the model is mis-specified. Under some regularity conditions, the model selection criteria that estimate the marginal likelihood are always applicable, since the marginal likelihood can be calculated regardless of the relationship between these two models.

However, the BPIC in (7.3) and PL_2 in (5.46) requires a situation that the model is correctly specified. In such a case, we have to use the BPIC in (7.4) and PL in (5.45), though the forms of equation are relatively complex.

8.5.6 Consistency

In Section 5.5.5, we discussed the consistency of the Bayesian information criteria. Under a certain condition, the Bayesian information criteria hold the consistency of model selection. The Bayes factor (Bayes factor(M_k, M_j) in (5.4)) is consistent in the sense that if M_k is the true model, then Bayes factor$(M_k, M_j) \to \infty$ with probability one as $n \to \infty$ (O'Hagan (1997)). Therefore, the Bayes factor computed by simulation approaches, including Laplace-Metropolis in (6.2), Harmonic mean in (6.3), Chib's estimator from Gibbs sampling in (6.8), Chib's estimator from MH sampling in (6.11) and other computing emthods that estimated the marginal likelihood are also consistent.

The similar augments of BIC might apply to the generalized Bayesian information criteria, GBIC and the Modified BIC, MBIC, in (7.12). However, these criteria are derived under a situation $\log \pi(\boldsymbol{\theta}) \neq O_p(1)$ and thus the best model should be defined carefully. Hereafter, we assume that $\log \pi(\boldsymbol{\theta}) = O_p(1)$ to simplify the augument.

Consider a situation that the specified family of parametric models contain the true model and that these models are nested. It is known that all the various forms of intrinsic Bayes factor can also be shown to be consistent (O'Hagan (1997)). On the other hand, the fractional Bayes factor is not consistent in this sense if the fraction b in (5.38) is held fixed and is independent of the sample size n. However, we assume that $b = n(1)/n$ and thus the fractional Bayes factor is also consistent because $b \to 0$ as $n \to \infty$ (O'Hagan (1997)).

To investigate the properties of the following criteria, the cross validated predictive densities CVPD in (5.40), the expected predictive likelihood approach PL_2 in (5.46), assume that the predictive distribution can be approximated as $f(z|\boldsymbol{X}_n) = f(z|\hat{\boldsymbol{\theta}}_{\mathrm{MLE}}) + O_p(n^{-1})$. We then define the best model as the predictive density that has the lowest Kullback–Leibler (1951) divergence from the true model. This framework is clearly the case treated in Section 5.5.5. Thus these criteria are also consistent.

TABLE 8.1: Comparison of various model selection criteria 1. PMELL: posterior mean of expected log-likelihood, ELL: expected log-likelihood, ML: Marginal likelihood, PL: Penalized likelihood, PS: Posterior simulation, PML: Pseudo marginal likelihood, NA: Not applicable. [a] Under a fixed value of fraction b, its computation is easy. [b] However, PBF tends to overfit. [c] Various estimation methods are covered, including M-estimation and other robust estimation procedures.

Criteria	Improper prior	Utility	Computation	Estimation	Mis-specification
BIC in (5.14)	Robust	ML	Easy	MLE	Applicable
GBIC in (5.16)	NA	ML	Complex	PL	Applicable
AIBF in (5.34)	Robust	PML	Intensive	Analytical, PL, PS	Applicable
GIBF in (5.35)	Robust	PML	Intensive	Analytical, PL, PS	Applicable
FBF in (5.38)	Robust	PML	Easy[a]	Analytical, PL, PS	Applicable
PBF in (5.39)	Robust	PML	Easy[b]	Analytical, PL, PS	Applicable
CVPD in (5.40)	Robust	PML	Intensive	Analytical, PL, PS	Applicable
PL in (5.45)	Robust	ELL	Complex	Analytical, PL, PS	Applicable
PL_2 in (5.46)	Robust	ELL	Easy	Analytical, PL, PS	NA
Laplace–Metropolis in (6.2)	NA	ML	Easy	PS	Applicable
Harmonic mean in (6.3)	NA	ML	Easy	PS	Applicable
Chib's estimator from Gibbs sampling in (6.8)	NA	ML	Relatively Complex	Gibbs sampling	Applicable
Chib's estimator from MH sampling in (6.11)	NA	ML	Complex	MH sampling	Applicable
BPIC in (7.3)	Robust	PMELL	Easy	Analytical, PS	NA
BPIC in (7.4)	Robust	PMELL	Complex	Analytical, PS	Applicable
DIC in (7.10)	Robust	ELL	Easy	Analytical, PS	Applicable
MBIC in (7.12)	Robust	Unclear	Easy	PL	Applicable
GIC in (7.20)	Robust	ELL	Complex	MLE, PL, Others[c]	Applicable
AIC in (8.13)	—	ELL	Easy	MLE	NA

Chapter 9

Bayesian model averaging

In the previous chapters, we have seen various types of Bayesian model selection criteria to select the best model among a set of candidate models $M_1, ..., M_r$. However, it is known that this approach ignores the uncertainty in model selection. To treat the model uncertainty, Bayesian model averaging (BMA) provides a coherent mechanism. The idea of BMA was developed by Leamer (1978), and has recently received a lot of attention in the literature, including Madigan and Raftery (1994), Raftery et al. (1997), Hoeting et al. (1999), Fernandez et al. (2001), Clyde and George (2004), Viallefont et al. (2001), Wasserman (2000), Wright (2008). This chapter describes the definition and practical implementations of BMA and related model averaging approaches.

9.1 Definition of Bayesian model averaging

We have seen that a main purpose of model selection is to select a "single" model that is regarded as the best among all candidate models. Once we select the best model, all subsequent decisions are made under the chosen model. In a real situation, however, the quality of the subsequent decisions after the model selection strictly depend on the selected model. Since picking up a single model ignores the uncertainty in model selection, one may obtain unstable model selection results, which comes from a randomness of observations. See also Leeb and Potscher (2003, 2005) who investigated several issues associated with inference after model selection. To incorporate model uncertainty into the decisions, a simple idea is to average a set of competing models. This approach is called model averaging and this method has been attracting many researchers and practitioners.

Consider a universe of r models $M_1, ..., M_r$. We have seen that the posterior probability of the model M_k for a particular data set $\boldsymbol{X}_n$ was given by

$$P(M_k|\boldsymbol{X}_n) = \frac{P(M_k)\displaystyle\int f_k(\boldsymbol{X}_n|\boldsymbol{\theta}_k)\pi_k(\boldsymbol{\theta}_k)d\boldsymbol{\theta}_k}{\displaystyle\sum_{j=1}^{r} P(M_\alpha)\int f_j(\boldsymbol{X}_n|\boldsymbol{\theta}_j)\pi_j(\boldsymbol{\theta}_j)d\boldsymbol{\theta}_j},$$

where the marginal likelihoods may be estimated by the asymptotic or simulation approaches.

In the Bayesian model averaging framework (Raftery et al. (1997) and Hoeting et al. (1999)), the predictive distribution for a future observation $\boldsymbol{z}$, $f(\boldsymbol{z}|\boldsymbol{X}_n)$ is defined as

$$f(\boldsymbol{z}|\boldsymbol{X}_n) = \sum_{j=1}^{r} P(M_j|\boldsymbol{X}_n) f_j(\boldsymbol{z}|\boldsymbol{X}_n), \tag{9.1}$$

with

$$f_j(\boldsymbol{z}|\boldsymbol{X}_n) = \int f_j(\boldsymbol{z}|\boldsymbol{\theta}_j)\pi_j(\boldsymbol{\theta}_j|\boldsymbol{X}_n)d\boldsymbol{\theta}_j, \quad j = 1, ..., r.$$

This predictive distribution $f(\boldsymbol{z}|\boldsymbol{X}_n)$ is an average of the predictive distributions under each of the models considered, weighted by their posterior model probability.

Let Δ be the quantity of our interest. Similarly to the estimation of the predictive distribution, model averaged estimates of quantities of interest, e.g., the mean and variance are obtained as

$$\begin{aligned} E\left[\Delta|\boldsymbol{X}_n\right] &= \sum_{j=1}^{r} P(M_j|\boldsymbol{X}_n)\Delta_j, \\ \mathrm{Var}\left[\Delta|\boldsymbol{X}_n\right] &= \sum_{j=1}^{r} \left[\mathrm{Var}\left[\Delta_j|\boldsymbol{X}_n, M_j\right] + \Delta_j^2\right] P(M_j|\boldsymbol{X}_n) - E\left[\Delta|\boldsymbol{X}_n\right]^2, \end{aligned}$$

with

$$\Delta_j = E\left[\Delta_j|\boldsymbol{X}_n, M_j\right], \quad j = 1, ..., r.$$

In favorable cases where the marginal likelihood can be derived analytically, the computation of the weight $P(M_k|\boldsymbol{X}_n)$, the posterior model probability might be straightforward under the stiation that the number of averaging models r is relatively small. Even when the analytical expression of the marginal likelihood $P(\boldsymbol{X}_n) = \int f_k(\boldsymbol{X}_n|\boldsymbol{\theta}_k)\pi_k(\boldsymbol{\theta}_k)d\boldsymbol{\theta}_k$ is not available, we can use the Bayesian computation approaches. However, under a situation where the number of averaging models r involved in the posterior model probability is relatively large, we often face the problem that the computation is practically infeasible. In such cases, a common approach is to resort to an MCMC algorithm, by which we run MCMC in the model space $\{M_1, \ldots, M_r\}$, like reversible jump MCMC (Green (1995)), the product space search (Carlin and Chib (1995)) and other algorithms. In the context of regression models, one can also employ George and McCulloch (1993, 1997), and Raftery et al. (1997).

There are many applications of BMA. Wright (2008) applied the BMA to the exchange rate forecasting. For the portfolio selection, the BMA approach was employed by Ando (2009b). Viallefont (2001) considered the variable selection and Bayesian model averaging in case-control studies. See also Hoeting et al. (1999) and Clyde and George (2004) a nice review of BMA approach and related topics.

9.2 Occam's window method

It is known that averaging over all the models generally provides better average predictive ability (Madigan and Raftery (1994)). However, the number of models in summation (9.1) often makes the implementation of Bayesian model averaging to be impractical. To reduce the computational amount, the Occam's window method was proposed by Madigan and Raftery (1994). It averages over a reduced set of parsimonious, data-supported models.

There are two basic principles underlying the Madigan and Raftery (1994)'s approach. The first principle is that if a model with less predictive ability should be excluded from a set of averaging models. The second principle is that if complex models which receive less support from the data than their simpler counterparts should be excluded from a set of averaging models. More formally, the following process will exclude these unlikely models.

Let us denote a set of likely models and unlikely models to be R and Q. An initial set of likely models to be included is $R_0 = \{M_1, ..., M_r\}$, from which the models that are unlikely are a posteriori excluded.

Firstly, the model M_k, which has a largest marginal likelihood score $P(M_k|\boldsymbol{y}) = \text{argmax}_j P(M_j|\boldsymbol{X}_n)$ is identified. Secondly, given the value of C, a set of unlikely models

$$Q_1 = \left\{ M_j ; \frac{P(M_k|\boldsymbol{X}_n)}{P(M_j|\boldsymbol{X}_n)} \geq C \right\}$$

is excluded from R_0. We then obtain the updated likely models $R_1 = \{M_j; M_j \notin Q_1\}$.

As a third step, we further exclude a set of unlikely models. Focusing on each of the models $M_j \in R_1$, if there exists a model $M_l \in R_1$ that satisfies $M_l \subset M_j$ and $\pi(M_l|\boldsymbol{y})/\pi(M_j|\boldsymbol{y}) \geq 1$, then such a model M_j is excluded. Then, from R_1, we obtain a set of unlikely models

$$Q_2 = \left\{ M_j ; M_l \subset M_j, M_j, M_l \in R_1, \frac{P(M_l|\boldsymbol{X}_n)}{P(M_j|\boldsymbol{X}_n)} \geq 1 \right\}.$$

The final set of likely models is $R_2 = \{M_j; M_j \in R_1, M_j \notin Q_2\}$ and the corresponding predictive distribution becomes

$$h(\boldsymbol{z}|\boldsymbol{X}_n) = \sum_{M_j \in R_2} P(M_j|\boldsymbol{X}_n) f_j(\boldsymbol{z}|\boldsymbol{X}_n)$$

which greatly reduces the number of models in the summation.

Although the Occam window approach for the Bayesian model averaging is useful, the selection of the size of Occam's razor is still unclear. We often want to evaluate the goodness of the predictive distributions $f(\boldsymbol{z}|\boldsymbol{X}_n)$ in (9.1), which depends on sampling density and the prior density parameter, and the number of models to be included in the predictive distribution. To select the value of C, Ando (2008b) proposed BPIC to evaluate the constructed models based on BMA. This approach will be explained below.

Recently, Ando (2008b) extended the Bayesian predictive information criterion Ando (2007) to cover the evaluation of the predictive distributions developed by the Bayesian model averaging approach. To determine the best predictive distribution among different statistical models, Ando (2008b) considered the maximization of the posterior mean of the expected log-likelihood:

$$\begin{aligned}\eta &= \int \left[\sum_{j=1}^{r} P(M_j|\boldsymbol{X}_n) \times \left\{ \int \log f_j(\boldsymbol{z}|\boldsymbol{\theta}_j)\pi_j(\boldsymbol{\theta}_j|\boldsymbol{X}_n)d\boldsymbol{\theta}_j \right\} \right] dG(\boldsymbol{z}) \\ &= \sum_{j=1}^{r} P(M_j|\boldsymbol{X}_n) \times \int \left\{ \int \log f_j(\boldsymbol{z}|\boldsymbol{\theta}_j)\pi_j(\boldsymbol{\theta}_j|\boldsymbol{X}_n)d\boldsymbol{\theta}_j \right\} dG(\boldsymbol{z}), \qquad (9.2)\end{aligned}$$

where $dG(\boldsymbol{z})$ is the Lebesgue measure with respect to a probability density $g(\boldsymbol{z})$ for the true model. Ando (2008b) developed the Bayesian predictive information criterion for the evaluation of the models constructed by the Bayesian model averaging approach:

$$\begin{aligned}\text{BPIC} &= -2\sum_{j=1}^{r} P(M_j|\boldsymbol{y}) \times \left\{ \int \log f_j(\boldsymbol{X}_n|\boldsymbol{\theta}_j)\pi_j(\boldsymbol{\theta}_j|\boldsymbol{X}_n)d\boldsymbol{\theta}_j - n\hat{b}_j \right\} \\ &= \sum_{j=1}^{r} P(M_j|\boldsymbol{y}) \times \text{BPIC}_j, \qquad (9.3)\end{aligned}$$

where $\hat{b}_j$ is the bias term of BPIC in (7.3) under the model M_j. The criterion is therefore expressed as a linear combination of the BPIC score for each of the models M_j, BPIC_j, weighted by the posterior model probabilities $P(M_j|\boldsymbol{X}_n)$.

9.3 Bayesian model averaging for linear regression models

Let us consider the set of linear regression models

$$\boldsymbol{y}_n = X_{jn}\boldsymbol{\beta}_j + \boldsymbol{\varepsilon}_{jn},$$

where $\boldsymbol{y}_n$ is a $n \times 1$ vector of observations on a variable that we want to predict, X_{jn} is a $n \times p_j$ matrix of predictors, $\boldsymbol{\beta}_j$ is a $p_j \times 1$ parameter vector, ε_{jn} is the error vector, the errors are independently, identically distributed with mean zero and variance σ^2.

For the model priors, the following prior specification is often used in literatures (e.g., Fernandez et al. (2001), Wright (2008)). For the coefficient vector $\boldsymbol{\beta}_j$, we take the natural conjugate g-prior specification. Conditional on σ^2, it is $N(\boldsymbol{0}, \phi\sigma^2(X_{jn}^T X_{jn})^{-1})$. For the prior σ^2, we use the improper prior that is proportional to $1/\sigma^2$.

Let $\boldsymbol{\theta}_j = (\boldsymbol{\beta}_j^T, \sigma^2)^T$. It is known that one can then calculate the marginal likelihood of the j-th model M_j analytically as

$$P(\boldsymbol{y}_n|M_j) = \int f_j(\boldsymbol{y}_n|X_{jn}, \boldsymbol{\theta}_j)\pi_j(\boldsymbol{\theta}_j)d\boldsymbol{\theta}_j \propto \frac{1}{2}\frac{\Gamma(n/2)}{\pi^{n/2}}(1+\phi)^{-p_j} S_j^{-n}$$

with $f_j(\boldsymbol{y}_n|X_{jn}, \boldsymbol{\theta}_j)$ is the likelihood and

$$S_j^2 = \boldsymbol{y}_n^T \boldsymbol{y}_n - \frac{\phi}{1+\phi}\boldsymbol{y}_n^T X_{jn}(X_{jn}^T X_{jn})^{-1} X_{jn}^T \boldsymbol{y}_n$$

see, e.g., Wright (2008).

Wright (2008) pointed out that the prior for σ^2 is an inverse gamma prior with parameter 0 and 0, and is improper. Usually, we cannot use improper priors for model-specific parameters. Due to the original nature of improper priors, they are unique only up to an arbitrary multiplicative constant. Thus, we have a difficulty in calculating the model posterior probabilities (Kass and Raftery, 1995). However, the improper prior of the variance parameter σ^2 can be employed, because this parameter is common to all of the set of models. Thus the posterior probabilities are not affected by the arbitrary multiplicative constant in this prior.

Under the equal prior model probabilities, $P(M_k) = 1/r$, $k = 1, ..., r$, the posterior model probability for the model M_k is given as

$$P(M_k|\boldsymbol{y}_n) = \frac{(1+\phi)^{-p_k} S_k^{-n}}{\displaystyle\sum_{j=1}^{r}(1+\phi)^{-p_j} S_j^{-n}}, \quad k = 1, ..., r.$$

Using the posterior model probability, one can construct the predictive mean and variance by using the formulas given in Section 9.1.

9.4 Other model averaging methods

For model averaging, we consider a universe of J models denoted by $M_1, ..., M_J$. An essential idea of model averaging is to construct the predictive distribution $p(\boldsymbol{z}|J)$ by combining the individual predictive distributions such that

$$p(\boldsymbol{z}|J) = \sum_{j=1}^{J} w(M_j) \times p(\boldsymbol{z}|M_j), \quad \sum_{j=1}^{J} w(M_j) = 1,$$

with optimal weights $w(M_j)$. This problem has been widely studied in the literature. The simplest approach is to use unweighted average across all models. However, it is obviously desirable that the models that perform well would be weighted higher than predictions from poorly performing models. Broadly speaking, we can distinguish three strands of model averaging methods.

9.4.1 Model averaging with AIC

Let $f(\boldsymbol{y}|\boldsymbol{\theta}_j, M_j)$ be the probability density function of $\boldsymbol{y}$ under model M_j with parameter $\boldsymbol{\theta}_j$ for $j = 1, ..., J$. In the frequentist approach, each model M_j is estimated by the maximum likelihood method. The predictive distribution of model M_j for a future observation $\boldsymbol{z}$ is given by $p(\boldsymbol{z}|M_j) = f(\boldsymbol{z}|\widehat{\boldsymbol{\theta}}_{j,\mathrm{MLE}}, M_j)$, where $\widehat{\boldsymbol{\theta}}_{j,\mathrm{MLE}}$ is the MLE of $\boldsymbol{\theta}$. Akaike (1979) and Kapetanios et al. (2006) use the Akaike information criterion (Akaike, 1974) to construct the weights

$$w(M_j) = \frac{\exp\{-0.5\,(\mathrm{AIC}_j - \mathrm{AIC}_{\min})\}}{\sum_{k=1}^{J} \exp\{-0.5\,(\mathrm{AIC}_k - \mathrm{AIC}_{\min})\}},$$

where $\mathrm{AIC}_{\min}$ is the minimum AIC score among the universe of J models.

As an alternative to AIC, Hansen (2007, 2008) proposed selecting the weights by minimizing a Mallows criterion for averaging across least squares estimates obtained from a set of models.

9.4.2 Model averaging with predictive likelihood

Recently, Eklund and Karlsson (2005) extended the standard approach of Bayesian model averaging by constructing the weights based on the predictive likelihood, instead of the standard marginal likelihood $\pi(\boldsymbol{y}|M_j)$. These authors pointed out that the use of predictive measure offers a greater protection against in-sample overfitting and improves the performance of out-of-sample forecasting. They applied this approach to forecast the Swedish inflation rate and showed that model averaging using predictive likelihood outperforms the standard Bayesian model averaging using the marginal likelihood.

However, in applying their procedure, Eklund and Karlsson (2005) partitioned the data into a training subsample and a hold-out subsample. The predictive likelihood is then estimated by using a cross validation procedure. Such a procedure is time consuming when the number of observations is large.

Ando and Tsay (2009) constructed the weights based on the expected log-predictive likelihood (5.41). As discussed before, the expected log-predictive likelihood measures the predictive performance of a statistical model. We use it here to construct the optimal weights for model averaging. An advantage of this new approach is that it does not require any cross-validation in the construction of weights for individual models.

Using the proposed approach, the optimal weights for model averaging are given by

$$w(M_j) = \frac{\exp\{\eta(M_j) - \eta_{\max}\} P(M_j)}{\sum_{k=1}^{J} \exp\{\eta(M_k) - \eta_{\max}\} P(M_k)}, \quad j = 1, ..., J,$$

where $\eta_{\max}$ is the maximum expected log-predictive likelihood score among the universe of J models. Although the weight $w(M_j)$ depends on the unknown quantities $\eta(M_j)$, we can estimate the expected log-predictive likelihood by using (5.45) or (5.46).

For model averaging, we refer to excellent textbooks by Burnham and Anderson (2002) and Claeskens and Hjort (2008).

Exercises

1. *Generate random samples* $\{(x_{1\alpha}, ..., x_{8\alpha}, y_\alpha); \alpha = 1, ..., n\}$ *from the true model* $y_\alpha = 0.1x_{1\alpha} + 0.2x_{2\alpha} + \varepsilon_\alpha$*, where the noises* ε_α *are generated from the normal with mean* 0 *and the standard deviation* $\sigma = 0.2$ *and* $x_{j\alpha}$ *are uniformly distributed within* $[-2, 2]$. *Then, using the BIC score, implement the Bayesian model averaging based on the size of Occam's razor* $C = 20$. *The* **R** *package* **BMA** *is useful to implement this model.*

2. *(Continued). Compare the result of Bayesian model averaging based on the size of Occam's razor* $C = 100$.

3. *Dataset, the effect of punishment regimes on crime rates, can be found in the* **R** *package* **MASS**. *Using the BIC score, implement the Bayesian model averaging based on the size of Occam's razor* $C = 20$.

4. *In Section 6.7.1, Bayesian analysis of the probit model has been implemented for the analysis of default data. Using the BIC score, implement the Bayesian model averaging of logit model. We can set the size of Occam's razor to be* $C = 20$. *The* **R** *package* **BMA** *is useful to implement this anlysis.*

5. *In Section 5.5.4, Bayesian analysis of the survival analysis has been implemented. Using the Bayesian model averaging of the gamma model, analyse the ovarian cancer survival data. You can use the BIC score to*

implement the analysis with the size of Occam's razor $C = 20$. *The* **R** *package* **BMA** *is useful to implement this anlysis.*

Bibliography

Abrevaya, J. 2001. The effects of demographics and maternal behavior on the distribution of birth outcomes. *Empirical Economics* 26: 247–257.

Aguilar, O. and West, M. 2000. Bayesian dynamic factor models and variance matrix discounting for portfolio allocation. *Journal of Business and Economic Statistics* 18: 338–357.

Aitkin, M. 1991. Posterior Bayes factor (with discussion). *Journal of the Royal Statistical Society* B53: 111–142.

Akaike, H. 1973. Information theory and an extension of the maximum likelihood principle. In *Proc. 2nd International Symposium on Information Theory*, ed. Petrov, B. N. and Csaki, F., 267–281. Budapest: Akademiai Kiado.

Akaike, H. 1974. A new look at the statistical model identification. *IEEE Transactions on Automatic Control* 19: 716–723.

Akaike, H. 1979. A Bayesian extension of the minimum AIC procedure of autoregressive model fitting. *Biometrika* 66: 237–242.

Albert, J. 2007. *Bayesian Computation with R.* New York: Springer.

Albert, J. H. and Chib, S. 1993. Bayesian analysis of binary and polychotomous response data, *Journal of the American Statistical Association* 88: 669–679.

Albert, J. H. and Chib, S. 2001. Sequential ordinal modeling with applications to survival data. *Biometrics* 57: 829–836.

Alizadeh, A. A., Eisen, M. B., Davis, R. E., et al. 2000. Distinct types of diffuse large B-cell lymphoma identified by gene expression profiling. *Nature.* **403**, 503–511.

Alon, U., Barkai, N., Notterman, D. A., et al. 1999. Broad patterns of gene expression revealed by clustering analysis of tumor and normal colon tissues probed by oligonucleotide arrays. *Proceedings of the National Academy of Sciences* 96: 6745–6750.

Alpaydin, E. and Kaynak, C. 1998. Cascading classifiers, *Kybernetika* 34: 369–374.

Ando, T. 2004. Nonlinear regression and multi-class classification based on radial basis function networks and model selection criteria. Ph.D. thesis, Graduate School of Mathematics, Kyushu University.

Ando, T. 2006. Bayesian inference for nonlinear and non-Gaussian stochastic volatility model with leverage effect. *Journal of the Japan Statistical Society* 36: 173-197.

Ando, T. 2007. Bayesian predictive information criterion for the evaluation of hierarchical Bayesian and empirical Bayes models, *Biometrika* 94: 443–458.

Ando, T. 2008a. Measuring the sales promotion effect and baseline sales for incense products: a Bayesian state space modeling approach. *Annals of the Institute of Statistical Mathematics* 60: 763-780.

Ando, T. 2008b. Bayesian model averaging and Bayesian predictive information criterion for model selection. *Journal of the Japan Statistical Society* 38: 243–257.

Ando, T., Konishi, S. and Imoto, S. 2008. Nonlinear regression modeling via regularized radial basis function networks *Journal of Statistical Planning and Inference* 138: 3616-3633.

Ando, T. 2009a. Bayesian factor analysis with fat-tailed factors and its exact marginal likelihood. *Journal of Multivariate Analysis* 100: 1717–1726.

Ando, T. 2009b. Bayesian portfolio selection using multifactor model and Bayesian predictive information criterion. *International Journal of Forecasting* 25: 550–566.

Ando, T. 2009c. Bayesian inference for the hazard term structure with functional predictors using Bayesian predictive information criterion. *Computational Statistics and Data Analysis* 53: 1925–1939.

Ando, T. and Konishi S. 2009. Nonlinear logistic discrimination via regularized radial basis functions for classifying high-dimensional data. *Annals of the Institute of Statistical Mathematics* 61: 331–353.

Ando, T. and Tsay, R. 2009. Predictive marginal likelihood for the Bayesian model selection and averaging. *International Journal of Forecasting* in press,

Andrieu, C., de Freitas, N. and Doucet, A. 2001. Robust full Bayesian learning for radial basis networks. *Neural Computation* 13: 2359–2407.

Andrews, D. F. and Mallows, C. L. 1974. Scale mixtures of normal distributions. *Journal of the Royal Statistical Society* B 36: 99–102.

Ardia, D. 2009. Bayesian estimation of a Markov-switching threshold asymmetric GARCH model with Student-*t* innovations. *Econometrics Journal* 12: 105–126

Barnard, J., McCulloch, R. and Meng, X., 2000. Modeling covariance matrices in terms of standard deviations and correlations, with application to shrinkage. *Statistica Sinica* 10: 1281–1311.

Banerjee, S., Carlin, B. P. and Gelfand, A. E. 2004. *Hierarchical Modeling and analysis of Spatial Data.* London: Chapman and Hall/CRC.

Barndorff-Nielsen, O. E. and Cox, D. R. 1989. *Asymptotic Techniques for Use in Statistics.* London: Chapman and Hall.

Barndorff-Nielsen, O. E. and Shephard, N. 2001. Econometric analysis of realised volatility and its use in estimating stochastic volatility models, *Journal of the Royal Statistical Society* B 64: 253–280.

Bauwens, L., Lubrano M. and Richard, J. F. 1999. *Bayesian Inference in Dynamic Econometric Models.* Oxford University Press.

Berg, A., Meyer, R. and Yu, J. 2004. Deviance information criterion comparing stochastic volatility models. *Journal of Business and Economic Statistics* 22: 107–120.

Berger, J. O. 1985. *Statistical Decision Theory and Bayesian Analysis.* New York: Springer.

Berger, J. O. and Pericchi, L. R. 1996. The intrinsic Bayes factor for linear models. In *Bayesian Statistics 5*, ed. J. M. Bernardo, J. O. Berger, A. P. Dawid and A. F. M. Smith, 25–44. Oxford: Oxford University Press.

Berger, J. O. and Pericchi, L. R. 1998a. Accurate and stable Bayesian model selection: the median intrinsic Bayes factor. *Sankhyā* B 60: 1–18.

Berger, J. O. and Pericchi, L. R. 1998b. Objective Bayesian Methods for Model Selection: Introduction and Comparison. In *Model selection* ed. P. Lahiri, 135–207. OH: Beachwood.

Berger, J. O., Ghosh, J. K. and Mukhopadhyay, N. 2003. Approximations to the Bayes factor in model selection problems and consistency issues. *Journal of Statistical Planning and Inference* 112: 241–258.

Bernardo, J. M. 1979. Reference posterior distributions for Bayesian inference. *Journal of the Royal Statistical Society* B 41: 113–147.

Bernardo, J. M. and Smith, A. F. M. 1994. *Bayesian theory.* Chichester: John Wiley.

Besag, J. York, J. and Mollie, A. 1991. Bayesian image restoration, with two applications in spatial statistics (with discussion). *Annals of the Institute of Statistical Mathematics* 43: 1–59.

Besag, J. and Higdon, D. 1999. Bayesian analysis of agricultural field experiments. *Journal of the Royal Statistical Society* B 61: 691–746.

Billio, M., Monfort, A. and Robert, C. P. 1999. Bayesian estimation of switching ARMA models. *Journal of Econometrics* 93: 229–255.

Bollerslev, T. 1986. Generalized autoregressive conditional heteroskedasticity. *Journal of Econometrics* 31: 307–327.

Box, G. E. P. 1976 Science and statistics. *Journal of the American Statistical Association* 71: 791–799.

Box, G. E. P. and Tiao, G. C. 1973. *Bayesian Inference in Statistical Analysis.* MA: Addison–Wesley.

Breiman, L., Friedman, J. H., Olshen, R. A., and Stone, C. J. 1984. *Classification and Regression Trees.* CA: Belmont, Wadsworth.

Brooks, S. P. and Gelman, A. 1997. General methods for monitoring convergence of iterative simulations. *Journal of Computational and Graphical Statistics* 7: 434–455.

Burnham, K. P. and Anderson, D. 2002. *Model Selection and Multi-Model Inference: A Practical Information-Theoretic Approach,* 2nd Edition. Springer Statistical Theory and Methods.

Candes, E. and Tao, T. 2007. The dantzig selector: Statistical estimation when p is much larger than n. *Annals of Statistics* 35: 2313–2351.

Carlin, B. P. and Chib, S. 1995. Bayesian Model choice via Markov chain Monte Carlo methods. *Journal of the Royal Statistical Society* B 57: 473–484.

Carlin, B. and Louis, T. 2000 *Bayes and empirical Bayes methods for data analysis.* New York: Chapman and Hall.

Celeux, G., Forbes, F., Robert, C., and Titterington, D. M. 2006. Deviance information criteria for missing data models, *Bayesian Analysis* 1: 651–674.

Chakrabartia, A. and Ghosh, J. K. 2006. A generalization of BIC for the general exponential family. *Journal of Statistical Planning and Inference* 136: 2847–2872.

Chao, J. C. and Phillips, P. C. B. 1998. Bayesian posterior distributions in limited information analysis of the simultaneous equation model using Jeffreys' prior, *Journal of Econometrics* 87: 49–86.

Chen, M.-H., Shao, Q.-M., and Ibrahim, J. G. 2000. *Monte Carlo Methods in Bayesian Computation.* New York: Springer-Verlag.

Chib, S. 1995. Marginal Likelihood from the Gibbs output. *Journal of the American Statistical Association* 90: 1313–1321.

Chib, S. and Jeliazkov, I. 2001. Marginal likelihood from the Metropolis-Hastings output. *Journal of the American Statistical Association* 96: 270–281.

Chib, S., Nardari, F. and Shephard, N. 2002. Markov chain Monte Carlo methods for stochastic volatility models. *Journal of Econometrics* 108: 281–316.

Chintagunta, P. K. and Prasad, A. R. 1998. An Empirical Investigation of the "Dynamic McFadden" Model of Purchase Timing and Brand Choice: Implications for Market Structure. *Journal of Business and Economic Statistics* 16: 2–12.

Chopin, N. and Pelgrin, F. 2004. Bayesian inference and state number determination for hidden Markov models: an application to the information content of the yield curve about inflation. *Journal of Econometrics* 123: 327–344.

Claeskens, G. and Hjort, N. L. 2008. *Model Selection and Model Averaging.* Cambridge: Cambridge University Press.

Clark. P. K. 1973. A Subordinated Stochastic Process Model with Finite Variance for Speculative Prices. *Econometrica* 4: 135–156.

Clarke, B. S. and Barron, A. R. 1994. Jeffreys' proir is asmptotically least favorable under entropy risk. *Journal of Statistical Planning and Inference* 41: 37–40.

Clyde, M. and George, E. I. 2004. Model uncertainty. *Statistical Science* 19: 81–94.

Clyde, M., DeSimone, H., and Parmigiani, G. 1996. Prediction via orthogonalized model mixing. *Journal of the American Statistical Association* 91: 1197–1208.

Congdon, P. 2001. *Bayesian Statistical Modelling.* New York: Wiley.

Congdon, P. 2007. Applied Bayesian Models New York: John Wiley & Sons.

Cox, D. R. 1972. Regression models and life-tables. *Journal of the Royal Statistical Society* B 34: 187–220.

Cox, D. R. and Hinkley, D.V. 1974. *Theoretical Statistics.* London: Chapman & Hall.

Cox, D. R. and Wermuth, N. 1996. *Multivariate Dependencies.* Chapman & Hall, London.

Craven, P. and Wahba, G. 1979. Smoothing Noisy Data with Spline Functions. *Numerische Mathematik* 31: 377–403.

Davison, A. C. 1986. Approximate predictive likelihood. *Biometrika* 73: 323–332.

de Boor, C. 1978. *A Practical Guide to splines.* Berlin: Springer.

Dellaportas, P., Forster, J. J. and Ntzoufras, I. 2002. On Bayesian model and variable selection using MCMC. *Statistics and Computing* 12: 27–36.

Dellaportas, P., Giudici, P. and Roberts, G. 2003. Bayesian inference for non-decomposable graphical Gaussian models. *Sankhya: The Indian Journal of Statistics* 65: 43–55.

Dempster, A. P. 1972. Covariance selection. *Biometrics* 28: 157–175.

Denison, D. G. T., Holmes, C. C., Mallick, B. K. and Smith, A. F. M. 2002. *Bayesian Methods for Nonlinear Classification and Regression.* New York: Wiley.

Denison, D. G. T., Mallick, B. K., Smith, A. F. M. 1998. Automatic Bayesian curve fitting. *Journal of the Royal Statistical Society* B 60: 333–350.

Deschamps, P. J. 2006. A flexible prior distribution for Markov switching autoregressions with Student-*t* Errors. *Journal of Econometrics* 133: 153–190.

DiCiccio, T. J., Kass, R. E., Raftery, A. E. and Wasserman, L. 1997. Computing Bayes factors by combining simulation and asymptotic approximations. *Journal of the American Statistical Association* 92: 903–915.

Dickey, J. 1971. The weighted likelihood ratio, linear hypotheses on normal location parameters. *Annals of Statistics* 42: 204–223.

DiMatteo, I. Genovese, C. R., Kass, R. E. 2001. Bayesian curve-fitting with free-knot splines. *Biometrika* 88: 1055–1071.

Donald, S. G. and Paarsch, H. J. 1993. Piecewise pseudo-maximum likelihood estimation in empirical models of auctions. *International Economic Review* 34: 121–148.

Drton, M. and Perlman, M. D. 2004. Model selection for Gaussian concentration graphs. *Biometrika* 91: 591–602.

Drèze, J. H. 1976. Bayesian limited information analysis of the simultaneous equations model. *Econometrica* 44: 1045–1075.

Drèze, J. H. and Morales, J. A. 1976. Bayesian full information analysis of simultaneous equations. *Journal of the American Statistical Association* 71: 329–354.

Edwards, D. M. 2000. *Introduction to Graphical Modelling* New York: Springer.

Efron, B. and Tibshirani, R. J. 1993. *An Introduction to the Bootstrap.* New York: Chapman and Hall.

Efron, B., Hastie, T., Johnstone, I. and Tibshirani, R. 2004. Least Angle Regression. *Annals of Statistics* 32: 407–499.

Eilers, P. H. C. and Marx, B. D. 1996. Flexible smoothing with B-splines and penalties (with discussion). *Statistical Science* 11: 89–121.

Eilers, P. H. C. and Marx, B. D. 1998. Direct generalized additive modeling with penalized likelihood. *Computational Statistics and Data Analysis* 28: 193–209.

Eklund, J. and Karlsson, S. 2005. Forecast combination and model averaging using predictive measures. Sveriges Riksbank Working Paper, vol. 191.

Engle, R. F. 1982. Autoregressive conditional heteroscedasticity with estimates of the variance of United Kingdom inflation. *Econometrica* 50: 987–1008.

Fama, E. and French, K. 1993. Common risk factors in the returns on stocks and bonds. *Journal of Financial Economics* 33: 3–56.

Fernandez, C., Ley, E. and Steel, M. F. J. 2001. Benchmark priors for Bayesian model averaging. *Journal of Econometrics* 100: 381–427.

Finley, A. O., Banerjee, S. and Carlin, B. P. 2007. spBayes: An R Package for Univariate and Multivariate Hierarchical Point-referenced Spatial Models. *Journal of Statistical Software* 19: 4.

Forster, J. J., McDonald, J. W. and Smith, P. W. F. 2003. Markov chain Monte Carlo exact inference for binomial and multinomial logistic regression models. *Statistics and Computing* 13: 169–177

Foster, D. P. and George, E. I. 1994. The risk ination criterion for multiple regression. *Annals of Statistics* 22: 1947–1975.

Fruhwirth-Schnatter 2001. Fully Bayesian analysis of switching Gaussian state space models. *Annals of the Institute of Statistical Mathematics* 53: 31–49.

Fujii, T. and Konishi, S. 2006. Nonlinear regression modeling via regularized wavelets and smoothing parameter selection. *Journal of Multivariate Analysis* 97: 2023–2033.

Gamerman, D. and Lopes, H. F. 2006. Markov Chain Monte Carlo: Stochastic Simulation for Bayesian Inference (2nd edition). London: Chapman & Hall/CRC Press.

Gelfand, A. E. and Dey, D. K. 1994. Bayesian model choice: asymptotics and exact calculations. *Journal of the Royal Statistical Society* B56: 510–514.

Gelfand, A. E. and Ghosh, S. K. 1998. Model choice: a minimum posterior predictive loss approach. *Biometrika* 85: 1–11.

Gelfand, A. E., Banerjee, S. and Gamerman, D. 2005. Spatial process modeling for univariate and multivariate dynamic spatial data. *Environmetrics* 16: 465–479.

Gelfand, A. E., Dey, D. K. and Chang, H. 1992. Model determination using predictive distributions with implementation via sampling-based methods (with discussion). In *Bayesian Statistics 4* ed. J. M. Bernardo, J. O. Berger, A. P. Dawid and A. F. M. Smith, 147-167. Oxford: Oxford University Press.

Gelfand, A. E., Kim, H-J., Sirmans, C. F. and Banerjee, S. 2003. Spatial modeling with spatially varying coefficient processes. *Journal of the American Statistical Association* 98: 387-96.

Gelman, A., Carlin, B., Stern, S. and Rubin, B. 1995. *Bayesian Data Analysis.* London: Chapman and Hall/CRC.

Gelman, A. and Meng, X. L. 1998. Simulating normalizing constants: From importance sampling to bridge sampling to path sampling. *Statistical Science* 13: 163–185.

Gelman, A. and Rubin, D. B. 1992. Inference from iterative simulation using multiple sequences. *Statistical Science* 7: 457–511.

Geman, S. and Geman, D. 1984. Stochastic relaxation, Gibbs distributions, and the Bayesian restoration of images. *IEEE Transactions on Pattern Analysis and Machine Intelligence* 6: 721–741.

George, E. I. and McCulloch, R. E. 1993. Variable selection via Gibbs sampling. *Journal of the American Statistical Association* 88: 881–889.

George, E. I. and McCulloch, R. E. 1997. Approaches for Bayesian variable selection. *Statistica Sinica* 7: 339–373.

Gerlach, R. and Tuyl, F. 2006. MCMC methods for comparing stochastic volatility and GARCH models. *International Journal of Forecasting* 22: 91–107

Geweke, J. F. 1989a. Bayesian inference in econometric models using Monte Carlo integration, *Econometrica* 57: 1317–1339.

Geweke, J. F. 1989b. Exact predictive densities for linear models with ARCH disturbances *Journal of Econometrics* 40: 63–86.

Geweke, J. F. (1992). Evaluating the accuracy of sampling-based approaches to calculating posterior moments. In *Bayesian Statistics 4* ed. J. M. Bernado et al. 169–193. Oxford: Clarendon Press.

Geweke, J. F. 1993. Bayesian Treatment of the Independent Student-t Linear Model. *Journal of Applied Econometrics* 8: 19–40.

Geweke, J. F. 1996. Variable Selection and Model Comparison in Regression. In *Bayesian Statistics 5*, ed. J.M. Bernardo, J.O. Berger, A.P. Dawid and A.F.M. Smith, 609–620 Oxford: Oxford University Press.

Geweke, J. F. 2005. *Contemporary Bayesian Econometrics and Statistics* New York: Wiley.

Geweke, J. F. and Singleton, K. J. 1980. Interpreting the likelihood ratio statistic in factor models when sample size is small. *Journal of the American Statistical Association* 75: 133–137.

Gilks, W. R., Richardson, S. and Spiegelhalter, D. J. 1996. *Markov Chain Monte Carlo in Practice.* New York: Chapman and Hall.

Giudici, P. and Green, P. J. 1999. Decomposable graphical Gaussian model determination. *Biometrika* 86: 785–801.

Golub, T. R., Slonim, D. K., Tamayo, P. et al. 1999. Molecular classification of cancer: class discovery and class prediction by gene expression monitoring. *Science* 286: 531–537.

Green, P. 1995. Reversible jump Markov chain Monte Carlo computation and Bayesian model determination, *Biometrika* 82: 711–732.

Green, P. J. and Silverman, B. W. 1994. *Nonparametric Regression and Generalized Liner Models.* London: Chapman & Hall/CRC.

Green, P. J. and Yandell, B. 1985. Semi-parametric generalized linear models. In *Generalized Linear Models* ed. Gilchrist, R. Francis, B. J. & Whittaker, J., Lecture Notes in Statistics 32: 44–55, Berlin: Springer.

Gupta, S. and Donald R. L. 2003. Customers as Assets. *Journal of Interactive Marketing* 17: 9–24.

Gupta, S., Hanssens, D., Hardie, B. et al. 2006. Modeling Customer Lifetime Value. *Journal of Service Research* 9: 139–155.

Hamilton, J. D. 1989. A new approach to the economic analysis of nonstationary time series and the business cycle. *Econometrica* 57: 357–384.

Han, C. and Carlin, B. P. 2001. Markov chain Monte Carlo methods for computing Bayes factors: a comparative review. *Journal of the American Statistical Association* 96: 1122–1132.

Hansen, B. E. 2007. Least Squares Model Averaging. *Econometrica* 75: 1175–1189.

Hansen, B. E. 2008. Least Squares Forecast Averaging. *Journal of Econometrics* 146: 342–350.

Härdle, W. 1990. *Applied Nonparametric Regression.* Cambridge: Cambridge University Press.

Hastie, T. and Tibshirani, R. 1990. *Generalized Additive Models.* London: Chapman & Hall/CRC.

Hastie, T., Tibshirani, R. and Friedman, J. 2009. The Elements of Statistical Learning: Data Mining, Inference, and Prediction (Second Edition). New York: Springer.

Hastings, W. K. 1970. Monte Carlo sampling methods using Markov chains and their application. *Biometrika* 57: 97–100.

Hoeting, J., Madigan, D., Raftery, A. and Volinsky, C. 1999. Bayesian model averaging. *Statistical Science* 14: 382–401.

Holmes, C. C. and Mallick, B. K. 1998. Bayesian radial basis functions of variable dimension. *Neural Computation* 10: 1217–1233.

Holmes, C. C. and Mallick, B. K. 2003. Generalized nonlinear modeling with multivariate free-knot regression splines. *Journal of the American Statistical Association* 98: 352–368.

Holmes, C. C. Denison, D. G. T. and Mallick, B. K. 2002. Accounting for model uncertainty in seemingly unrelated regressions. *Journal of Computational and Graphical Statistics* 11: 533–551.

Hosmer, D. W. and Lemeshow, S. 1989. *Applied Logistic Regression.* New York: Wiley-Interscience.

Hurvich, C. M., Simonoff, J. S. and Tsai, C.-L. 1998. Smoothing parameter selection in nonparametric regression using an improved Akaike information criterion. *Journal of the Royal Statistical Society* B60: 271–293.

Ibrahim, J. G., Chen, M. H. and Sinha, D. 2007. *Bayesian Survival Analysis.* Springer-Verlag.

Imai, K. and van Dyk, A. D. 2005. A Bayesian analysis of the multinomial probit model using marginal data augmentation. *Journal of Econometrics* 124: 311–334.

Imoto, S. and Konishi, S. 2003. Selection of smoothing parameters in B-spline nonparametric regression models using information criteria. *Annals of the Institute of Statistical Mathematics* 55: 671–687.

Jacquier, E., Nicholas, G. P. and Rossi, P. E. 2004. Bayesian Analysis of Stochastic Volatility Models with Fat-tails and Correlated Errors. *Journal of Econometrics* 122: 185–212.

Jensen, J. L. and Petersen, N. V. 1999. Asymptotic normality of the maximum likelihood estimator in state space models. *Annals of Statistics* 27: 514–535.

Jeffreys, H. 1946. An Invariant Form for the Prior Probability in Estimation Problems. *Proceedings of the Royal Society of London* A 196: 453–461.

Jeffreys's, H. 1961. *Theory of Probability.* Oxford: Oxford University Press.

Jobson, J. D. and Korkie, B. 1980. Estimation for Markowitz efficient portfolios. *Journal of the American Statistical Association* 75: 544–554.

Kadane, J. B. and Dickey, J.M. 1980. Bayesian decision theory and the simplification of models. In *Evaluation of Econometric Models* ed. Kmenta, J. and Ramsey, J., 245-268. New York: Academic Press.

Kadane, J. B. and Lazar, N. A. 2004. Methods and criteria for model selection. *Journal of the American Statistical Association* 99: 279–290.

Kapetanios, G., Labhard, V. and Price, S. 2006. Forecasting using predictive likelihood model averaging. *Economics Letters* 91: 373–379.

Kass, R. E., Tierney, L. and Kadane, J. B. 1990. The validity of posterior expansions based on Laplace's method. In *Essays in Honor of George Barnard*, ed. S. Geisser, J. S. Hodges, S. J. Press and A. Zellner, 473–488. Amsterdam: North-Holland.

Kass, R. E. and Raftery, A. 1995. Bayes factors. *Journal of the American Statistical Association* 90: 773–795.

Kass, R. E. and Wasserman, L. 1995. A reference Bayesian test for nested hypotheses and its relationship to the Schwarz criterion. *Journal of the American Statistical Association* 90: 928–934.

Khan, J., Wei, J. S., Ringner, M., et al. 2001. Classification and diagnostic prediction of cancers using gene expression profiling and artificial neural networks. *Nature Medicine* 7: 673–679.

Kim, C. J. and Nelson, C. R. 1998. Business cycle turning points: a new coincident index, and tests of duration dependence based on a dynamic factor model with regime switching. *Review of Economics and Statistics* 80: 188–201.

Kim, C. J. and Nelson, C. R. 1999. *State-Space Models with Regime Switching: Classical and Gibbs Sampling Approaches with Applications.* The MIT Press.

Kim, S., Shephard, N. and Chib, S. 1998. Stochastic volatility: likelihood inference comparison with ARCH models. *Review of Economic Studies* 65: 361–393.

Kitagawa, G. 1987. Non-Gaussian state-space modeling of nonstationary time series. *Journal of the American Statistical Association* 82: 1032–1063.

Kitagawa, G. 1996. Monte Carlo filter and smoother for Gaussian nonlinear state space models. *Journal of Computational and Graphical Statistics* 5: 1–25.

Kitagawa, G. and Gersch, W. (1996) *Smoothness Proirs Analysis of Time Series.* Lecture Notes in Statistics 116. Springer.

Kleibergen, F. R., Zivot, E. 2003. Bayesian and classical approaches to instrumental variable regression. *Journal of Econometrics* 114: 29–72.

Kleibergen, F. R., Van Dijk, H.K. 1998. Bayesian simultaneous equations analysis using reduced rank structures. *Econometric Theory* 14: 701–743.

Knorr-Held, L. and Rue, H. 2002. On block updating in Markov random field models for disease mapping. *Scandinavian Journal of Statistics* 29: 597–614.

Koenker, R. 2005. *Quantile Regression.* Econometric Society Monograph Series, Cambridge University Press.

Koenker, R. and Bassett, G. S. 1978. Regression quantiles. *Econometrica* 46: 33–50.

Koenker, R. and Geling, O. 2001. Reappraising medfly longevity: A quantile regression survival analysis. *Journal of the American Statistical Association* 96: 458–468.

Konishi, S. and Kitagawa, G. 1996. Generalised information criteria in model selection. *Biometrika* 83: 875–890.

Konishi, S. and Kitagawa, G. 2003. Asymptotic theory for information criteria in model selection — functional approach. *Journal of Statistical Planning and Inference* 114: 45–61.

Konishi, S. and Kitagawa, G. 2008. *Information Criteria and Statistical Modeling.* Springer.

Konishi, S., Ando, T. and Imoto, S. 2004. Bayesian information criteria and smoothing parameter selection in radial basis function networks. *Biometrika* 91: 27-43.

Koop, G. 2003. *Bayesian Econometrics.* New York: Wiley.

Koop, G., Poirier, D. J. and Tobias, J. L. 2007. *Bayesian Econometric Methods.* Cambridge University Press.

Kullback, S. and Leibler, R. A. 1951. On information and sufficiency. *Annals of Mathematical Statistics* 22: 79–86.

Lancaster, T. 2004. *An Introduction to Modern Bayesian Econometrics*. Blackwell Publishing.

Lang, S. and Brezger, A. 2004. Bayesian *P*-Splines. *Journal of Computational and Graphical Statistics* 13: 183–212.

Lanterman, A. D. 2001. Schwarz, Wallace, and Rissanen: Intertwining themes in theories of model selection. *International Statistical Review* 69: 185–212.

Lauritzen, S. L. 1996. *Graphical Models.* New York: Oxford University Press.

Leamer, E. E. 1978. *Specification Searches: Ad Hoc Inference with Non-Experimental Data.* New York: Wiley.

Lee, P. M. 2004 *Bayesian Statistics – An Introduction.* London: Arnold.

Lee, S.-Y. 2007. *Structural Equation Modelling: A Bayesian Approach.* John Wiley & Sons.

Lee, Y. and Lee, C. K. 2003. Classification of multiple cancer types by multicategory support vector machines using gene expression data. *Bioinformatics* 19: 1132–1139.

Leeb, H. and Potscher, B.M. 2003. The finite sample distribution of post-model-selection estimators and uniform versus non-uniform approximations. *Econometric Theory* 19: 100–142.

Leeb, H. and Potscher, B.M. 2005. Model selection and inference: facts and fiction. *Econometric Theory* 21: 21–59.

Lewis, S. M. and Raftery, A. E. 1997. Estimating Bayes factors via posterior simulation with the Laplace-Metropolis estimator. *Journal of the American Statistical Association* 92: 648–655.

Liang, F., Truong, Y. and Wong, W. 2001. Automatic Bayesian model averaging for linear regression and applications in Bayesian curve fitting. *Statistica Sinica* 11: 1005–1029.

Liang, F., Paulo, R., Molina, G., Clyde, M. A. and Berger, J. O. 2008. Mixtures of g Priors for Bayesian Variable Selection. *Journal of the American Statistical Association* 103: 410–423.

Linhart, H. and Zucchini, W. 1986. *Model Selection.* New York: Wiley.

Liu, J. S. 1994. *Monte Carlo Strategies in Scientific Computing.* New York: Springer.

Lopes, H. F. and West, M. 2004. Bayesian model assessment in factor analysis. *Statistica Sinica* 14: 41–67.

MacKay, D. J. C. 1992. A practical Bayesian framework for backpropagation networks. *Neural Computation* 4: 448–72.

Madigan, D. and Raftery, A. E. 1994. Model selection and accounting for model uncertainty in graphical models using Occam's window, *Journal of the American Statistical Association* 89: 1535–1546.

Mallick, B. K., 1998. Bayesian curve estimation by polynomial of random order. *Journal of Statistical Planning and Inference* 70: 91–109.

Markowitz, H. 1952. Portfolio selection. *Journal of Finance* 7: 77–91.

Matsui, S., Araki, Y. and Konishi, S. 2008. Multivariate regression modeling for functional data, *Journal of Data Science* 6: 313–331.

McCullagh, P. 2002. What is a statistical model? (with discussion). *Annals of Statistics* 30: 1225–1310.

McCullagh, P. and Nelder, J. A. 1989. *Generalized Linear Models.* Chapman & Hall/CRC.

McCulloch, R. E. and Rossi, P. E. 1992. Bayes factors for nonlinear hypotheses and likelihood distributions, *Biometrika* 79: 663–676.

McCulloch, R. E. and Rossi, P. E. 1994. An exact likelihood analysis of the multinomial probit model. *Journal of Econometrics* 64: 207–240.

McCulloch, R. E., Polson, N. G., Rossi, P. E. 2000. A Bayesian analysis of the multinomial probit model with fully identified parameters. *Journal of Econometrics* 99: 173–193.

Meng, X. L. and Wong, W. H. 1996. Simulating ratios of normalizing constants via a simple identity: a theoretical exploration. *Statistica Sinica* 6: 831–860.

Metropolis, N., Rosenbluth, A. W., Rosenbluth, M. N., Teller, A. H. and Teller, E. 1953. Equations of state calculations by fast computing machine. *Journal of Chemical Physics* 21: 1087–1092.

Meyer, R. and Yu, J. 2000. BUGS for a Bayesian Analysis of Stochastic Volatility Models. *Econometrics Journal* 3: 198–215.

Mitchell, T.J. and Beauchamp, J.J. 1988. Bayesian variable selection in linear regression (with discussion). *Journal of the American Statistical Association* 83: 1023–1036.

Moody, J. 1992. The *effective* number of parameters: an analysis of generalization and regularization in nonlinear learning systems. In *Advances in Neural Information Processing System 4*, ed. J. E. Moody, S. J. Hanson and R. P. Lippmann, 847–854. San Mateo: Morgan Kaufmann.

Nakatsuma, T. 1998. A Markov-Chain Sampling Algorithm for GARCH Models. *Studies in Nonlinear Dynamics and Econometrics* 3: 107–117.

Nakatsuma, T. 2000. Bayesian Analysis of ARMA-GARCH Models: A Markov Chain Sampling Approach. *Journal of Econometrics* 95: 57–69.

Neal, R. M. 1996. *Bayesian Learning for Neural Networks.* Lecture Notes in Statistics 118. New York: Springer-Verlag.

Nelder, J. A. and Wedderburn, R. W. M. 1972. Generalized linear models. *Journal of the Royal Statistical Society Series* A 135: 370–384.

Newton, M. A. and Raftery, A. E. 1994. Approximate Bayesian inference by the weighted likelihood bootstrap (with discussion). *Journal of the Royal Statistical Society* B 56: 3–48.

Nobile, A. 1998. A hybrid Markov chain for the Bayesian analysis of the multinomial probit model. *Statistics and Computing* 8: 229–242.

O'Hagan, A. 1995. Fractional Bayes factors for model comparison (with discussion). *Journal of the Royal Statistical Society* B 57: 99–138.

O'Hagan, A. 1997. Properties of intrinsic and fractional Bayes factors. *Test* 6: 101–118.

O'Sullivan, F., Yandell, B. S. and Raynor, W. J. 1986. Automatic smoothing of regression functions in generalized linear models. *Journal of the American Statistical Association* 81: 96–103.

Park, T. and Casella, G. 2008. The Bayesian Lasso. *Journal of the American Statistical Association* 103: 681–686.

Pastor, L. 2000. Portfolio selection and asset pricing models. *Journal of Finance* 55: 179–223.

Pastor, L. and Stambaugh, R. F. 2000. Comparing asset pricing models: an investment perspective. *Journal of Financial Economics* 56: 335–381.

Patrick, R. 1982. An Extension of Shapiro and Wilk's W Test for Normality to Large Samples, *Applied Statistics* 31: 115–124.

Pauler, D. 1998. The Schwarz criterion and related methods for normal linear models. *Biometrika* 85: 13–27.

Percy, D. F. 1992. Predictions for seemingly unrelated regressions, *Journal of the Royal Statistical Society* B 54: 243–252.

Perez, J. M. and Berger, J. O. 2002. Expected-posterior prior distributions for model selection. *Biometrika* 89: 491–512.

Phillips, D. B. and Smith, A. F. M. 1995. Bayesian Model Comparison Via Jump Diffusions. In *Practical Markov Chain Monte Carlo in Practice* ed. W.R. Gilks, S. Richardson and D.J. Spiegelhalter, 215–239. London: Chapman & Hall.

Pitt, M. and Shephard, N. 1999. Filtering via simulation: Auxiliary particle filter. *Journal of the American Statistical Association* 94: 590–599.

Plummer, M. 2008. Penalized loss functions for Bayesian model comparison. *Biostatistics* 9: 523–539.

Pole, A., West, M. and Harrison 2007. *Applied Bayesian forecasting and times series analysis.* Chapman & Hall.

Pollack, J. R., Perou, C. M., Alizadeh, A. A. et al. 1999. Genome-wide analysis of DNA copy-number changes using cDNA microarrays. *Nature Genetics* 23: 41–46.

Press, S. J. 2003. *Subjective and objective Bayesian statistics: principles, models, and applications.* New York: Wiley.

Press, S. J. and Shigemasu, K. 1989. Bayesian inference in factor analysis, In: *Contributions to probability and statistics*, ed. L. Gleser, M. Perleman, S.J. Press, A. Sampson, 271–287. New York: Springer-Verlag.

Press, S. J. and Shigemasu, K. 1999. A note on choosing the number of factors. *Communications in Statistics, Theory Methods* 28: 1653–1670.

Quinn, K. M. and Martin, A. D. 2002. An Integrated Computational Model of Multiparty Electoral Competition. *Statistical Science* 17: 405–419.

Raftery, A. E. and Lewis, S. M. 1992. One long run with diagnostics: Implementation strategies for Markov chain Monte Carlo. *Statistical Science* 7: 493–497.

Raftery, A. E., Madigan, D. and Hoeting, J. A. 1997. Bayesian model averaging for linear regression models. *Journal of the American Statistical Association* 92: 179–191.

Ramsay, J. O. and Silverman, B. W. 1997. *Functional data analysis.* New York: Springer.

Rao, C. R. and Wu, Y., 2001. On model selection (with Discussion), In *Model Selection* ed. by P.Lahiri 1–64 IMS Lecture Notes - Monograph Series 38.

Richard J.F. and Steel M.F.J. 1988. Bayesian analysis of systems of seemingly unrelated regression equations under a recursive extended natural conjugate prior density. *Journal of Econometrics* 38: 7–37.

Rios Insua, D. and Müller, P. 1998. Feedforward neural networks for nonparametric regression. In *Practical Nonparametric and Semiparametric Bayesian Statistics*, ed. D. K. Dey, P. Müller and D. Sinha, 181–191. New York: Springer Verlag.

Ripley, B. D. 1987. *Stochastic Simulation.* New York: Wiley.

Robert, C. 2001. *Bayesian Choice.* Springer Verlag.

Robert, C. P. and Titterington, D. M. 2002. Discussion of a paper by D. J. Spiegelhalter, et al. *Journal of the Royal Statistical Society* B 64: 621–622.

Rossi, P., Gilula, Z. and Allenby, G. 2001. Overcoming scale usage heterogeneity: a Bayesian hierarchical approach, *Journal of the American Statistical Association* 96: 20–31.

Rossi, P., Allenby, G. and McCulloch, R. 2005. *Bayesian Statistics and Marketing.* John Wiley & Sons.

Roverato, A. 2002. Hyper inverse Wishart distribution for non-decomposable graphs and its application to Bayesian inference for Gaussian graphical models. *Scandinavian journal of statistics* 29: 391–411.

Santis, F. D. and Spezzaferri, F. 2001. Consistent fractional Bayes factor for nested normal linear models. *Journal of Statistical Planning and Inference* 97: 305–321.

Schwarz, G. 1978. Estimating the dimension of a model. *Annals of Statistics* 6: 461–464.

Seber, G. A. F. 1984. *Multivariate Observations.* New York: Wiley.

Sha, N., Vannucci, M., Tadesse, M. G., et al. 2004. Bayesian variable selection in multinomial probit models to identify molecular signatures of disease stage. *Biometrics* 60: 812–819.

Sharpe, W. F. 1964. Capital asset prices: A theory of market equilibrium under conditions of risk. *Journal of Finance* 19: 425–442.

Sheather, S. J. and Jones, M. C. 1991. A reliable data-based bandwidth selection method for kernel density estimation. *Journal of the Royal Statistical Society* B 53: 683–690.

Shephard, N. 2005. *Stochastic Volatility: Selected Readings*, Oxford: Oxford University Press.

Shibata, M. and Watanabe, T. 2005. Bayesian analysis of a Markov switching stochastic volatility model, *Journal of Japan Statistical Society* 35: 205–219.

Silverman, B. W. 1985. Some aspects of the spline smoothing approach to nonparametric regression curve fitting (with Discussion). *Journal of the Royal Statistical Society* B 47: 1–52.

Silverman, B. W. 1986. *Density Estimation for Statistics and Data Analysis.* Chapman & Hall.

Sin, C.-Y. and White, H. 1996. Information criteria for selecting possibly misspecified parametric models. *Journal of Econometrics* 71: 207–225.

Sivia, D. S. 1996. *Data Analysis: A Bayesian Tutorial.* Oxford: Oxford University Press.

Smith, M. and Kohn, R. 1996. Nonparametric regression using Bayesian variable selection. *Journal of Econometrics* 75: 317–343.

Smith, A. F. M. and Gelfand, A. E. 1992. Bayesian statistics without tears: a sampling-resampling perspective. *American Statistician* 46: 84–88.

Smith, A. F. M. and Spiegelhalter, D. J. 1980. Bayes factors and choice criteria for linear models. *Journal of the Royal Statistical Society* B 42: 213–220.

So, M., Lam, K. and Li, W. 1998. A stochastic volatility model with Markov switching. *Journal of Business and Economic Statistics* 16: 244–253.

Spiegelhalter, D. J., Best, N. G., Carlin, B. P., and van der Linde, A. 2002. Bayesian measures of model complexity and fit (with discussion and rejoinder). *Journal of the Royal Statistical Society* B 64: 583–639.

Stone, C. J. 1974. Cross-validatory choice and assessment of statistical predictions (with discussion). *Journal of the Royal Statistical Society Series* B36: 111–147.

Stone, M. 1979. Comments on model selection criteria of Akaike and Schwarz. *Journal of the Royal Statistical Society* B 41 276–278.

Takeuchi, K. 1976. Distribution of information statistics and criteria for adequacy of models (*in Japanese*). *Mathematical Sciences* 153: 12–18.

Tanizaki, H. 2004. On Asymmetry, Holiday and Day-of-the-Week Effects in Volatility of Daily Stock Returns: The Case of Japan. *Journal of the Japan Statistical Society* 34: 129–152.

Taylor, S. J. 1982. Financial returns modelled by the product of two stochastic processes – A study of the daily sugar prices 1961–75, In *Time Series Analysis: Theory and Practice 1*, ed. Anderson, O. D., 203–226, Amsterdam: North-Holland.

Tibshirani, R. 1996. Regression shrinkage and selection via the lasso. *Journal of the Royal Statistical Society Series* B58: 267–288.

Tierney, L. 1994. Markov chains for exploring posterior distributions (with discussion). *Annals of Statistics* 22: 1701–1762.

Tierney, L. and Kadane, J. B. 1986. Accurate approximations for posterior moments and marginal densities. *Journal of the American Statistical Association* 81: 82–86.

Tierney, L., Kass, R. E. and Kadane, J. B. 1989. Fully exponential Laplace approximations to expectations and variances of nonpositive functions. *J. Am. Statist. Assoc* 84: 710–6.

Tsay, R. S. 2002. *Analysis of Financial Time Series.* New York: Wiley.

van der Linde, A. 2005. DIC in variable selection. *Statistica Neerlandica* 59: 45–56.

van Dyk D. A. and Meng, X. L. 2001. The art of data augmentation. *Journal of Computational and Graphical Statistics* 10: 1–50.

Veer, L. and Jone, D. 2002. The microarray way to tailored cancer treatment. *Nature Medicine* 8: 13–14.

Viallefont, V., Raftery, A. E. and Richardson, S. 2001. Variable selection and Bayesian model averaging in case-control studies. *Statistics in Medicine* 20: 3215–3230.

Verdinelli, I. and Wasserman, L. 1995. Computing Bayes factor using a generalization of the Savage-Dickey density ratio. *Journal of the American Statistical Association* 90: 614–618.

Viallefont, V., Raftery, A. E. and Richardson, S. 2001. Variable selection and Bayesian model averaging in case-control studies. *Statistics in Medicine* 20: 3215–3230.

Volinsky, C. T. and Raftery, A. E. 2000. Bayesian information criterion for censored survival models. *Biometrics* 56: 256–262.

Wasserman, L. 2000. Bayesian model selection and model averaging *Journal of Mathematical Psychology* 44: 92–107.

White, H. 1982. Maximum Likelihood Estimation of Misspecified Models. *Econometrica* 50: 1–25.

Whittaker, E. 1923. On a new method of graduation. *Proc. Edinburgh Math. Soc.* 41: 63–75.

Whittaker, J. 1990. *Graphical Models in Applied Multivariate Statistics.* Chichester: Wiley.

Wong, F., Carter, C. K. and Khon, R. 2003. Efficient estimation of covariance selection models. *Biometrika* 90: 809–830.

Wright, J. H. 2008. Bayesian model averaging and exchange rate forecasts *Journal of Econometrics* 146: 329–341.

Yu, J. 2005. On leverage in a stochastic volatility model. *Journal of Econometrics* 127: 165–178.

Yu, K. and Moyeed, R. A. 2001. Bayesian quantile regression *Statistics & Probability Letters* 54: 437–447.

Yu, K. and Stander, J. 2007. Bayesian analysis of a Tobit quantile regression model. *Journal of Econometrics* 137: 260–276.

Zellner, A. 1962. An efficient method of estimating seemingly unrelated regression equations and tests for aggregation bias. *Journal of the American Statistical Association* 57: 348–368.

Zellner, A. 1971. *An Introduction to Bayesian Inference and Econometrics*. Wiley.

Zellner, A. 1977. Maximal data information prior distributions, In: *New developments in the applications of Bayesian methods* ed. A. Aykac and C. Brumat, 211–232. Amsterdam: North-Holland.

Zellner, A., 1996. Models, prior information, and Bayesian analysis *Journal of Econometrics* 75: 51–68.

Zellner, A. 2006. S. James Press and Bayesian Analysis. *Macroeconomic Dynamics* 10: 667–684.

Zellner, A. and Chen, B. 2001. Bayesian modeling of economies and data requirements. *Macroeconomic Dynamics* 5: 673–700.

Zellner, A. and Chetty, V. K. 1965. Prediction and decision problems in regression models from the Bayesian point of view. *Journal of the American Statistical Association* 60: 608–616.

Zellner, A. and Min, C. K. 1995. Gibbs sampler convergence criteria. *Journal of the American Statistical Association* 90: 921–927.

Zellner, A., Bauwens, L., Van Dijk, H.K. 1988. Bayesian specification analysis and estimation of simultaneous equation models using Monte-Carlo integration, *Journal of Econometrics* 38: 39–72.

Zou, H. 2006. The Adaptive Lasso and Its Oracle Properties. *Journal of the American Statistical Association* 101: 1418–1429.

Index